An Introduction to
Spectroscopy for Biochemists

An Introduction to

Edited by

ACADEMIC PRESS
A Subsidiary of Harcourt Brace Jovanovich, Publishers
London New York Toronto Sydney San Francisco

Spectroscopy for Biochemists

S. B. BROWN

Reader in Biochemistry

University of Leeds

1980

AP

ACADEMIC PRESS INC. (LONDON) LTD.
24/28 Oval Road
London NW1

United States Edition published by
ACADEMIC PRESS INC.
111 Fifth Avenue
New York, New York 10003

British Library Cataloguing in Publication Data

An introduction to spectroscopy for biochemists.
1. Biological chemistry—Technique
2. Spectrum analysis
I. Brown, Stanley B
574.1′92′028 QP519.9.S6 79-41632

ISBN 0-12-137080-1

Printed in Great Britain at the Alden Press, Oxford

Contributors

PETER BAYLEY, Biophysics Division, National Institute for Medical Research, Mill Hill, London, NW7 1AA.

S. B. BROWN, Department of Biochemistry, University of Leeds, 9 Hyde Terrace, Leeds LS2 9LS.

P. GANS, Department of Inorganic and Structural Chemistry, School of Chemistry, University of Leeds, Leeds LS2 9JT.

A. J. GEDDES, Department of Biophysics, University of Leeds, Leeds LS2 9JT.

R. JONES, Department of Physics, University of Lancaster, Lancaster, LA1 4YB.

G. R. PENZER, British Council (Cairo), c/o F.C.O. King Charles Street, London SW1A 2AH.

E. J. WOOD, Department of Biochemistry, University of Leeds, 9 Hyde Terrace, Leeds LS2 9LS.

Preface

Over the past 15 years there have been major developments in the application of spectroscopic methods to systems of biological and medical interest. A quick scan through any biochemical periodical will show the extent to which spectroscopic techniques such as nuclear magnetic resonance, electron spin resonance or circular dichroism have become standard tools in many biochemical investigations. Increased sensitivity has led to techniques such as atomic absorption becoming routine methods for analysis in clinical biochemistry.

Although there are numerous excellent books and review articles dealing with specific spectroscopic techniques at an advanced level, there is a need for a single volume appropriate to the non-specialist biochemist or clinician. This book is an attempt to meet that need. It should be appropriate to the undergraduate science or medical student who wishes to appreciate the role played by spectroscopy in the development of modern biochemistry. It is also intended for those engaged in biochemically related research as an aid to an understanding of the literature and also in helping to create an awareness of the great potential of spectroscopic methods in solving specific problems.

The majority of the contributors to the book have direct experience in the teaching of spectroscopy to the non-specialist. Each has attempted to present, so far as possible, a non-mathematical, yet rigorous account of a particular technique. Emphasis has been placed on a wide survey of examples of the application and scope of the various methods, rather than on complete and critical explanations of the techniques themselves. References to more advanced and detailed works are given at the end of each chapter. The literary styles of the various chapters differ somewhat as is inevitable in a multi-author work. However, the format is consistent throughout and it is hoped that the differences in style may enhance rather than detract from the readability of the book. So far as possible, common themes have been emphasized among the examples chosen in the different contributions. For example haems and haemoproteins appear in almost every chapter. This also serves to illustrate the complementary nature of the various approaches.

In selecting spectroscopic techniques for consideration in an introductory text, it is inevitable that some of the less widely applied methods are neglected. The most notable omission in the present text is that of Mössbauer spectroscopy. Conversely, the technique of atomic absorption which the "pure" spectroscopist will probably consider as marginal has been included. The criterion throughout has been that of general applicability to biochemical systems.

Finally a chapter on mass spectrometry is included. Although it is not a spectroscopic technique as normally defined, it is of increasing importance in biochemistry and clinical chemistry and is often complementary to the other techniques discussed.

Leeds,
April 1979

Stanley B. Brown

Acknowledgements

All the contributors are grateful for the support, encouragement and guidance given by their families and colleagues at various stages in the preparation of the book. The editor wishes to express appreciation to the authors for their prompt and willing cooperation and to various final-year biochemistry students at the University of Leeds for help in assessing the level of the text. Finally, particular thanks are due to Mrs E. A. Brown for reading the manuscript and preparation of the index.

Contents

1 Introduction to Spectroscopy

S. B. Brown

Introduction

In its modern usage, the term spectroscopy includes a large group of techniques which differ widely in their mode of application and the information they reveal. As a useful working definition, we may state that spectroscopy is a study of the interaction of electromagnetic radiation with matter, although we shall not adhere strictly to this definition, otherwise the technique of mass spectroscopy would be excluded.

Several factors have led to the branching of the spectroscopic technique along different directions. Most significant is the order of magnitude of the energies involved (see p. 2) but additional factors (e.g. the presence of a magnetic field) and instrumental considerations are also important. Thus the various branches of spectroscopy, ultraviolet (UV) infrared (IR), nuclear magnetic resonance (NMR) etc. have been developed. Because the different branches require distinct instrumentation and yield different kinds of information, to a large extent they are usually considered independently of one another. However, there is a common basis for all spectroscopic techniques and it is the purpose of this brief introductory chapter to consider these fundamental concepts before the detailed discussion of each branch in the following chapters.

Spectroscopic techniques have been applied in investigation of all states of matter although, in biochemistry, we are concerned chiefly with solutions and solids and only to a minor degree with the gas phase. There is a wealth of information which may be derived from careful application of spectroscopic methods to systems of interest. The interpretation of this information may be at one extreme on a precise mathematical basis (this is physics) or at the other extreme on a purely empirical level (biochemical systems often fall into this category). Thus physicists and physical chemists often choose simple systems so that they may attempt to interpret spectra in the precise mathematical terms of the quantum theory. On the other hand, the organic chemist and the biochemist find that the complex systems in which they are most interested can often be treated only in a non-mathematical, semi-empirical manner. Theoretical progress is usually made by the movement of these extreme approaches toward one another, but the direction of greatest progress in recent years has been the tremendous improvement in instrumentation. It is largely for this reason that it is only in the past 15 years or so that spectroscopy has begun to be widely applied to biochemical systems.

Energy

Spectroscopy is intimately concerned with changes in energy states. For biochemical purposes only molecular energy states are important, i.e. with one exception (see Chapter 8), we are not concerned with free atoms, gaseous atomic ions or free subatomic particles. It is important to appreciate from the outset, the various subsections into which the total energy of a molecule can be classified, since it is from these subdivisions that the various branches of spectroscopy arise. The energy of a molecule may change by virtue of changes in its translational energy (change in position), electronic energy, vibrational energy, rotational energy, electronic orientation energy and nuclear orientation energy. Although strictly, these terms are mutually dependent, for many purposes it is valid to treat them independently. The first term is unimportant in spectroscopy and whilst the exact significance of the others will become obvious later, it is important at this stage to realize that the orders of magnitude of the various terms are quite different. Thus changes in electronic energy are much greater than changes in vibrational or rotational energy and changes in nuclear or electronic orientation energy are comparatively minute.

The basis of all spectroscopy is the concept known as the "quantization of energy" which is manifested in the application of quantum theory to atoms and molecules. The concept is usually illustrated by reference to electronic energy when we see that electrons cannot possess any arbitrary energy but are constrained to a strictly limited number of possible energy values, known as energy levels. (The terms energy level and energy state are often used interchangeably.) Energy values between these levels are not possible energy states for electrons in that molecule. This situation is generally represented by an energy level diagram such as that depicted in Fig. 1.1. Similar considerations apply to the other molecular energy terms.

Ultimately the extent to which energy levels may be precisely specified is determined by a rule known as the Heisenberg uncertainty principle and depends on the time for which a molecule occupies the energy state. For most

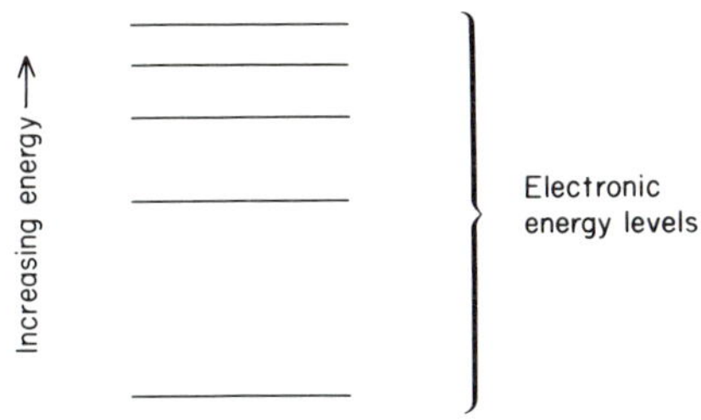

FIG. 1.1 Hypothetical energy level diagram for electrons.

practical purposes, however, it is usual to consider energies as precise quantities which may be accurately measured and used to label energy states.

The question arises, why does the total energy of practical systems (gases, liquids, solutions, etc.) appear to be able to assume any value, i.e. why in macroscopic terms do we observe a continuum of energy and not a series of jumps? The answer is that such a series of jumps does in fact occur, but the gaps are so small that we actually observe an apparent continuum. Therefore, although at any instant each molecule is constrained to a particular composite energy level (composed of all the various energy terms), there are a large number of such levels and it is a thermodynamic necessity that the molecules are distributed among the various levels. Since the increment between the levels (per molecule) is minute in macroscopic terms, by suitable change in the distribution of molecules among the levels, we may obtain any desired energy within the limits of practical measurement. It is for these reasons that the laws of thermodynamics are valid, and we can use quantities such as free energy, enthalpy, etc. without reference to quantum theory. Spectroscopy, on the other hand involves, not the total gross energy of a set of molecules, but the individual energy levels available to each molecule.

Complex Energy Levels

Before considering how spectra arise, it is necessary to define two commonly used spectroscopic terms—ground state and excited state. The ground state may, for practical purposes, be defined as the state corresponding to the lowest possible energy level which the system may occupy and in which more and more molecules are found as the temperature is reduced towards absolute zero. A molecule is said to be in an excited state when it occupies other (higher) energy levels. We may therefore refer to the ground electronic state, the first excited electronic state, etc.

Up to now we have considered the various energy terms in isolation from one another. Thus in Fig. 1.1 we showed only single, well-separated electronic energy levels and ignored vibrational levels. Clearly, however, for any molecule in a given electronic level, several vibrational levels are possible and for any one of these vibrational levels, several rotational levels are possible. This breakdown may be continued through all the lower energy processes. The total number of molecular energy levels is thus very large and a complete energy level diagram would be extremely complex. Fortunately for the spectroscopist, since the various molecular energy terms are well separated in terms of energy, for practical purposes we may restrict our consideration to processes adjacent to one another in the energy spectrum, e.g. electronic and vibrational or vibrational and rotational. This situation is illustrated in Fig. 3.2 (p. 76) in which ground and first excited electronic states are shown separated into their vibrational components.

Electromagnetic Radiation

With the exception of mass spectroscopy, all the techniques discussed in this book involve the interaction of electromagnetic radiation with matter. In the cases of NMR spectroscopy and electron paramagnetic resonance spectroscopy (EPR), this interaction takes place in an applied magnetic field.

A discussion of the precise physical significance of electromagnetic radiation is beyond the scope of this book, but the term derives from the fact that an electromagnetic wave consists of an electric field always at right angles to a magnetic field both being at right angles to the direction of propagation of the wave. For our purposes the chief property of an electromagnetic wave is that it is a source of energy. Figure 1.2 illustrates the basic features of such a wave. It is usual to show only a single sine wave which may represent either the electric or the magnetic vector but not both. The distance travelled by the wave between two repeating points is called the wavelength (λ), while the number of such repeating units or cycles occurring per second is called the frequency (ν). It is clear that the velocity of the wave (c) is given by

$$c = \nu\lambda \tag{1.1}$$

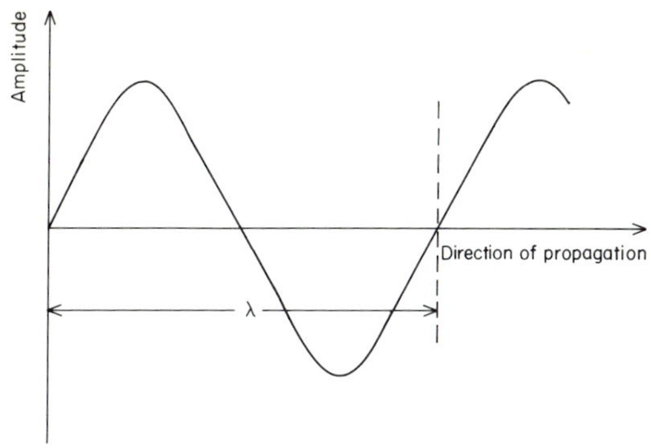

FIG. 1.2 The characteristic form of an electromagnetic wave. The curve may be taken to represent either the electric or the magnetic vector (see the text).

Since the velocity of all electromagnetic radiation is constant in any particular medium, and since light is an electromagnetic radiation, c is equal to the velocity of light.

A fundamental property of electromagnetic radiation is that it can behave as though it exists as discrete packets of energy known as photons or quanta. The energy (E) of a single quantum is determined by the frequency of the radiation according to the equation

$$E = h\nu = hc/\lambda \tag{1.2}$$

where h is a universal constant (Planck's constant). In considering the interaction of radiation with matter the smallest quantity of energy that may be involved is a single quantum and energy changes may occur only by absorption or emission of whole quanta, i.e. the energy change is equal to $nh\nu$ where n is an integer.

It has been found using natural or artificial means that it is possible to obtain electromagnetic radiation over a wide range of frequencies. This continuum of frequencies is called the electromagnetic spectrum. For reasons which are largely historical, various names have been given to different regions of the spectrum. These are shown in Fig. 1.3 along with the corresponding frequency, wavelength and energy ranges.

Radiation is called monochromatic when it consists of a single frequency and polychromatic when it contains a variety or spectrum of frequencies. In practice it is possible to obtain radiation which is almost monochromatic. Even more important, using modern instruments, it is possible to vary the frequency of the monochromatic radiation obtained almost continuously over large regions of the spectrum.

In defining a particular point in the spectrum, it is equally valid to refer to

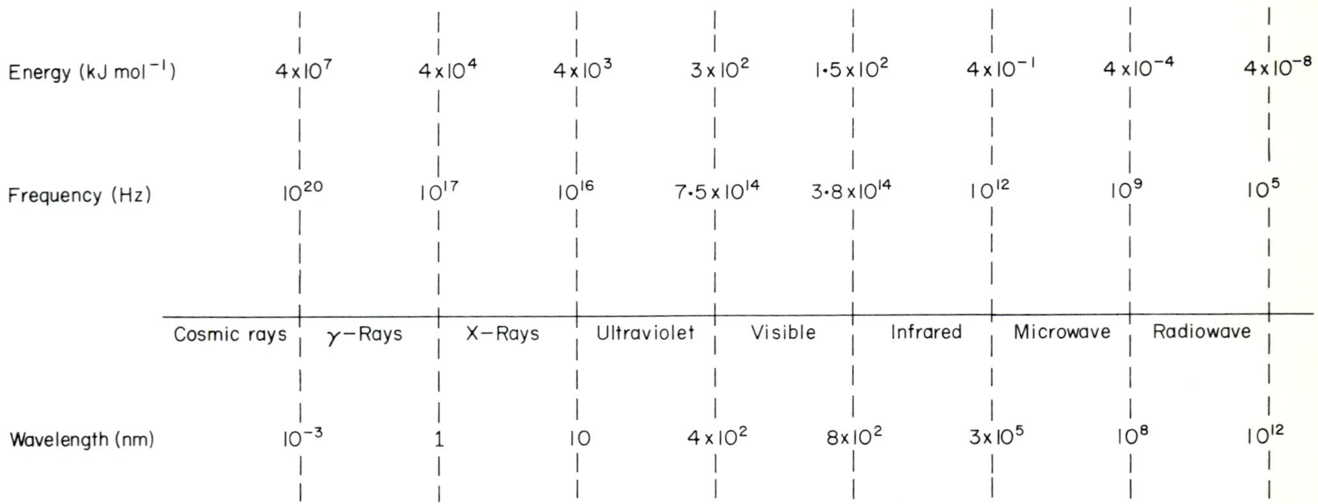

FIG. 1.3 The electromagnetic spectrum.

energy, wavelength or frequency. A further term which is sometimes used is wavenumber defined by

$$\bar{\nu} = 1/\lambda \tag{1.3}$$

Again the particular quantity which is used and the units in which it is expressed are largely a combination of history and convenience. Wavelength is the usual quantity in discussing the visible and UV region. The nanometer (nm) is now the recommended unit but millimicron (mμ) and Ångstrom (Å) were commonly used until recently. In the IR region, wavenumber expressed in cm^{-1} has been preferred whilst hertz (cycles per second) and multiples thereof (e.g. megahertz) are used for lower energy regions. It is curious that energy units are rarely invoked. The relationship between these quantities and units is shown in Fig. 1.3. They may of course be readily interrelated using eqns (1.2) and (1.3). The energy equivalent to these values is also shown in Fig. 1.3 in kJ mol^{-1} assuming that one quantum interacts with one molecule. Since the energy of chemical bonds is of the order of 10–10^2 kJ mol^{-1} it is immediately apparent that at low energies, e.g. in the microwave region, there is no question of electromagnetic radiation being powerful enough to disrupt chemical bonds. Not until the UV–visible region does the energy of the radiation approach the order of that required for chemical reactions. In this region, light-induced reactions such as those involved in photosynthesis become possible. At higher energies (X-rays) disruption of molecules is much more likely whilst at very high energies (γ-rays and cosmic rays) chemical modification and consequent tissue damage are inevitable from prolonged exposure. Table 1.1 gives values of some of the constants necessary for conversion using eqns (1.1), (1.2) and (1.3).

TABLE 1.1 Some useful constants

Symbol	Quantity	Value in SI units	Value in other units
c	velocity of light *in vacuo*	$2{\cdot}998 \times 10^{8}$ m s^{-1}	$2{\cdot}998 \times 10^{10}$ cm s^{-1}
h	Planck's constant	$6{\cdot}626 \times 10^{-34}$ J s	$6{\cdot}626 \times 10^{-27}$ erg s
e	electronic charge	$1{\cdot}602 \times 10^{-19}$ C	$4{\cdot}803 \times 10^{-10}$ esu
N	Avogadro's Number	$6{\cdot}022 \times 10^{23}$ mol^{-1}	—
R	gas constant	$8{\cdot}314$ J mol^{-1} K^{-1}	$1{\cdot}987$ cal deg^{-1} mol^{-1}

Transitions, Absorption and Emission

A transition is said to occur when the energy of a molecule changes from one state to another, e.g. from the ground state to the first excited state. It is usual to represent such transitions on the energy level diagram by a vertical arrow between the levels involved, the direction of the arrow indicating whether the molecule has gained or lost energy. Several such transitions are shown in Fig. 1.4, which is an electronic energy level diagram with the vibrational levels omitted. How may such transitions arise? There are several answers to this question but we are concerned chiefly with spectroscopic transitions, i.e. transitions due to the interaction of the molecule with electromagnetic radiation.

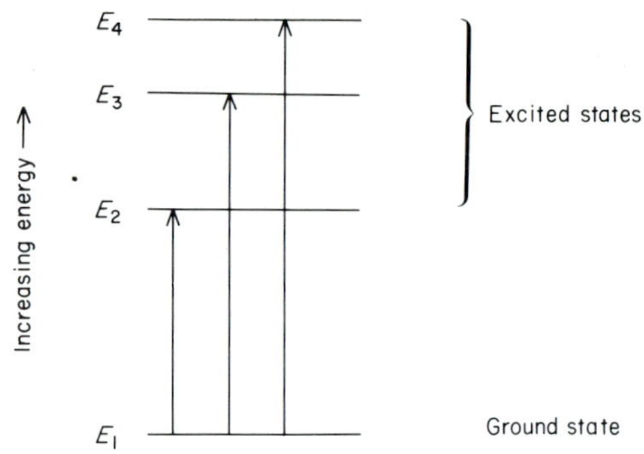

FIG. 1.4 Transitions illustrating absorption of radiation.

Two fundamental processes, absorption and emission, characterize the interaction of electromagnetic radiation with matter. Absorption is said to occur when existing radiation causes an increase in energy of a system with which it interacts. This energy transfer is manifested as a decrease in intensity of the radiation. Emission occurs when electromagnetic radiation is produced due to a decrease in energy of a system corresponding to a transition from an excited state to a lower energy state. Although absorption is always accompanied by an increase in molecular energy, a decrease in molecular energy does not necessarily produce emission since the energy may be dissipated in other ways (see p. 11).

Absorption is readily observable by irradiating a sample and comparing the intensity of the incident and emergent radiation. To observe emission, molecules or atoms must first be raised to an excited energy level. One way in which this may be accomplished is by supplying electrical energy, e.g. in a sodium or mercury vapour lamp. Electrically stimulated emission is unimportant for biochemical purposes and is essentially the province of the physicist. Another way of enabling molecules to achieve an excited energy state is to use existing electromagnetic radiation, i.e. absorption. Part of the energy may then be lost by an emission process when the phenomenon of induced emission or fluorescence occurs. This is the basis of an important technique and is considered in detail in Chapter 3.

Spectra

Suppose a molecule in the ground state, energy E_1 (Fig. 1.4) is subjected to electromagnetic radiation. Quantization tells us that if this molecule is to absorb radiation, it must increase its energy to at least the next lowest energy level, E_2, i.e. by an amount equal to $(E_2 - E_1)$. As we have seen, since the energy of a quantum is equal to $h\nu$, there will be only one point in the entire spectrum at which quanta have exactly the right energy to promote the transition $E_1 \rightarrow E_2$. The frequency of this transition is given by

$$E_2 - E_1 = h\nu \tag{1.4}$$

If the molecule is irradiated with radiation which contains a component at the transition frequency, a transition *may* occur and radiation will be absorbed. Measurement of the wavelength or frequency at which this absorption occurs yields, directly, the energy of the transition.

In Fig. 1.4, the transitions $E_1 \rightarrow E_3$ and $E_1 \rightarrow E_4$ may also be possible. Hence, we might expect three transitions each at different and precisely defined energies. After passage of radiation through the system, a plot of intensity of the emergent radiation versus wavelength might be expected to appear as in Fig. 1.5(a). In the case where Fig. 1.4 refers to electronic energy levels it is more usual to plot directly the amount of radiation absorbed versus wavelength, as shown in Fig 1.5(b). Such a plot, which as we shall see is greatly oversimplified, is termed an absorption spectrum.

For various reasons, some of which are considered in later parts of this book, instead of getting single sharp *lines* of negligible width as in Fig. 1.5(b), absorption *bands* of finite width are observed. In addition, the amount of radiation absorbed varies enormously and depends on many factors (e.g. see pp. 27–42). The three transitions of Fig. 1.4 might therefore be expected to give a spectrum such as that shown in Fig. 1.5(c). However, the true situation is yet more complex. This is because we have ignored the vibrational energy levels into which each of the electronic energy levels is subdivided (see p. 4). For each parent transition shown in Fig. 1.5(c), there are therefore many subsidiary transitions which will vary slightly in wavelength. Instead of the single bands shown in Fig. 1.5(c), we might therefore expect clusters of bands as shown in Fig. 1.5(d). This phenomenon is known as fine structure. The extent to which fine structure is observable depends upon an important property called *resolution*, which in turn depends partly upon intrinsic proper-

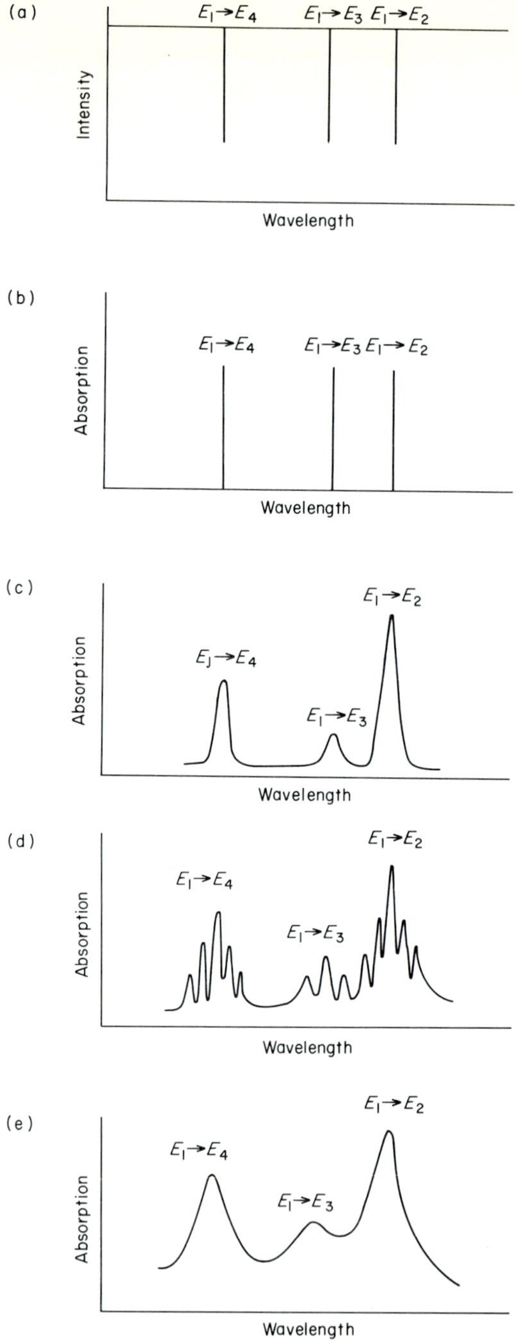

FIG. 1.5 Diagram to illustrate the components of an electronic spectrum. (For details, see the text.)

ties of the system and partly upon instrumental considerations. This is considered in detail in the following chapters. In the example of the electronic transitions presently under consideration however, it is unlikely that such fine structure could be observed, particularly if the system involves molecules in solution. In this case the observed spectrum represents the envelope of the various transitions and would appear as in Fig. 1.5(e). This plot is now typical of the form observed for electronic spectra of molecules in solution.

In a similar way, transitions between other types of energy level give rise to other branches of spectroscopy. Table 1.2 shows the processes from which these various branches arise, the region of the electromagnetic spectrum used to stimulate the appropriate transitions and the parameters used in defining the spectra.

Relaxation Processes

The actual process of absorption of radiation is very fast. If there were no means by which molecules in the excited state could return to the ground state, then all the molecules in a system under study would quickly reach the excited state and no further absorption could be observed. However, since we can continuously observe spectra, it is clear that there are ways in which molecules in excited states may lose energy. Such events are called *relaxation processes*.

There are several kinds of relaxation process and a detailed discussion is reserved for the appropriate sections of the following chapters. In general, thermal relaxation is likely to occur by transfer of energy by molecular or submolecular motion. The efficiency of such thermal relaxation depends upon many factors such as the size, shape and rigidity of the molecule in which the transition has occurred, the solvent, the presence of other solute molecules, etc. It is easiest to regard relaxation processes as involving a competition between thermal energy loss and loss in other ways. These include re-emission of radiation (leading to fluorescence) and chemical utilization of the excitation energy resulting in a photochemical reaction. The average *lifetime* of an excited state therefore depends upon all of these factors.

It is possible to obtain very useful structural and kinetic information from careful study of relaxation processes and excited state lifetimes. This is particularly true for the techniques of fluorescence (Chapter 3) and NMR spectroscopy (Chapter 6).

Type of Information Available in Biochemical Systems

The past 15 years have seen tremendous advances in the application of spectroscopy to systems of biochemical interest. As will be seen from the

TABLE 1.2 Transitions and associated spectroscopic techniques

Energy term producing transition	Region	Usual defining quantity and units	Associated spectroscopic techniques
Outer orbital electrons	UV and visible	Wavelength (nm)	UV and visible spectroscopy Optical rotatory dispersion Circular dichroism Fluorescence spectroscopy
Molecular vibrations and rotations	IR	Wavenumber (cm^{-1}) or frequency (s^{-1} or Hz)	IR spectroscopy Raman spectroscopy
Electronic orientation	Microwave (applied field)	Frequency (Hz)	EPR (ESR) Spectroscopy
Nuclear orientation	Radiowave (applied field)	Frequency (Hz)	NMR Spectroscopy

following chapters, spectroscopic techniques are extremely versatile in providing information about such systems. This information may refer to small molecules, to macromolecules or to interactions between the two.

One of the most important applications of spectroscopy has been in the determination of molecular conformation. In this context, the development of the spectroscopic approach has depended to a large extent on comparison with X-ray crystallographic data. In many instances, the X-ray structure remains the ultimate against which spectroscopic techniques are measured. Nevertheless, spectroscopic methods have advantages over X-ray crystallography and in many cases can provide information not available from the X-ray structure. One such advantage is that of speed, since spectroscopic techniques are often the quickest and most convenient means of obtaining specific pieces of structural information. Sometimes this is at a relatively superficial level as in the determination of the degree of exposure of certain residues in proteins by electronic absorption spectroscopy (see pp. 57–61), but it can also be at a very precise atomic level as in the determination of the conformation of certain substrates by NMR spectroscopy (see p. 264). Spectroscopic methods also have the virtue that they are non-destructive and can often be carried out under conditions approaching those pertaining in living systems. Indeed, recent developments in NMR spectroscopy suggest that valuable results can now be obtained by direct measurements on living cells.

A particular advantage of the spectroscopic approach is that it can be applied in solution and thereby yields dynamic rather than static information. It is therefore possible to determine rates of reaction and the lifetimes of transient species. There are many examples of the use of spectroscopy in determining kinetic data in the following chapters.

2 Ultraviolet and Visible Spectroscopy

S. B. Brown

Introduction

Vision

Perhaps the oldest form of spectroscopy is represented by the ability of the eye to distinguish different colours. The human eye is sensitive to a comparatively narrow range of the electromagnetic spectrum defined as the visible region. For practical purposes this may be taken as the wavelength range 400–700 nm or 7×10^{16}–4×10^{16} Hz expressed in terms of frequency. It is now conventional to express all UV and visible spectra in terms of wavelength, the accepted unit of which is the nanometer (nm). Radiation at the high-wavelength (low-energy) extreme of the visible region appears red whilst that at the low-wavelength (high-energy) extreme appears blue. As we have seen in Chapter 1 the regions of the electromagnetic spectrum immediately adjacent to the high- and low-energy extremes of the visible region are known as the ultraviolet (UV) and infrared (IR) regions respectively. Thus a quantum of UV radiation possesses more energy than one in the visible region whilst the reverse is true for a quantum of IR radiation.

The human eye cannot detect light absorbed by an object, but only light which that object reflects. When viewed in white light, therefore, objects which absorb red light appear blue and objects which absorb blue light appear red. It is important to remember this complementarity when dealing with solutions since, for example, a blue solution will have an absorption band in the red region of the spectrum around 600–700 nm.

Electronic transitions

Spectra in the UV and visible region arise from transitions involving electrons (see Chapter 1). More specifically, the electrons concerned are those in the outer orbitals of atoms and molecules, sometimes referred to as "valency electrons". Because chemical reactions involve changes in the energies of these same electrons, it follows that the energies of chemical bonds are of the same order of magnitude as the energies associated with electronic transitions giving rise to UV and visible spectra. The relationship is seen more explicitly in the fact that light in the UV and visible region may sometimes stimulate chemical reactions which do not occur in the dark. This gives rise to the phenomena of photochemistry and photobiology. Conversely, chemical reactions in the dark may give rise to the emission of light. This phenomenon is known as chemiluminescence and is quite distinct from fluorescence (see Chapter 3). Chemiluminescence also finds biological applications as, for example, in the firefly and in iridescent deep-sea animals. It is no coincidence that the human eye is sensitive to the region of the electromagnetic spectrum with energies comparable to those in chemical bonds, since vision arises from the stimulation of biochemical reactions by light entering the eye.

As we have already seen in Chapter 1, electronic transitions can give rise to several branches of spectroscopy, each of which finds useful applications in biochemical studies. In turn, electronic absorption spectroscopy of molecules has become subdivided into two distinct groups. High-resolution electronic spectroscopy is concerned with the precise evaluation of spectroscopic data in terms of the quantum theory. For this reason, the systems studied are usually limited to di- or triatomic molecules in the gas phase. Low-resolution electronic spectroscopy is concerned with a more empirical relationship between spectra and molecular structure and the quantitative determination of absorbing species (chromophores). It is this latter branch which is of value to organic chemists and biochemists and only this area will be considered further in this book.

Spectrophotometry

Undoubtedly, the most widespread use of UV and visible spectroscopy in biochemistry is in the quantitative analysis of certain chromophores, known as spectrophotometry. The great majority of enzyme assays and determinations of metabolites are carried out in this way.

The Beer–Lambert law

All spectrophotometric methods depend on the application of two basic laws which, combined together, are known as the Beer–Lambert law. The first of these, Lambert's law, states that *the fraction of light absorbed by a transparent medium is independent of the incident light intensity, and each successive layer of the medium absorbs an equal fraction of the light passing through it.*

The law may be expressed mathematically as

$$\log_{10}\frac{I_0}{I}=k\,l$$

where I_0 is the intensity of the incident light, I is the intensity of the transmitted light, l is the path length, and k is a constant for the medium.

The second law, Beer's law, states that *the amount of light absorbed is proportional to the number of molecules of the chromophore, through which the light passes.* It is clear that Beer's law implies that the constant k, is proportional to the concentration of the chromophore. The mathematical expression of the combined Beer–Lambert law is

$$\log_{10}\frac{I_0}{I}=\varepsilon\,c\,l$$

where c is the concentration and ε is a constant. Since ε is independent of concentration and of the path length, it is a property of the chromophore itself. The term $\log_{10} I_0/I$ is called the *absorbance* (A) and is synonymous with the now outdated terms optical density and extinction. When the path length is measured in centimetres and the concentration in terms of molarity, the constant ε is known as the molar absorption coefficient, previously called the molar extinction coefficient or molar absorptivity. In biochemical studies, concentrations are often measured in terms of millimolar units and correspondingly a millimolar coefficient is defined. The simple relationship

$$A=\varepsilon\,c\,l \tag{2.1}$$

can be readily used for the determination of concentration, provided ε and l are known. Examples of the use of this relationship and the consequences of deviations from Beer's law are discussed on pp. 54–57. Although absorbance is the quantity most often used, the intensity is sometimes described in terms of *transmission*, which is defined as I/I_0.

Atomic and Molecular Orbitals

In Chapter 1, we introduced the concept of a spectroscopic transition arising from the acquisition of energy by an electron in moving from one atomic orbital to another of higher energy (Fig. 1.4). For molecules, the basic concept is identical, but instead of dealing with atomic orbitals it is often useful to think in terms of molecular orbitals, i.e. energy levels intrinsically associated with the molecule rather than with its component atoms. A detailed treatment of molecular orbital theory is beyond the scope of this book, but the following qualitative treatment may help in the understanding of the origin of molecular spectra.

For organic molecules, several types of molecular orbital may be distinguished. In saturated compounds, such as methane, ethane, etc., carbon atoms are linked by "single bonds" or σ bonds. The only types of molecular orbitals here are σ orbitals at lowest energy levels and σ^* orbitals at higher energy levels. In the ground state, each σ orbital is usually occupied by a pair of electrons and the σ^* orbitals are not occupied. The occupied σ orbitals are known as *bonding orbitals* whilst the unoccupied σ^* orbitals are known as *antibonding orbitals*. In compounds which contain double bonds, such as ethylene, in addition to σ orbitals, other orbitals known as π orbitals (bonding) are occupied by a pair of electrons whilst the higher energy π^* orbitals (antibonding) are not occupied in the ground state. A double bond may therefore be considered in terms of a pair of electrons in a σ orbital and another pair in a π orbital. A further type of orbital, known as an *n orbital* occurs in molecules containing an atom such as O or N which possesses a lone pair of electrons. Again, the n orbitals contain a pair of electrons, but in this case these electrons may not be involved in chemical bonding and n orbitals are therefore known as *non-bonding orbitals*. In general the energies of these types of orbital increase in the sequence $\sigma < \pi < n < \pi^* < \sigma^*$. In systems containing *conjugated* or *delocalized* double bonds, the energy difference between the π and π^* orbitals can become quite small. From this simple sequence, therefore, it is possible to see that in saturated molecular groupings (e.g. the hydrocarbon chains of fatty acids) the only possible transition is $\sigma \rightarrow \sigma^*$ and this would be at high energy or low wavelength. This explains why such compounds are colourless and do not show any significant absorption above 200 nm. In unsaturated systems, however, and particularly in delocalized systems such as benzene or porphyrins, the $\pi \rightarrow \pi^*$ transition is of sufficiently small energy (long wavelength) to produce an absorption band in the near UV region (benzene) or even in the visible region (porphyrins).

Nature of Electronic Transitions

Types of transition

We have seen that spectra arise from transitions of electrons from a low energy state (usually the ground state) to a higher energy state (the excited state). In general a movement or displacement of electronic charge must occur if an interaction with the electromagnetic field and hence spectral effects are to be observed. With unpolarized light, the interaction with the electromagnetic field is independent of directional effects (i.e. the symmetry or asymmetry) of the chromophore. However, with molecules which are intrinsically asymmetric, the extent of absorption depends upon the way in which the light is polarized. This is the basis of ORD and CD and is considered in detail in Chapter 5. In statistical terms there are many possible transitions combining the types of molecular orbital considered above. Table 2.1 shows some of the commonly observed types of transition in organic molecules, the kinds of bonds they are associated with and the wavelength range in which the spectroscopic bands appear. For experimental reasons it is very difficult to measure spectra below 190 nm and, in practice, this means that, for organic compounds, observable spectra arise only from unsaturated molecules. In addition to the organic groups we have already considered, certain other types of electronic transition are of importance in biology. For example, compounds which are ionic in nature such as transition metal salts, display characteristic spectra which essentially reflect the atomic orbitals on the central metal atom. The wide variation in colour of the transition metals is due to transitions between d atomic orbitals on the metal known as d–d transitions. Up to now we have considered only electronic transitions between energy levels in the same atom or molecular grouping. However, an electron may also be transferred from one atom or chemical group to another, for example in the complexes formed by iodine in aromatic solvents and in metalloproteins such

TABLE 2.1 Common transitions in organic molecules

Group	Example	λ_{max} (nm)	ε (M^{-1} cm^{-1})	Assignment
Alkane	hexane	150	—	$\sigma \rightarrow \sigma^*$
Isolated olefin	ethylene	163	15 000	$\pi \rightarrow \pi^*$
Carbonyl	$(CH_3)_2CO$	188	1 860	$n \rightarrow \sigma^*$
		277	12	$n \rightarrow \pi^*$
Nitro	CH_3NO_2	275	15	$n \rightarrow \pi^*$
Hydroxyl	CH_3OH	177	200	$n \rightarrow \sigma^*$
Sulphide	C_2H_5—S—C_2H_5	195	4 600	$n \rightarrow \sigma^*$
Disulphide	C_2H_5—S—S—C_2H_5	194	5 500	$n \rightarrow \sigma^*$
		250	380	$n \rightarrow \sigma^*$
Phenyl	benzene	256	230	$\pi \rightarrow \pi^*$
Carboxyl	CH_3COOH	204	45	$n \rightarrow \pi^*$
Amino	$(C_2H_5)_2NH$	195	2 500	$n \rightarrow \sigma^*$
Amide	CH_3CONH_2	205	160	$n \rightarrow \pi^*$

(The λ_{max} value is the wavelength corresponding to the absorption peak. The values given above are only approximate since solvent effects are ignored—see p. 26)

as haemocyanin (see p. 42). These are often referred to as charge transfer transitions.

Selection rules

Although it is not easy to predict precisely which transition may occur in any given molecule, some general guidelines may be given. In particular, restrictions, known as selection rules, limit the occurrence of certain transitions. Transitions which are inconsistent with the selection rules are said to be *forbidden* whilst those which are acceptable in terms of the rules are said to be *allowed*. Some selection rules involve the application of symmetry arguments to the ground and excited state of the electron but consideration of these is outside the scope of the present discussion. Among the most important selection rules is one which does not allow a change in the spin quantum number during the electronic excitation process. With the exception of free radicals (most of which are unstable and of little interest in terms of the use of electronic absorption spectroscopy in biology) and transition metals (which will be discussed later) most biologically important compounds contain spin-paired electrons, i.e. every electron is paired with another electron of opposite spin. During excitation an electron becomes "separated" from its previous partner, but this selection rule demands that the spins of the two electrons remain unchanged. The normal spin-paired states are called *singlet states*, whilst, less usual situations where electrons are unpaired are called *triplet states*. Thus singlet–singlet transitions are allowed and are common. Triplet–triplet transitions are also allowed but are less common because triplet states are rare. However triplet–singlet transitions are forbidden. This selection rule is illustrated diagrammatically in Fig. 2.1.

Sometimes transitions are observed which are formally forbidden. This occurs because of the interaction between electronic and vibrational energy levels. An example is the n→π^* transition of the carbonyl group, shown in

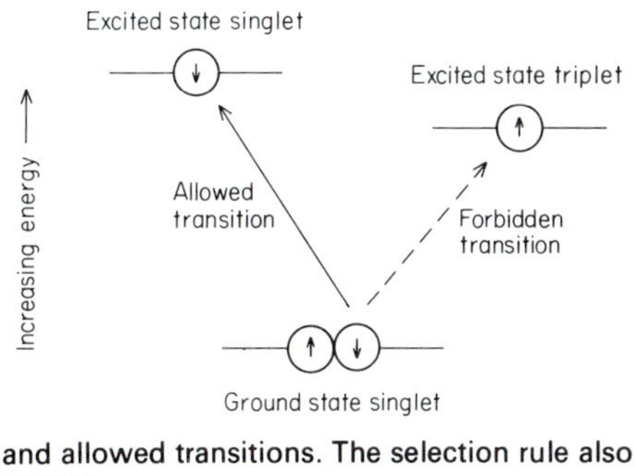

FIG. 2.1 Forbidden and allowed transitions. The selection rule also applies to the corresponding emission processes.

Table 2.1. However, such transitions are usually of low intensity with molar absorption coefficients of about 10–100 M^{-1} cm^{-1}.

Intensity of absorption

The theoretical factors governing the intensities of absorption of radiation during electronic transitions will not be considered here, In general terms we may say that the intensity depends on the nature of the charge displacement or *transition moment* associated with the transition. For the present purposes it is best to treat intensities in empirical terms. Thus intensities of d–d transitions are relatively low, whilst those of π–π^* transitions in aromatic compounds are relatively high. Further details of the typical intensities of various chromophores are given on pp. 27–42.

Electronic relaxation processes

We have seen in Chapter 1 that the transition from the ground state to the excited state is so fast as to be regarded as instantaneous and that after this "instantaneous" event, the lifetime of the excited state depends upon several factors. Since excited states are intrinsically unstable, the electron will return to the ground state by the easiest route by a so-called relaxation process. There are three ways in which the electronic excitation energy can be lost and the ground state attained. The most common of these is loss by transfer of the energy thermally to the surroundings (e.g. the solvent) by vibrations, rotations and collisions. Another possibility is that part of the energy may be lost by emission, leading to the phenomena of fluorescence and phosphorescence. Finally, in certain cases, the energy of the excited state may be sufficient to activate a chemical process in which case a photochemical reaction occurs. There may be a competition between these three processes and their relative importance depends very much upon the structural nature of the electronic excited state and its environment. Because these relaxation processes are of fundamental importance in fluorescence spectroscopy, a detailed consideration is given in Chapter 3.

Conjugated Systems

Conjugated systems occur when double and single bonds alternate along a chain of carbon atoms which may or may not be cyclic. The result of such a molecular arrangement is the *delocalization* of the double bonds and a spreading of electron density across the whole molecule. The effect on the molecular orbitals is best illustrated by reference to the simplest conjugated molecule butadiene. The difference between this molecule and two isolated ethylene molecules is depicted in Fig. 2.2. Two ethylene molecules would possess two π orbitals of equal energy and two π^* orbitals of equal energy. The effect of conjugation in butadiene is to split the orbitals such that one of the resulting orbitals (π_1) is at a higher energy and the other (π_2) is at a lower energy than the π orbital of ethylene. The same effect occurs for the π^* antibonding orbitals, to yield new orbitals π^*_1 and π^*_2. The most important result of this consideration is that several π–π^* type transitions are now possible and in particular the transition π_1–π^*_1 now appears at lower energy (higher wavelength) than the

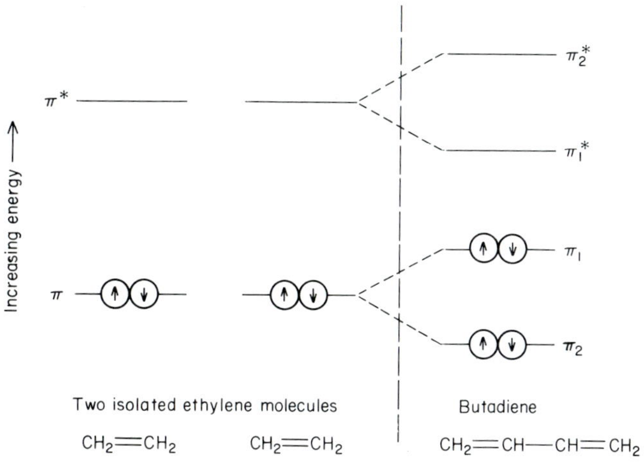

FIG. 2.2 Energy level diagram showing effect of delocalization.

π–π^* transition of ethylene. This simple example may be generalized. As the number of conjugated double bonds increases the wavelength of the lowest energy π–π^* transition increases, i.e. the absorption band moves closer to the visible region. This trend is illustrated in Table 2.2 where n is the number of double bonds per molecule. It may be seen that for $n > 8$, the absorption band extends into the visible region implying that these compounds are coloured. The β carotenes containing the equivalent of 11 double bonds per molecule, represent an example of this type of system (see p. 33).

Cyclic conjugated systems follow a similar pattern as shown in Table 2.3. Benzene, which is a completely delocalized molecule, shows a π–π^* absorption band at 254 nm whereas the corresponding band for naphthalene occurs at 275 nm (Table 2.3). The most highly conjugated systems in biology occur in the porphyrins (see p. 35) which contain the equivalent of 11 double bonds per molecule.

TABLE 2.2 Spectral parameters of conjugated polyenes. General formula $H(CH{=}CH)_nH$

n	λ_{max} (nm)	ε (M^{-1} cm^{-1})	Solvent
1	163	15 000	Hexane
2	217	21 000	Hexane
3	268	56 000	Hexane
4	304	—	Hexane
5	334	118 000	Hexane
6	364	138 000	Iso-octane
7	390	—	Iso-octane
8	410	—	Iso-octane
10	447	—	Iso-octane

TABLE 2.3 Spectral parameters of aromatic conjugated systems

Compound	Structure	λ_{max} (nm)	ε (M^{-1} cm^{-1})	Solvent
Benzene		254	204	Hexane
Styrene	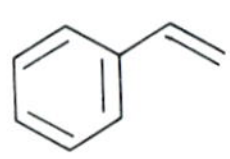	282	750	Ethanol
Naphthalene	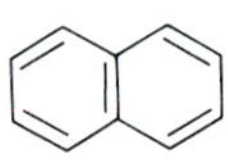	275	5 620	Ethanol
Anthracene	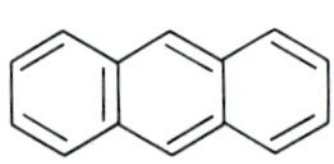	355	7 940	Cyclohexane
Tetracene	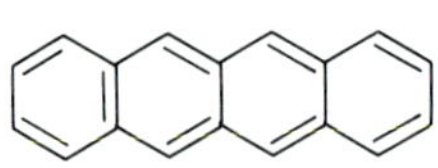	445	10 000	Benzene

The intensities of the π–π^* transitions in conjugated compounds also follow a well-defined pattern. In both the straight-chain and cyclic molecules, absorption coefficients increase markedly with the number of double bonds as shown in Tables 2.2 and 2.3. The result of this trend is that when sufficient double bonds have been introduced such that absorption in the visible range is possible, molar absorption coefficients have very high values often of the order of 100 000 or even greater. Such molecules as the carotenoids and porphyrins are therefore valuable as biological pigments.

Spectrophotometers

The modern UV and visible spectrophotometer is familiar to all biologists. Instruments may be of the single-beam or double-beam type although the former is now used only for simple absorption measurements at a single wavelength. In the single-beam spectrophotometer, the instrument must first be calibrated or "zeroed" with a reference solution in the light beam. This is then replaced by the sample and a reading of the absorbance is taken. In the double-beam spectrophotometer, the light beam is automatically passed through both the sample and reference solution so that a direct reading of the *difference* between sample and reference solution is obtained. The detailed construction of the various commercial instruments available varies considerably, but the overall principle of operation is the same and is illustrated in the block diagram in Fig. 2.3.

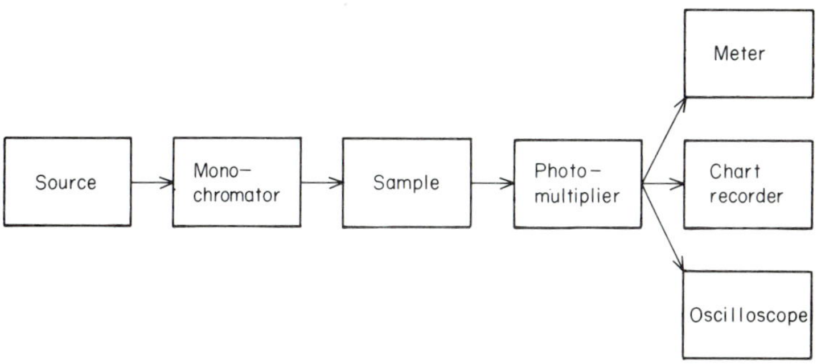

FIG. 2.3 Block diagram of the essential components of an ultraviolet and visible spectrophotometer.

Light source

The source is required to yield a high output of polychromatic light over a wide range of the spectrum. For the visible region, tungsten filament bulbs are invariably used and have a usable wavelength range from about 330 nm to 850 nm. For the UV region, deuterium lamps are usually used and have a range of approximately 200–450 nm. Although useful biochemical measurements can be made in the visible range, many enzyme assays and structural determinations require UV measurements. Most modern instruments in biochemical teaching and research laboratories are therefore equipped with both tungsten and deuterium sources which can be readily interchanged.

Monochromator

In order to make precise spectral measurements, light embracing a very narrow wavelength range of the electromagnetic spectrum and which may be termed *monochromatic* must be used. On simpler types of instrument, light filters used singly or in combination may be used. However, such filters yield a relatively wide wavelength band and are unsuited to accurate measurements. The more sophisticated spectrophotometers use one of two types of monochromator. In older types of instrument, such as the Unicam SP800, a fused-silica prism is used to disperse the light. However, in modern instruments diffraction gratings are used except where the highest accuracy is required when prism dispersion remains the method of choice. In Fig. 2.3, light proceeding out of the monochromator may therefore be considered to be essentially monochromatic.

Sample compartment

The monochromatic light then passes through the sample, which is contained in the familiar spectrophotometric cell or *cuvette*. Such cuvettes are accurately manufactured to a specified path length which is often 10 mm. However, cuvettes with path lengths covering the range 1–100 mm are readily available. The standard type of 10 mm cuvette has an internal volume of about 3 ml, whilst the volumes of cuvettes of other path lengths vary proportionately. For biochemical studies, where materials may be in very short supply, microcells may be used. These are cells which are usually of 10 mm path length, but present a much reduced area of cross-section to the light beam. They therefore have the advantage of using less solution without a corresponding decrease in path length, but their use requires careful alignment in the light beam. Other types of cell known as flow cells permit the continuous passage of solution across the instrument light path. Such cells are useful for continuous monitoring of eluants from columns or gradients.

It is of course essential that the cuvette wall is completely transparent to light of the wavelength being used. For the visible region, glass or certain types of plastic cuvette may be used. However, glass absorbs in the UV region, where fused silica (quartz) cells are necessary.

For accurate measurement of intensities and for all kinetic work, thermostatting of the cuvette compartment is essential. This is usually accomplished with the aid of a cuvette "block" through which a thermostatically controlled liquid is circulated or alternatively by means of electrical heating.

Photomultiplier and recorder

Light may or may not be absorbed by passage through the sample. This is determined by a quantitative measurement of the amount of light passing through the instrument by means of a photomultiplier which uses the photoelectric effect and electronic amplification to produce an output voltage. This output may be displayed or recorded in several ways. Some instruments simply employ a voltmeter, calibrated in appropriate units. In more sophisticated instruments a chart recorder is used. Although not absolutely essential, chart recorders can operate continuously during spectral measurements and are very much more convenient than meters for measuring the variation of absorbance with wavelength or for carrying out kinetic measurements. Alternatively, some modern instruments are equipped with a digital output accessory which enables a numerical print-out of the absorbance and wavelength variation to be obtained.

Other Practical Considerations

In order to obtain accurate and meaningful spectra, solutions must be completely free of all solid material. If this is not the case light scattering may occur which will distort the spectra obtained. For such measurements, solutions which are cloudy or appear colloidal must not be used. On the other hand, for the purely empirical measurements of the difference in absorbance at a given wavelength such as are required for enzyme assays, a certain degree of turbidity may be tolerated. Also, direct measurements of turbidity are useful in determination of the cell density of bacterial suspensions. These may be obtained in exactly the same way as absorption measurements, except that the decrease in intensity after passage of the light beam through the suspension is primarily due to scattering effects rather than absorption (see p. 45). The surfaces of the cuvettes must be kept scrupulously clean and dry otherwise large errors in absorbance measurement could occur. In general, reference solutions should contain all the ingredients of the sample cell except the chromophore under study. Thus for measurements of the spectrum produced by the reaction of dithionite with haemoglobin in phosphate buffer, the reference solution should contain dithionite and phosphate buffer at the same concentration as in the sample.

Solvents

Although much work of a biochemical nature is carried out in aqueous medium, the identification and measurement of metabolites and even some enzyme assays require non-aqueous solvents. The spectrum of most chromophores is solvent dependent to a small extent. Both the intensity of absorption and the wavelength of the absorption peak may change significantly, but not grossly. Presumably, solvent-induced changes reflect small changes in the energy levels caused by direct interaction between chromophores and solvent molecules. Typical solvent dependence data are shown in Table 2.4 which lists the variation in millimolar absorption coefficient and peak wavelength of protoporphyrin IX dimethyl ester in various solvents. Such solvent shifts can be utilized in the determination of macromolecular structure as indicated on pp. 57–61.

For spectra in the UV region, it is important to choose solvents which do not absorb a significant amount of light at the wavelength under study. A general guide to the minimum wavelength which can be used with a number of common solvents is given in Table 2.5.

TABLE 2.4 Solvent dependence of spectrum of protoporphyrin IX dimethylester

Solvent	λ_{max} (millimolar absorption coefficient in parenthesis)				
Chloroform	407(171)	505(14·2)	541(11·6)	575(7·4)	630(5·4)
Ether	404(158)	503(14·8)	536(11·9)	576(6·6)	633(6·6)
Dioxan	406(164)	504(14·7)	538(11·6)	575(6·9)	631(5·6)
Pyridine	409(163)	506(14·9)	541(11·9)	576(7·5)	631(5·5)

(Data taken from K.M. Smith, (Ed.) (1975). "Porphyrins and Metalloporphyrins", 872. Elsevier, Amsterdam.)

Calibration of spectrophotometers

It is sometimes necessary to check both the wavelength calibration and the absorbance accuracy of a spectrophotometer. For this purpose, some machines are supplied with solid filters which produce peaks and troughs at known wavelengths. However it is often desirable to make up a standard solution of known absorbance and peak wavelength (λ_{max}) values. Potassium chromate and potassium dichromate are commonly used for this purpose. The values of λ_{max} and absorbance necessary for the calibration of a spectrophotometer using these materials are shown in Table 2.6.

TABLE 2.5 Minimum wavelengths for use of common solvents

Solvent	Wavelength (nm) 10 mm cells	1 mm cells
Water	191	187
Cyclohexane	195	190
Hexane	201	197
Ethanol	204	198
Methanol	203	198
Butan-1-ol	210	204
Ether	210	203
Chloroform	237	223
Pyridine	305	295
Acetone	330	320

TABLE 2.6 Data for wavelength and absorbance calibration (25°C)

Chromophore	λ (nm)	ε (M^{-1} cm^{-1})
Potassium dichromate in 0·005 M H_2SO_4	235 (min)	3 676
	257 (max)	4 272
	313 (min)	1 436
	350 (max)	3 151
Potassium chromate in 0·05 M KOH	275 (max)	3 676
	313 (min)	211
	375 (max)	4 815

Naturally Occurring Chromophores

Before discussing examples of the applications of UV and visible spectroscopy in biochemistry, it is useful to consider some commonly occurring chromophores and how their spectra arise. Since special techniques are required to investigate spectra in the far UV region, only those chromophores with absorption above about 210 nm will be considered in detail here. For many chromophores, especially those which are coloured, measurement of the electronic absorption spectrum has been an important tool in determination of their structure.

Amino acids and proteins

Two absorption bands of the carboxyl groups of amino acids are known, neither of which is of significant value in biochemistry. The strongest band (probably due to an n→σ* transition) occurs in the region of 175–190 nm, a part of the spectrum which, as has already been mentioned, is difficult to study. The other band due to an n→π* transition occurs in a more accessible region around 205–210 nm but is of very low intensity. For general practical purposes therefore, the spectra of amino acids are determined by the nature of the side chains or R groups. Similarly in proteins, the peptide-bond absorption peak occurs at around 190 nm. Although some progress has been made in the study of proteins in this region, wide application of such studies has not yet been achieved. However, investigation of the peptide bond absorption is potentially of great value because it may reflect conformational details of the protein.

To a first approximation, the spectral characteristics (i.e. λ_{max} and ε values) of the R groups in proteins and in the free amino acids are the same and we may consider them together. Spectral details of the amino acids which have significant absorption above 210 nm are listed in Table 2.7. None of these amino acids has an absorption extending into the visible region and, in the absence of additional chromophores, proteins are therefore colourless. Figure 2.4 shows spectra of the aromatic amino acids in the range 230–310 nm. Tyrosine is alone among these aromatic acids in possessing an acid–base function in the side chain. As might be expected, the spectrum of the undissociated form (pH 6) is significantly different from that of the dissociated form (pH 13). This difference can be used to determine conformational information about proteins (see p. 61).

It should also be noted from Table 2.7 and Fig. 2.4 that the absorption coefficient for phenylalanine is lower than those for tryptophan and tyrosine by an order of magnitude. As a result, the contribution of phenylalanine to the spectrum of a protein is often insignificant compared with that of tyrosine and tryptophan. In general, therefore, protein absorption is maximal at about 280 nm. Almost all proteins display absorption in this region, although of course,

TABLE 2.7 Spectral details of amino acids in neutral aqueous solution

Amino Acid	Chromophore	λ_{max} (nm)	ε (M^{-1} cm^{-1})	Assignment
Cysteine	S^-	235	3 200	n→σ*
Cystine	—S—S—	250	320	n→σ*
Phenylalanine	phenyl	257	220	π→π*
Tyrosine	phenolic	274	1 440	π→π*
		222	7 900	π→π*
Tryptophan	indole	280	5 050	π→π*
		219	34 000	π→π*
Histidine	imidazole	211	6 300	π→π*

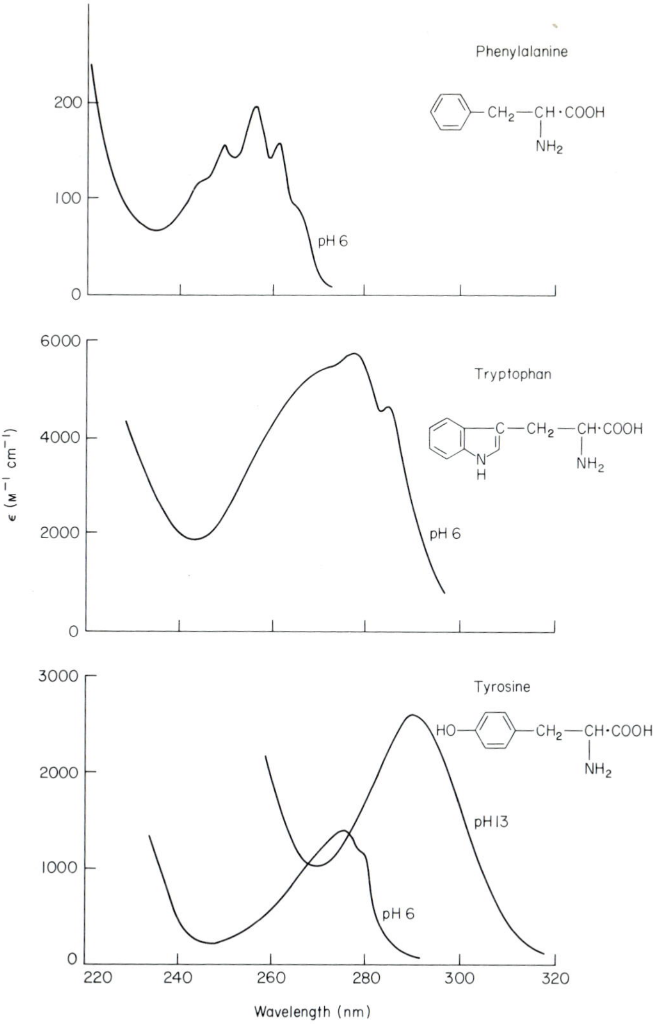

FIG. 2.4 Ultraviolet spectra of the aromatic amino acids.

there is a very large variation depending upon the number of aromatic residues per protein molecule. The spectrum of a typical protein, chymotrypsin, is shown in Fig. 2.5. Absorption at 280 nm is often used as a rough protein assay, for example in following the elution of a sample from a chromatographic column.

Although the intensity of absorption bears a superficial relationship to the number of aromatic residues, the precise spectra of proteins cannot be pre-

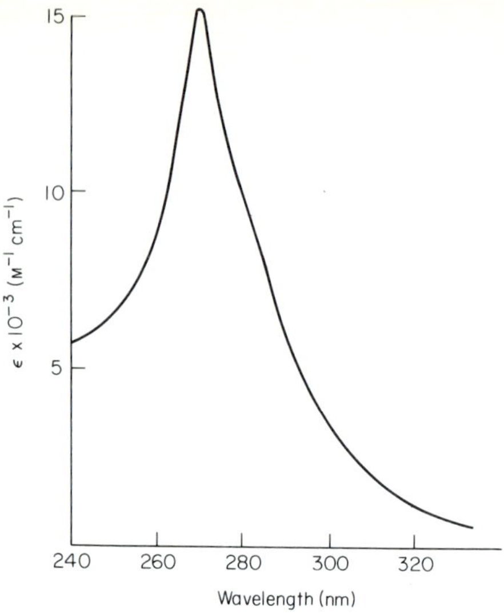

FIG. 2.5 Ultraviolet spectrum of chymotrypsin.

dicted as a summation of the absorption of the component amino acids. It follows that the observed spectra reflect, to some extent, the varying environments of the amino acid residues within the protein, which depend upon its structure and conformation. Examples of the use of spectral measurement of the aromatic residues in determining information on the conformation of proteins are considered on pp. 57–65.

With the exception of these aromatic amino acids, Table 2.7 shows that only cysteine, cystine and histidine absorb above 210 nm. However, the absorptions of both histidine and cystine are very weak compared with those of tryptophan and tyrosine in the same region.

Some proteins, such as the haemoproteins and copper proteins, contain strongly absorbing non-polypeptide chromophores which impart characteristic colours. For these proteins the ratio between the absorption coefficient at the visible region maximum and the 280 nm region maximum is a characteristic property of the protein. It is useful as a purity index since these proteins can be purified to a constant ratio of absorbances at the two maxima.

Nucleosides and nucleic acids

Both the pyrimidine and the purine ring systems display extensive delocalization and are associated with $\pi \rightarrow \pi^*$ transitions giving rise to absorption in the region of 240–275 nm. Table 2.8 shows the spectral characteristics for a number of bases and nucleosides. (The structures of the bases are shown in Fig. 2.6.) The nucleosides, containing a sugar residue, are much more water soluble than the free bases. For any given series, Table 2.8 shows that the spectra of the free base and the corresponding nucleoside are quite similar, to be expected, since only the purine or pyrimidine base component is a chromophore around 240–275 nm. The spectra of the nucleotides (nucleoside phosphates) are almost identical to those of the corresponding nucleosides.

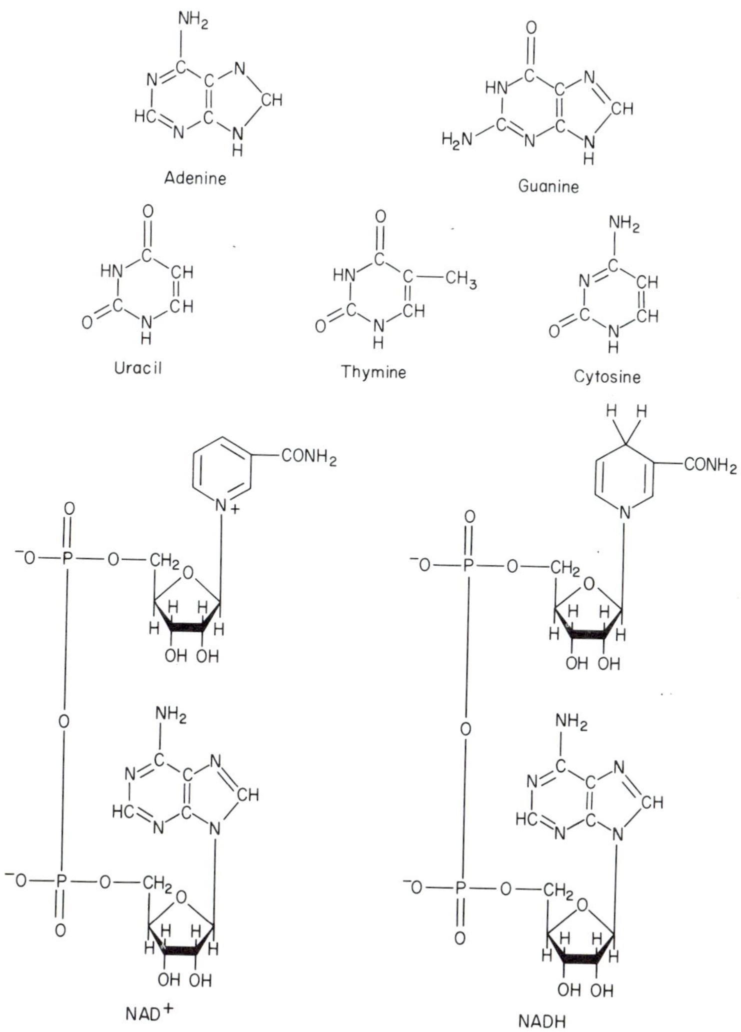

FIG. 2.6 Structures of bases and nucleotides.

TABLE 2.8 Spectral characteristics of bases and nucleosides in neutral solution

		λ_{max} (nm)	ε (M^{-1} cm^{-1})	pH
Base				
Adenine	(Pu)[a]	261	13 300	7·0
Guanine	(Pu)	246	10 700	7·0
Uracil	(Py)[b]	276	8 150	7·2
		259	8 100	
Thymine	(Py)	265	7 890	7·2
Cytosine	(Py)	267	6 130	7·2
Nucleoside				
Adenosine	(Pu)	260	14 900	6·4
Guanosine	(Pu)	253	13 700	6·0
Uridine	(Py)	262	10 100	7·2
Thymidine	(Py)	267	9 650	7·2
Cytidine	(Py)	271	9 100	7·2

[a] Pu = purine
[b] Py = pyrimidine
Data taken from Beaven, G.H., Holiday, E.R. and Johnson, E.A. (1955). "The Nucleic Acids", Vol. I, (Eds Chargaff, E. and Davidson, J.N.), 493–545. Academic Press, New York and London.

As would be expected because of protonation of the base nitrogen atoms, spectra of these compounds are highly pH dependent and it is essential to take account of pH in making spectral measurements. Typical variations with pH are shown for the base cytosine in Fig. 2.7. Because of the absorption of their component nucleotides, all samples of DNA and RNA absorb strongly in the region 240–270 nm. with a peak close to 260 nm. The spectrum of a typical DNA sample is shown in Fig. 2.8 at three pH values. As with the component nucleotides, it may be seen that spectra of DNA and RNA are pH dependent.

The variation in absorption coefficient of DNA or RNA per residue is comparatively small and hence the absorption at 260 nm can be used as a reliable measure of nucleic acid concentration. However, the spectrum of a nucleic acid cannot be accurately expressed as the simple sum of the absorptions from the component nucleotides. In general, the total absorbance of a nucleic acid is significantly less than its component nucleotides. This phenomenon is known as the hypochromism of nucleic acids and arises because of the effect of the neighbouring residues upon one another. The effect is discussed further on pp. 65–67.

Nucleotides are also important as vitamins or coenzymes, but in these molecules there is often a second base which contributes an additional chromophore. For example, the coenzymes nicotinamide adenine dinucleotide (NAD) and nicotinamide adenine dinucleotide phosphate (NADP), function

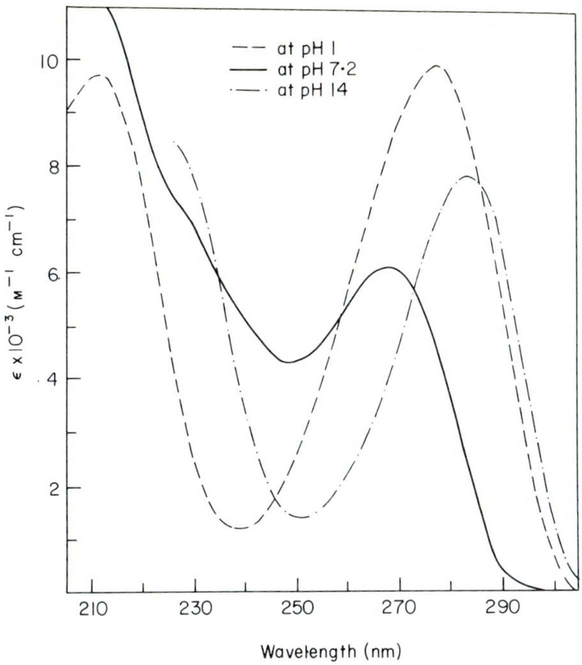

FIG. 2.7 pH Dependence of cytosine spectra. (Spectra taken from G. H. Beavan, E. R. Holiday and E. A. Johnson. (1955). *In* "The Nucleic Acids", Vol. I, (Eds E. Chargaff and J. N. Davidson), 493–545. Academic Press, New York and London.)

in the action of many dehydrogenases. These molecules contain a nicotinamide group in addition to adenine (Fig. 2.6). Upon reduction, to NADH and NADPH, the structure of the nicotinamide moiety changes substantially (Fig. 2.6) with the consequent production of an absorption peak at 340 nm. A very large number of enzyme reactions can be conveniently followed by monitoring the change in absorbance of 340 nm (see p. 47).

Carotenoids

Carotenoid is the general name given to a very large class of red, yellow and orange plant pigments. These compounds are characterized chemically by long carbon chains containing many conjugated double bonds, such as occur in β-carotene (Fig. 2.9). This is responsible for their characteristic spectra, presumably due to π–π^* transitions. Molar absorption coefficients are high with values often in excess of 10^5. As indicated earlier (p. 23) an increase in the number of ethylenic linkages in the carotenoids leads to an increase in absorption coefficient and a red shift.

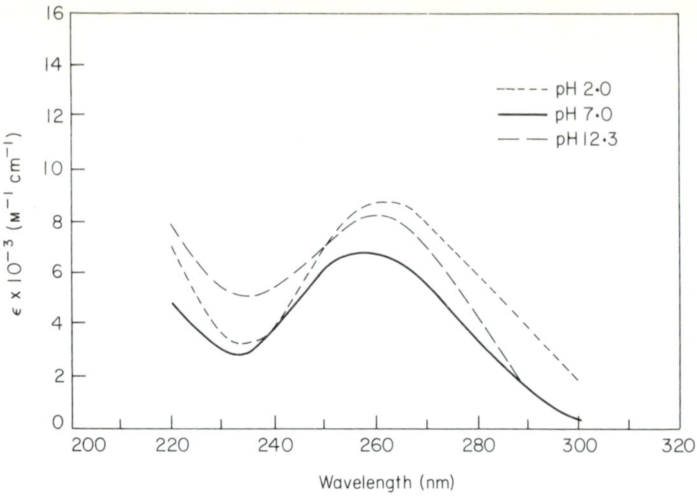

FIG. 2.8 pH Dependence of spectrum of calf thymus DNA. (Spectra taken from G. H. Beavan, E. R. Holiday and E. A. Johnson, (1955). *In* "The Nucleic Acids", Vol. I, (Eds. E. Chargaff and J. N. Davidson), 493–545. Academic Press, New York and London.)

The large number of alternating double bonds leads to the possibility of many stereoisomers. These can be distinguished spectroscopically as indicated in Fig. 2.9 which refers to three stereoisomers of β-carotene.

The spectra of carotenoids characteristically contain three sub-maxima as shown in Fig. 2.9. There is an extensive literature dealing with the identification of carotenoid pigments based on visible absorption spectra.

Flavinoids

Almost all of the brilliant shades of red and blue found in flowers and fruit are due to the presence of flavinoid compounds known as anthocyanins. These are glycosides of pigments called anthocyanidins. As with most other chromophores of biological importance, the anthocyanidins are highly conjugated systems. The names of these compounds (e.g. pelargonidin, cyanidin, delphinidin) are derived from the plants from which they are isolated. The structures and spectral characterization of three major anthocyanidins are shown in Table 2.9. Other flavinoid types are also important in plant pigmentation. For example flavones are responsible for the ivory and cream colours of many flowers.

The absorption and emission spectra of riboflavin and derivatives are considered in Chapter 3.

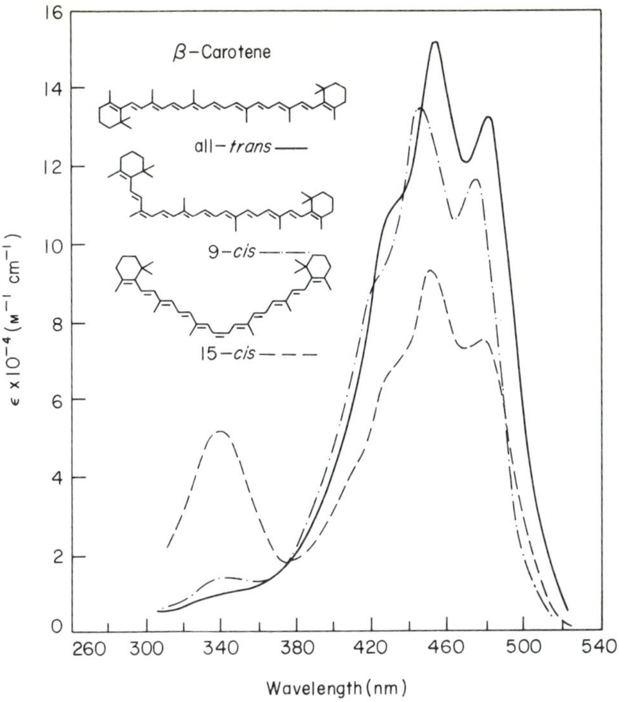

FIG. 2.9 Spectra of three stereoisomeric β-carotenes. (Spectra taken from H. H. Jaffe and M. Orchin, (1962). "Theory and Applications of Ultraviolet Spectroscopy" 233. Wiley, New York.)

Tetrapyrroles

The tetrapyrroles and their derivatives which include the porphyrins, haems, chlorophylls and bile pigments carry out an enormous range of functions in biology. One of the reasons for this versatility is the high degree of delocalization of these molecules and the subtle interactions of the delocalized systems with metal atoms and other molecular groupings. All of these compounds have intense UV and visible spectra with values for molar absorption coefficients approaching 2×10^5 or even higher. Because the biological functions often are associated with perturbation in the delocalized system, electronic absorption spectra of tetrapyrroles are of great value in studies of their reactions. The spectra of tetrapyrrolic derivatives are complex and a detailed account is beyond the scope of this chapter. Only the chief features are considered below.

TABLE 2.9 Spectral properties of anthocyanidins in Methanol/HCl

Name	Structure	Colour	λ_{max} (nm)
Pelargonidin	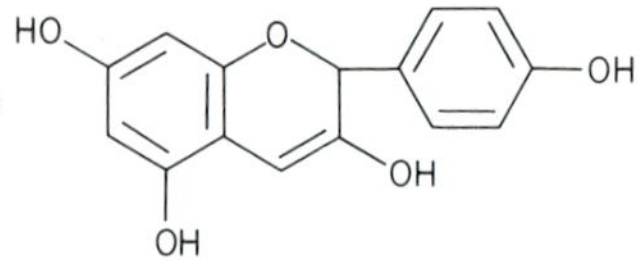	Orange-red	520
Cyanidin	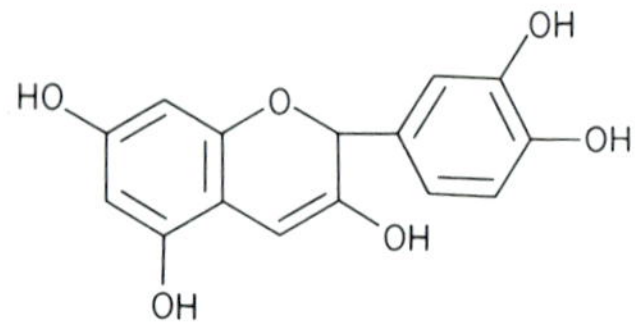	Magenta	535
Delphinidin	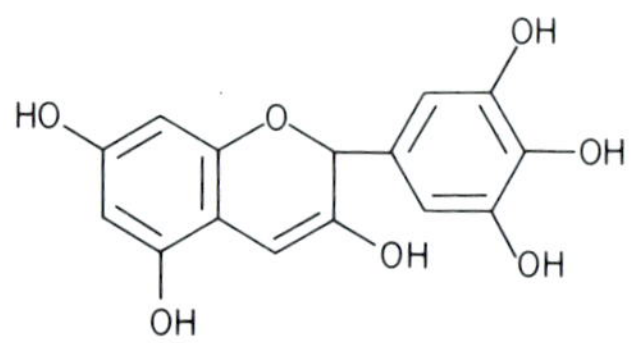	Mauve	545

(Data taken from Harborne, J.B. (1965). *In* "Chemistry and Physiology of Plant Pigments", (Ed. T. W. Goodwin), 252. Academic Press, London and New York.)

Tetrapyrroles may be divided into two classes, the macrocyclic tetrapyrroles e.g. porphyrins, chlorophyll and the linear tetrapyrroles or bile pigments.

The most common naturally occurring porphyrin derivative (excluding chlorophyll) is protohaem (the iron complex of protoporphyrin IX—Fig. 2.10) which is the prosthetic group of haemoglobin, myoglobin, catalase, cytochromes and other proteins. It may be seen that delocalization extends throughout the macrocycle and this gives rise to an intense $\pi \rightarrow \pi^*$ transition at around 400 nm. This band usually called the "Soret band", is found in all porphyrin derivatives including haemoproteins and vitamin B_{12} (a cobalt tetrapyrrole complex). In the porphyrinogens (Fig. 2.10), which are formed during the biosynthesis of porphyrins, the conjugation does not extend throughout the macrocycle. Consequently these compounds lack a Soret band and behave as four isolated pyrrole units. In the case of protoporphyrin, delocalization extends into the side chains, since the vinyl groups are conju-

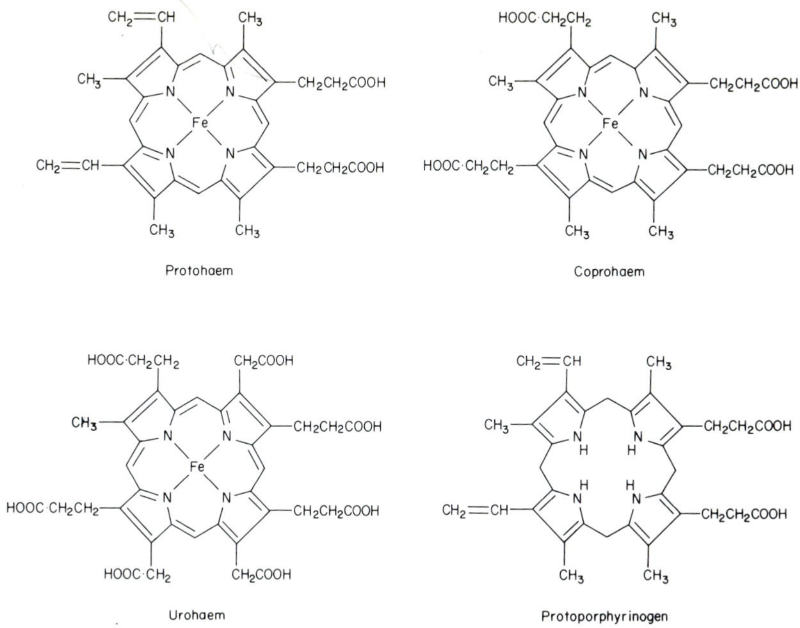

FIG. 2.10 Structures of porphyrin derivatives.

gated with the macrocycle. As might be expected, the Soret band for protohaem occurs at slightly longer wavelengths than that for coprohaem or urohaem, intermediates which are formed during haem biosynthesis and which lack this additional conjugation.

Besides the Soret band, several characteristic longer wavelength bands also occur. These are also sensitive to changes in the environments of the porphyrin and particularly to the presence of a metal atom. The spectrum of protoporphyrin IX in chloroform is shown in Fig. 2.11. The four bands between 450 nm and 650 nm in descending order of intensity are very characteristic of naturally occurring porphyrins. It should be noted that the intensity of these bands is an order of magnitude less than that of the Soret band. The spectra of metal-free porphyrins is extremely pH dependent. Upon insertion of an iron atom, only two bands are observed in the 450–650 nm region. Figure 2.12 shows spectra of various haemoglobin derivatives in the neighbourhood of the Soret band and also at longer wavelengths. It may be seen that the spectrum is very sensitive to changes in the iron-bound ligand. This type of change in the Soret region is extremely useful in studying the structure and function of haems and haemoproteins. Absorbance maxima and absorption coefficients for haemoglobin derivatives are shown in Table

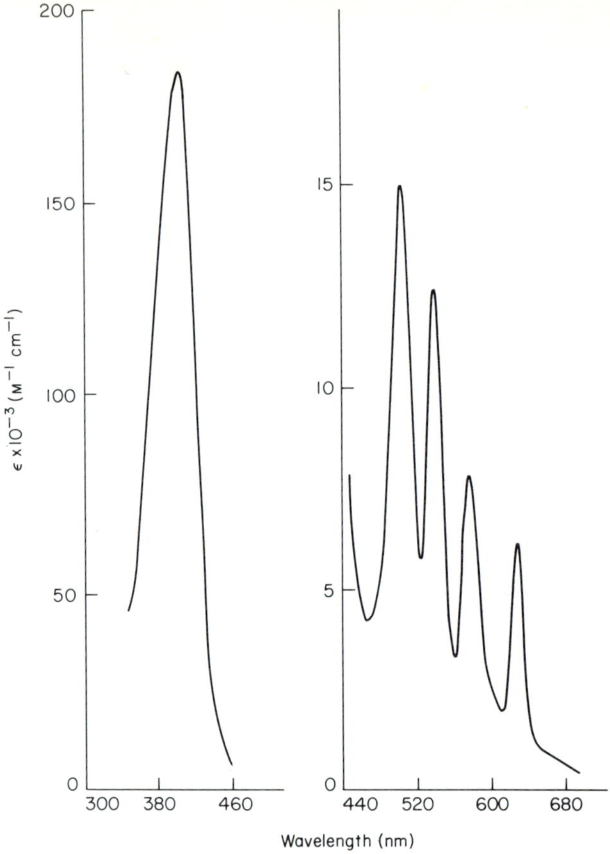

FIG. 2.11 Spectrum of protoporphyrin IX dimethylester in chloroform.

2.10. Because of the stability of the carbonmonoxy form, its spectrum is often used for the determination of haemoglobin.

A further example of the use of the Soret band is shown in Fig. 2.24 in which the association of two haem moieties to form a dimeric molecule results in large changes in the shape and intensity of the absorption band. Another advantage of studies in this region is that, due to the very high intensity of the Soret band, it is possible to work with extremely small amounts of material. For example, an easily measurable absorption could be obtained from a 1·0 μM haem solution in a 10 mm cuvette.

Chlorophylls (Fig. 2.13) may be considered as porphyrin derivatives in which one of the pyrrole rings has been reduced. However, it should be noted that delocalization still continues throughout the macrocyclic system. Chloro-

TABLE 2.10 Spectral data for human haemoglobin derivatives

Derivative	λ_{max} (nm)	ε_{mM} (M^{-1} cm^{-1} $\times 10^{-3}$)[a]
Oxyhaemoglobin	415	125
	542	14·4
	577	15·3
Deoxyhaemoglobin	430	133
	555	13·1
Carbonmonoxyhaemoglobin	419	194
	539	14·2
	569	14·3
Cyanmethaemoglobin	420	116
	543	11·1

[a] Absorption coefficients refer to one haemoglobin subunit.
(Data taken from M.R. Waterman, (1978). *In* "Methods in Enzymology", vol. LII, (Eds. S. Fleischer and L. Packer), p. 460. Academic Press, London and New York.)

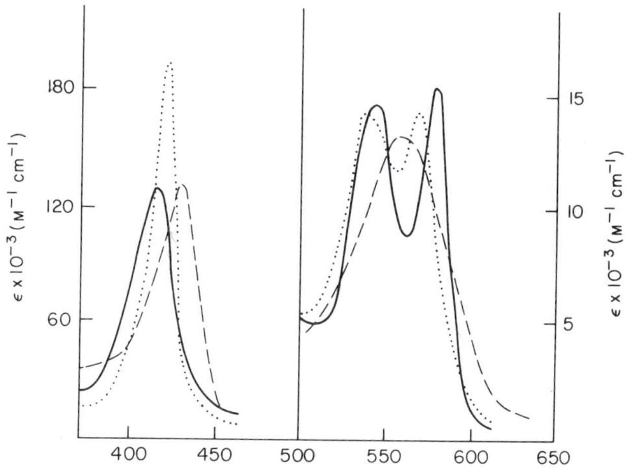

FIG. 2.12 Absorption spectra of haemoglobin derivatives. Oxyhaemoglobin (~), deoxyhaemoglobin (– – –), carbonmonoxyhaemoglobin (:·:.:).
(Spectra taken from M. R. Waterman, (1978). *In* "Methods in Enzymology", Vol. LII, (Eds S. Fleischer and L. Packer), 16. Academic Press, London and New York.)

phylls therefore show a Soret band. In addition, the chlorophylls contain a fused cyclopentanone ring as shown (Fig. 2.13) and a phytyl group which makes little or no contribution to the visible spectrum. The spectrum of chlorophyll *a* is shown in Fig. 2.14. It may be seen that although the reduction

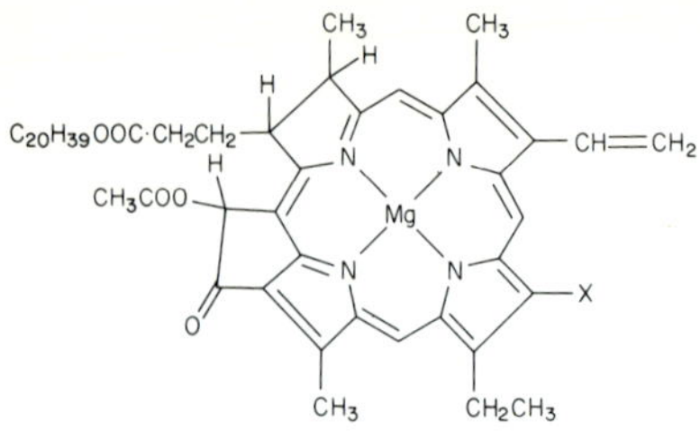

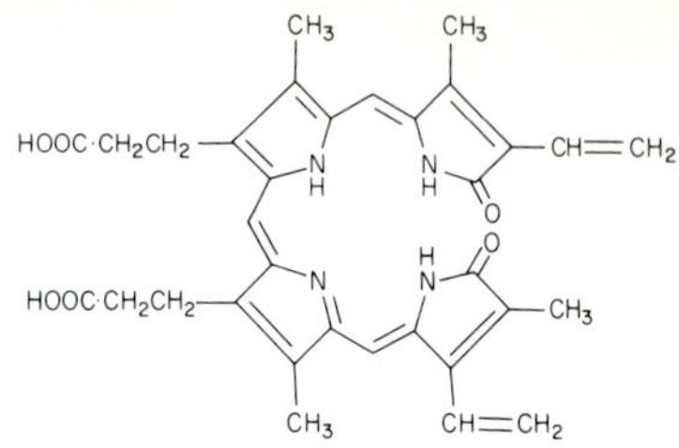

FIG. 2.13 Structures of chlorophylls and biliverdin.

of one of the pyrrole rings has led to a decrease in the Soret band, there is a significant increase in absorption in the region of 650 nm, which is responsible for the green colour of chlorophyll. The spectrum of magnesium-free chlorophyll *a* (known as pheophytin *a*) is similar to that shown in Fig. 2.14, which shows that removal of magnesium does not lead to major changes in the absorption.

The second important class of tetrapyrroles includes the bile pigments or bilins which are typified by biliverdin (Fig. 2.13). Several different subgroups of bilins are known, depending on the state of reduction of the molecule. Although biliverdin is a highly conjugated molecule, it lacks the cyclic nature of the delocalization in porphyrins. The spectra of bile pigments therefore do not show a Soret band, but nevertheless they have relatively intense UV and visible absorption bands. The mammalian bile pigments appear to have no biological function other than that of providing a pathway for the elimination of unwanted haem. However, certain bacteria and algae contain bile pigments linked to proteins (the phycobiliproteins) which act as photosynthetic accessory pigments, replacing the chlorophyll *b* of higher plants.

Metalloproteins

In the haemoproteins, the strong absorption bands are due primarily to the porphyrin rather than the metal. However, there are many other metalloproteins containing transition metals and which have absorption bands in the UV and visible region. In general, these spectra are due to charge transfer transitions involving the metal atom and a neighbouring group (see p. 19) and not to d–d transitions within the metal, which are very much weaker.

This effect is illustrated in Fig. 2.15 for the iron–sulphur protein rubredoxin. In the oxidized form (high spin Fe^{3+}), transitions intrinsic to the metal atom are extremely weak. The strong visible band around 480 nm is due to an $S \rightarrow Fe^{3+}$ charge transfer transition, i.e. the electron moves from the sulphur atom in the ground state to the iron atom in the excited state. On the other

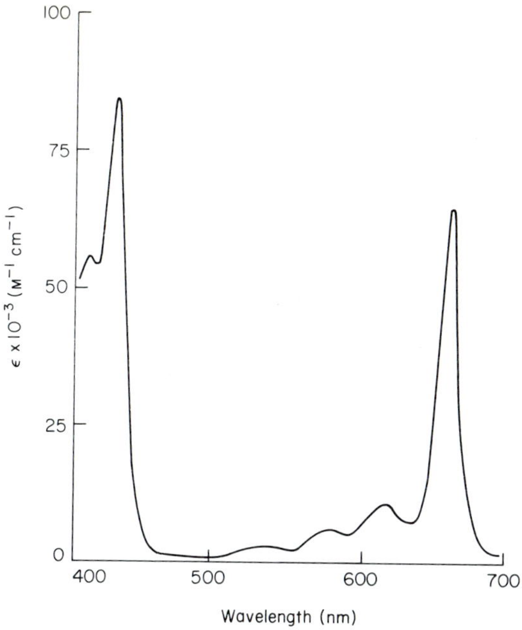

FIG. 2.14 Spectrum of chlorophyll *a* in ether.
(Spectrum taken from A. S. Holt, (1976). *In* "Chemistry and Physiology of Plant Pigments", (Ed. T. W. Goodwin), 14. Academic Press, London and New York.)

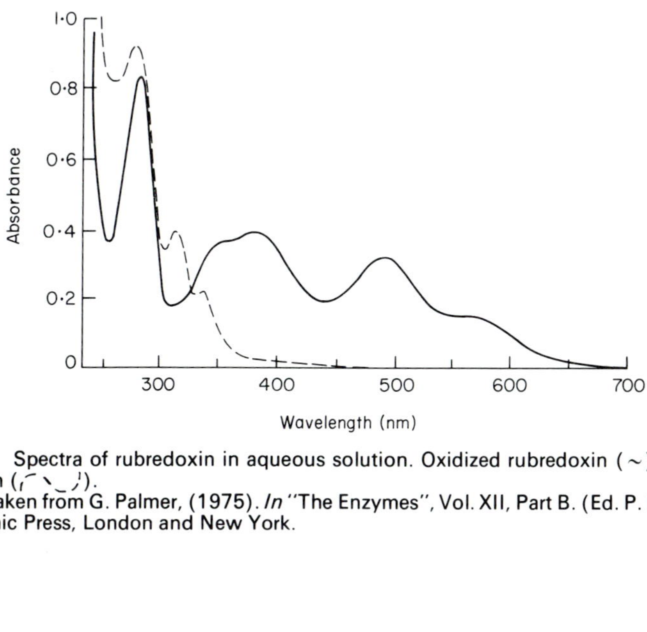

FIG. 2.15 Spectra of rubredoxin in aqueous solution. Oxidized rubredoxin (~), reduced rubredoxin (- - -).
(Spectra taken from G. Palmer, (1975). *In* "The Enzymes", Vol. XII, Part B. (Ed. P. D. Boyer), 7. Academic Press, London and New York.

hand, the reduced form is virtually colourless with no bands above 350 nm. Presumably, by decreasing the positive charge on the iron atom (i.e. reduction) the transfer of an electron from sulphur to iron becomes energetically more difficult and no charge transfer band is seen. Spectra of this type are used a great deal in studying iron–sulphur proteins.

A similar effect is seen in the oxygen transporting protein haemocyanin in which the substrate oxygen molecule binds to copper atoms within the protein. Figure 2.16 shows that the oxy-species exhibits an absorption band at 570 nm giving haemocyanin its characteristic blue colour. This is thought to be due to a charge transfer transition from the copper atom to a histidine residue. On the other hand the deoxy-species is colourless, lacking the 570 nm band.

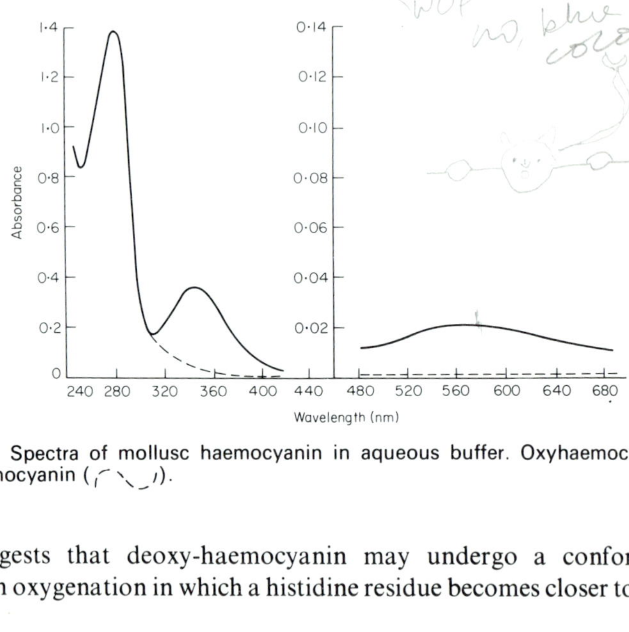

FIG. 2.16 Spectra of mollusc haemocyanin in aqueous buffer. Oxyhaemocyanin (~), deoxyhaemocyanin (---).

This suggests that deoxy-haemocyanin may undergo a conformational change on oxygenation in which a histidine residue becomes closer to a copper atom.

Applications of Spectrophotometry: Determination of Concentration

The familiar spectrophotometric method represents the most widely used technique in biochemistry for determination of concentration. It is based upon eqn (2.1) from which an unknown concentration can readily be determined from an absorbance measurement, provided ε is known.

Routine determination of concentration

For the routine estimation of concentration of a particular species in a series of samples, it is usual to construct a calibration curve at the time the samples are being measured, rather than rely upon literature values for ε. The calibration graph of absorbance versus concentration must embrace the concentration range expected in the samples. Concentrations of test samples may then simply be read off the graph, after measurement of their absorbance values. In this way, any deviations from Beer's law or specific and local effects such as non-linearity of the spectrophotometer output, can be compensated for. Often a species whose concentration is required does not absorb in an accessible region of the UV and visible spectrum. However, it is usually possible to form a derivative of the species, in solution, which does have a convenient absorption. Provided the agent producing the derivative is added in excess, and the reaction is rapid, the concentration of the derivative will equal that of the original species. Two examples of this method will be given.

The determination of acetoacetate and/or acetone is useful in diagnosis of ketotic patients. The total concentration of these metabolites in blood or urine is conveniently determined by a spectrophotometric method. Acetoacetate is converted to acetone catalytically by addition of aniline and the acetone so produced, together with any originally present, is treated with salicylaldehyde in the presence of strong alkali to give a red colour (dihydroxybenzylacetone). A series of standard acetone solutions (in the range 0–8 mg per 100 ml) is taken through the reaction procedure together with a blank. Absorbances at 520 nm are then measured in a 1 cm cell. The blank reading is subtracted from each of the absorbances of the standard solutions and a calibration graph such as that shown in Fig. 2.17 is constructed. The identical procedure is then repeated with samples, after which the acetone concentration can be read off the calibration graph, as shown in Fig. 2.17.

The determination of glucose in blood can be carried out by oxidation with glucose oxidase followed by determination of the resulting peroxide with peroxidase in the presence of a suitable oxygen acceptor, i.e. in this case there are two derivatization steps. *o*-Dianisidine dihydrochloride is used as it is quantitatively oxidized by peroxide to a derivative which absorbs at 436 nm. The reagents are often supplied in a package as a "test-kit". Glucose concentrations can then be readily determined by constructing a calibration graph as described above.

Mixtures of chromophores

The components of a mixture, all of which absorb, can be determined quantitatively, provided they exhibit different absorption coefficients at cer-

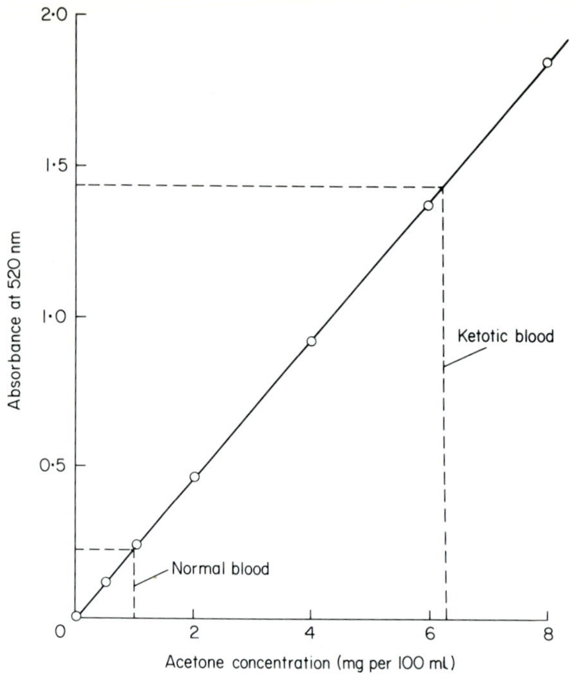

FIG. 2.17 Spectrophotometric determination of acetone concentration. (Data obtained from Pye-Unicam Spectrophotometric Methods, Pye-Unicam Ltd., York St., Cambridge, CB1 2PX.)

tain wavelengths. For a mixture of n species, the concentration of each may be determined if the absorbance of the mixture is measured at n different wavelengths and the absorption coefficient of each species at each wavelength is known. For the simple case of a binary mixture of species **B** and C, we may write,

$$A_1 = \varepsilon_1^B[\mathbf{B}] + \varepsilon_1^C[\mathrm{C}] \qquad (2.2)$$

and

$$A_2 = \varepsilon_2^B[\mathbf{B}] + \varepsilon_2^C[\mathrm{C}] \qquad (2.3)$$

where A_1 and A_2 are the measured absorbance values at two wavelengths 1 and 2 (for unit pathlength) and ε_1^B, ε_2^B and ε_1^C, ε_2^C are the molar absorption coefficients of **B** and C at the appropriate wavelengths. Provided that the appropriate absorption coefficients are known and A_1 and A_2 are measured, these simultaneous equations are readily solved to yield the concentrations of **B** and C. A simple example illustrates the method. A solution containing both NAD^+ and NADH had absorbances in a 1 cm cuvette of 1·560 at 260 nm and 0·498 at 340 nm. The molar absorption coefficients of NAD^+ and NADH are

18 000 and 15 000 respectively at 260 nm and zero and 6220 respectively at 340 nm. Using eqns (2.2) and (2.3) we may write

$$1{\cdot}560 = 15\,000\,[\mathrm{NADH}] + 18\,000[\mathrm{NAD}^+]$$

and

$$0{\cdot}498 = 6220\,[\mathrm{NADH}]$$

from which $[\mathrm{NADH}] = 8 \times 10^{-5}$M and $[\mathrm{NAD}^+] = 2 \times 10^{-5}$M.

In principle the components of more complex mixtures may similarly be determined, although this seldom proves necessary.

Light scattering

As we have seen a spectrophotometer measures the loss in intensity on passage of a beam of light through a sample. In the determination of concentration above, we have assumed that this loss is due entirely to absorption of radiation by the chromophore and that loss by scattering is negligible. However, since light scattering increases with the size of the particles in the sample, it is a factor which should be borne in mind when making measurements on biological samples. In general, for solutions of molecules of low molecular weight, scattering may be ignored. Even for macromolecules such as proteins, scattering may be regarded as negligible for measurements made above 250 nm, provided that clear solutions are used. However, for samples containing large particles such as viruses or even whole cells, light scattering must be taken into account.

Where scattering is significant but not overwhelming, a correction can be made to the measured absorbance and a true absorbance calculated. For example, Fig. 2.18 shows the spectrum of *E. coli* phage T7 which has a molecular weight of 5×10^7. The peak at 260 nm is due primarily to the DNA absorption, but also contains a contribution due to Rayleigh scattering. This contribution can be readily calculated since the magnitude of the scattering varies as λ^{-4}. At wavelengths above about 300 nm the measured "absorbance" is due entirely to scattering. The contribution of scattering at 260 nm can therefore be determined by extrapolation using the λ^{-4} law as shown in Fig. 2.18. The true absorbance due to DNA can then be determined and the DNA content calculated.

There are also situations where light scattering is the predominant factor producing loss of intensity when a light beam is passed through a sample. In such cases, where the solutions appear turbid it is not easy to make meaningful measurements of the absorption spectrum. However, in this situation, measurements of the absorbance are, in effect, measurements of the *turbidity* of the solution, which reflects the number of particles per unit volume scattering the incident light. The measurement of turbidity, which can be carried out in the usual manner on any spectrophotometer is useful in the determination

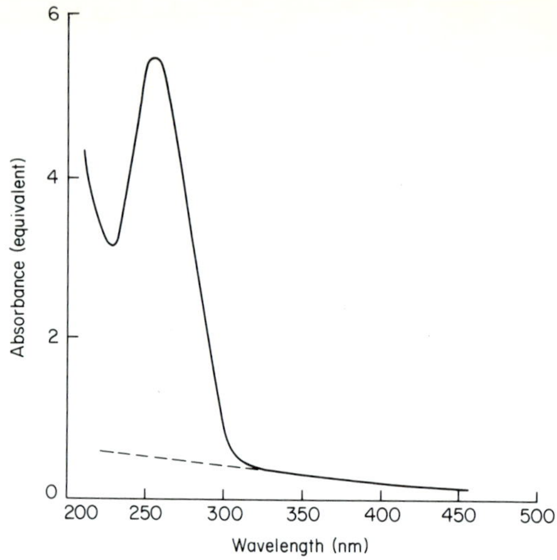

FIG. 2.18 Spectrum of *Escherichia coli* phage T7 showing Rayleigh scattering correction (----).

of the concentration of cells in a bacterial suspension. Although the analogous dependence to Beer's law is observed at low cell concentrations, significant deviations occur at higher concentrations and it is usual to draw a calibration curve. Such a graph is shown in Fig. 2.19 for cell suspensions of *Klebsiella aerogenes*. The cell density can thereby be determined very simply and this is useful in constructing growth curves for bacteria.

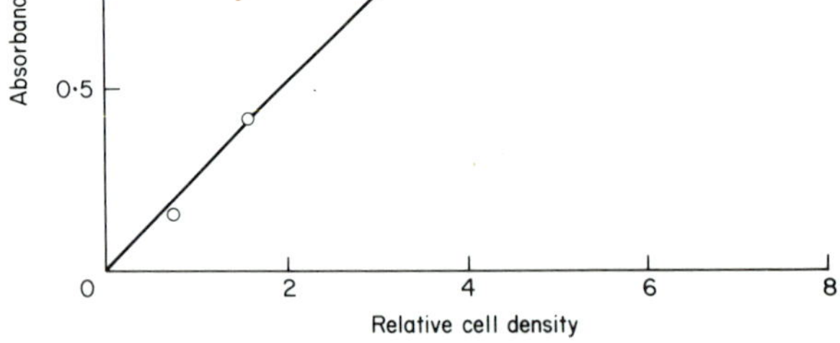

FIG. 2.19 Variation of turbidity with concentration for suspensions of *Klebsiella aerogenes*.

Applications of Spectrophotometry: Enzyme Assays

The great majority of enzyme assays are carried out via a spectrophotometric estimation of change in concentration of either a substrate or a product. Usually the change in absorbance with time at a specific wavelength is followed on a chart recorder. Under the conditions appropriate for measurement of initial rates, the resultant trace is a straight line, the slope of which readily yields the enzyme activity. Although it is sometimes convenient to express rates in terms of absorbance change per unit time, it is better, where possible, to express the rate in the more correct units of concentration per unit time (e.g. M s^{-1}).

There are many experimental variations in the measurement of enzyme activities by the spectrophotometric method. Usually, the assay is actually carried out in the spectrophotometer cuvette. In these circumstances it is essential to have the cuvette initially at the appropriate reaction temperature and then thermostatically maintained at that temperature for the duration of the assay. Large errors can be introduced into enzyme kinetic studies by inadequate temperature control.

Often the reaction of interest is coupled to another enzymic reaction in order to provide a suitable absorption change. In such cases, it is essential that the secondary reaction is fast, so that the overall rate reflects only the rate of the enzyme reaction of interest. This is usually accomplished by ensuring that the auxiliary enzyme is added in excess. An example is the estimation of glutamic-oxalacetate transaminase (GOT), which builds up in the blood following myocardial infarction. The enzyme may be assayed in serum by coupling with the malate dehydrogenase (MDH) reaction and estimating NADH. The system may be represented by the reactions

$$\text{aspartate} + \text{1-oxoglutarate} \underset{}{\overset{\text{GOT}}{\rightleftharpoons}} \text{glutamate} + \text{oxalacetate}$$

$$\text{oxalacetate} + \text{NADH} \overset{\text{MDH}}{\rightleftharpoons} \text{malate} + \text{NAD}^+$$

The decrease in absorbance at 340 nm, is directly proportional to the decrease in NADH concentration which, in turn, reflects the rate of formation of oxalacetate in the GOT reaction, provided that MDH is present in excess.

A further example is the assay of haem oxygenase, a microsomal enzyme responsible for the conversion of haem to the bile pigment biliverdin. The assay is usually carried out in the presence of excess biliverdin reductase which rapidly converts the biliverdin to bilirubin. Haem oxygenase is then determined by measuring the increase in absorbance at about 460 nm where bilirubin has a characteristic absorption maximum.

Applications of Spectrophotometry: Fast Reaction Techniques

The special techniques for following fast reactions are not usually considered in terms of spectrophotometry, but since the spectrophotometric method is almost always used as a method of analysis and in view of the growing importance of these techniques in biochemistry, a brief description is relevant here.

Biochemical relevance of fast reaction techniques

In order to carry out meaningful kinetic measurements, the components of the system to be studied must be effectively mixed. A rule underlying all kinetic studies is that the mixing time should be short compared with the reaction time, i.e. mixing should be effectively "instantaneous". If this was not the case, instead of measuring the rate of the chemical transformation, the measurement would relate to the rate of the mixing process. For conventional kinetic measurements (e.g. the majority of enzyme assays), reaction times are of the order of minutes or longer and this criterion can be fulfilled simply by shaking the components together. The shortest time in which mixing can be achieved using conventional techniques is about 0·5 s, as, for example, when initiating a reaction by delivering the contents of a syringe into a rapidly stirring reaction mixture. The effective lower limit on meaningful measurement of reaction times is therefore about 5 s. Special methods known as *fast reaction techniques* are required to study reactions occurring in a shorter time. Many important biochemical reactions are fast within this definition. Indeed most enzyme-catalysed reactions *in vivo* fall into this category. Of course, enzyme preparations can always be diluted so as to give a reaction rate in a measurable range but such dilutions may reduce the enzyme activity to far below its value *in vivo* with the necessity for extrapolation of any results obtained. Other reactions such as conformational changes, binding of metal ions and other smaller molecules to proteins and binding of hormones to receptors may also often fall into the category of fast reactions. It should also be emphasized that although an overall reaction may be relatively slow, individual steps of the reaction such as protonation or electron transfer may be very fast. Finally, fast reaction techniques have proved valuable in the direct study of transient species, such as enzyme–substrate complexes and in the investigation of pre-steady state kinetics.

Continuous flow technique

In general, finding a technique to follow a reaction sufficiently quickly presents no serious problems, since a spectrophotometer coupled to an oscilloscope can give an accurate response in less than 1 μs. The problem is therefore one of achieving adequate rapid mixing. One solution is provided by the *Flow techniques*, of which there are several variants. The principle is illustrated by the simple *Continuous flow* apparatus shown in Fig. 2.20. Steady streams of solutions A and B are passed along the appropriate tubes into the mixing chamber C. The geometry of the mixing chamber is extremely important in ensuring rapid mixing. The "mixed" solution then passes along tube D during which time reaction takes place.

Observation of the solutions may be carried out at the point E, in effect by having the tube D enclosed in a simple spectrophotometer. Provided the flow rate is constant, for any element of reaction mixture proceeding along D, a fixed amount of reaction will have taken place by the time E is reached, so that a steady reading of absorbance, and hence concentration of reactant or product, should be obtained at E. The reaction time is then readily calculated from a knowledge of the flow rate and geometry of the tube D. Results for different reaction times may be obtained by varying the flow rate and measuring the corresponding absorbance at E. Steady flow may be achieved either by gravity-feeding of tubes A and B from larger reservoirs or alternatively by applying suction along tube D. Using this technique, reaction times down to about 0·05 s may be obtained. The continuous flow method was first used to study the rate of oxygenation of haemoglobin where the availability of large amounts of reactants presents no problem. However, it is of very limited general use in biochemistry, because of the copious amounts of solution required.

Quenched flow technique

A modification of the continuous flow method which overcomes this problem is the *quenched flow method*, illustrated in Fig. 2.20. In this technique, flow is initiated and maintained by two motor-driven syringes containing reactants A and B. As with the continuous flow method, the reactants are mixed in the mixing chamber C and the reacting solution proceeds along the tube D. At the same time a quench solution, Q, is driven by the same motor along tube F to mix with the reacting solution at a second mixing chamber, E. The quench solution stops the reaction at E, which is therefore analogous to the observation point in the continuous flow method. Analysis may then be made at leisure and although this need not be carried out spectrophotometrically, very often this is the method of choice. The flow rate, and hence the reaction time, may be varied by simply altering the gearing on the drive motor. The most appropriate quench solution must, of course, depend upon the reaction being followed, but for enzyme reactions the use of acid or alkali is usually effective.

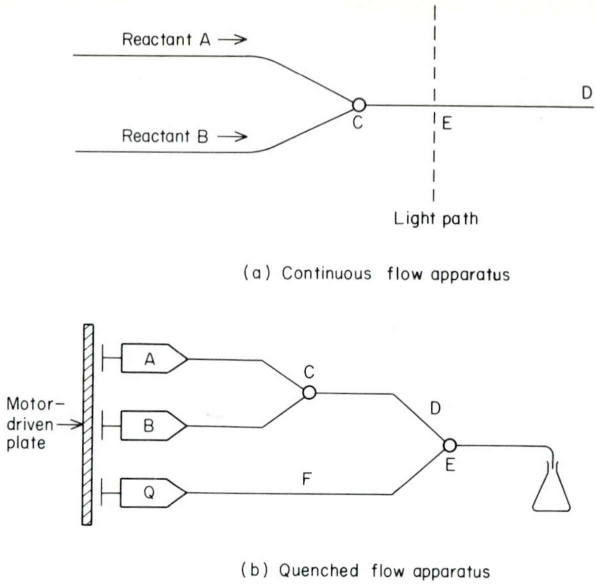

FIG. 2.20 Schematic diagram of (a) continuous flow and (b) quenched flow apparatus.

An example of the application of this technique in a study of xanthine oxidase is considered pp. 307–309.

Stopped flow technique

By far the most useful flow technique in biochemistry is the *stopped flow method.* A schematic representation of a typical stopped flow apparatus is illustrated in Fig. 2.21. The syringes A and B, typically of about 25 ml capacity, are reservoir syringes from which solution can be passed to the smaller drive syringes C and D by operation of valves 1 and 2 respectively. During the filling of the drive syringes, valves 3 and 4 are closed. When solution passes through valves 3 and 4, it also passes through a mixing chamber G into a cell F, which has optical windows and is set into a simple spectrophotometer. Syringe E is a stop syringe whose plunger is fully inserted into the barrel before the measurement begins. Valve 5 is present to enable syringe E to expel spent solution to waste and is always closed during a measurement. Before making a measurement the cell F and tubes leading to the stop syringe and to waste must be full, either with water, or with spent solution from a previous run. The whole apparatus must be scrupulously purged of air bubbles. To initiate a measurement, valves 1 and 2 are closed, valves 3 and 4 are opened, and syringes C and D are then rapidly driven forward, usually by means of a hydraulic ram with a pressure bypass. Solution rapidly flows out of C and D, through valves 3 and 4 into the mixing chamber and out into cell F. Solution previously in F is expelled into syringe E, whose plunger retreats rapidly until the stop is reached. At this stage, there is no further outlet for the solution and newly mixed reactants are thereby "stopped" in the cell F. To avoid shattering the apparatus it is essential to use a bypass on the ram to release the hydraulic pressure. With a suitably designed apparatus the processes from mixing to stopping can take place in about 1 ms. Observation of the change in absorption of the newly mixed solution with time is then made with the spectrophotometer which utilizes an oscilloscope instead of a chart recorder. The oscilloscope may be of the type which can store a transient signal or a synchronized photograph may be taken. It can be calibrated to generate directly a graph of absorbance (or transmission, see p.52) against time. Triggering of the oscilloscope takes place via a microswitch H, which is activated by the stop syringe plunger.

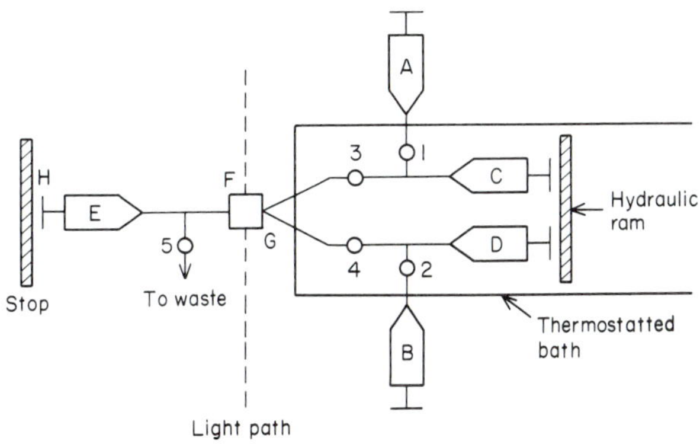

FIG. 2.21 Schematic diagram of stopped-flow apparatus.

Constant temperature is achieved by immersion of the drive syringes in a thermostatted bath. Many variations of the apparatus are possible and, in particular, the various syringes need not be in one plane as in Fig. 2.21.

An example of a typical oscilloscope trace is shown in Fig. 2.22 for the reaction of a haem derivative with hydrogen peroxide, resulting in the degradation of the haem to form a bile pigment. The trace represents a graph of transmission (which can be readily converted to concentration) versus time, from which the rate constant for the reaction can be determined. In this case, there is actually a very fast formation of a peroxide–haem complex before degradation occurs, but this process is too fast to be resolved, even by the stopped flow method. A further example is considered on p. 229.

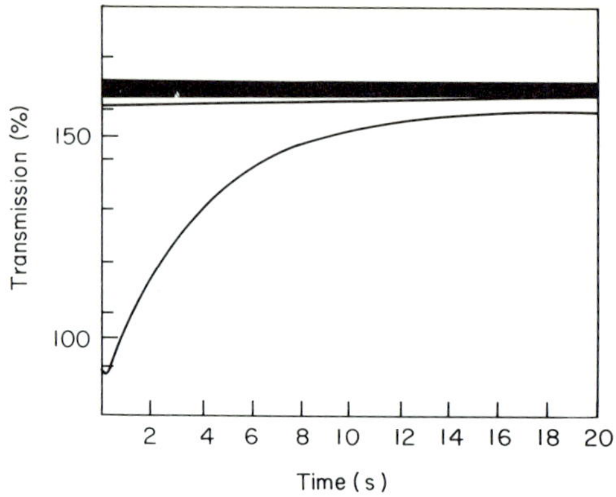

FIG. 2.22 Stopped flow trace for reaction of deuteroferrihaem with H_2O_2. (Taken from P. Jones, K. Prudhoe and T. Robson, (1975). *Biochem. J.* **135**, 361–365.)

Relaxation techniques: the temperature jump method

Even the most efficient rapid-flow mixing device is unable to achieve adequate mixing in less than about 1 ms and many important biochemical reactions therefore remain outside the range of study of the flow techniques. The problem of how to measure the rates of reaction which occur much faster than the time in which reactants can be mixed appears to be an intractable one. Its solution, at least in favourable cases, is an ingenious one, provided by the relaxation techniques, devised by Eigen and others. The method applies only to reversible reactions near equilibrium. At the beginning of the measurement the reactants are already mixed and at equilibrium. The system is then rapidly given a perturbation, the nature of which is such as to change the equilibrium constant. Immediately after the perturbation, therefore, the components are thoroughly mixed together at concentrations appropriate to the old equilibrium rather than the new equilibrium. The reactants then react together until concentrations appropriate to the new equilibrium are reached. With suitable instrumentation, the absorption change in the solution immediately following the perturbation may be monitored, thus providing the basic kinetic requirement, a variation in concentration with time.

Although, in principle, several ways of perturbing the equilibrium are possible, the only method applicable to biochemical systems is rapidly to induce an increase in temperature, i.e. cause the system to undergo a *temperature jump* (T-jump). This is done by means of a high-voltage electrical discharge across the solution. A temperature jump of up to 10°C can be achieved in about 1 μs, which means that meaningful rate measurements may be made down to about 10 μs. Test solutions must be conducting but this criterion is almost always attained in biochemical systems e.g. through buffer ions. It is important to note that the temperature at which reaction takes place is actually the higher temperature following the temperature jump. Thus, to make measurements at 25° with a T-jump of 5°, solutions must be pre-thermostatted at 20°C. A T-jump cell is essentially a silica cuvette (usually 1 cm pathlength) fitted with appropriate electrodes to permit passage of the electrical discharge. Any desired T-jump to the maximum may be obtained simply by varying the discharge voltage. The oscilloscope arrangement is similar to that for the stopped flow method. A few minutes after the T-jump, if the cell has been left in the instrument, its temperature will have fallen to its previous value and another experiment may be carried out immediately. Many repeat measurements may therefore be carried out very easily and quickly. The oscilloscope trace from a T-jump experiment always has an exponential form according to the equation,

$$A = A_0\, e^{-t/\tau}$$

when A_0, A are the absorbances at time zero and time t and τ is a constant, characteristic of the reaction rate, known as the *relaxation time*. For a simple enzyme–substrate binding reaction of the type,

$$E + S \underset{k_2}{\overset{k_1}{\rightleftharpoons}} ES$$

it is readily shown that

$$\tau^{-1} = k_1\,([E]+[S]) + k_2 \tag{2.4}$$

where [E] and [S] are the equilibrium concentrations of E and S respectively. Thus, after carrying out temperature jump measurement at a series of ([E]+[S]) values, various τ values may be determined and a graph of τ^{-1} versus ([E]+[S]) constructed. From the slope and intercept of the straight line obtained, both forward and reverse rate constants and the equilibrium constant (k_1/k_2) may be determined. Figure 2.23 shows results from T-jump experiments for the binding of oxygen to myoglobins and the monomeric haemoglobin from *Chironomus thummi*. The plots according to eqn (2.4) are linear showing that the binding is a simple, one-step process. Data are shown for native sperm whale myoglobin and also a cobalt-substituted sperm whale myoglobin. For the latter species, Fig. 2.23 shows a clearly measurable intercept and slope from which $k_1 = 3 \times 10^7\ \text{M}^{-1}\ \text{s}^{-1}$, $k_2 = 5{\cdot}2 \times 10^3\text{s}^{-1}$ and the stability constant of the haem-O_2 complex $= k_1/k_2 = 5{\cdot}77 \times 10^3\ \text{M}^{-1}$.

Although the lines for native sperm whale myoglobin and for the haemoglobin have an easily measurable slope, the intercepts are too small to permit accurate determination. This is a very common occurrence in T-jump studies. In such cases, k_1 can be determined accurately from the slope. An accurate value for k_2 is then determined by using an independently measured value of the stability constant and the measured k_1 value.

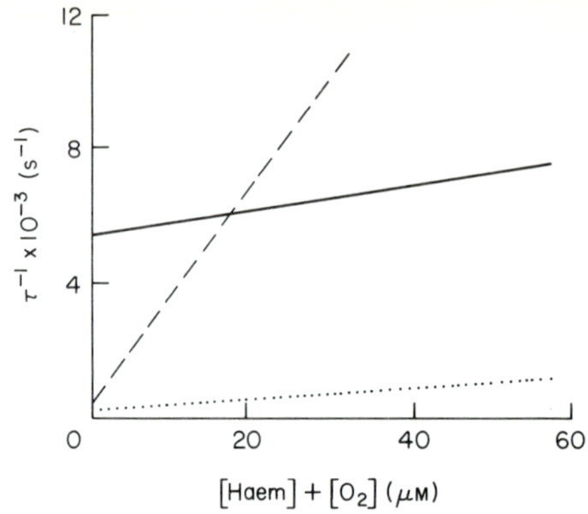

FIG. 2.23 Data from T-jump experiments for oxygen binding to myoglobin and haemoglobin derivatives. Sperm whale myoglobin (·········), cobalt-substituted sperm whale myoglobin (~), *Chironomus thummi* haemoglobin (-----). (Data taken from M. Brunori, (1978). *In* "Methods in Enzymology", Vol. LIV, (Eds S. Fleischer and L. Packer), 73. Academic Press, London and New York.)

Deviations from Beer's Law

There are many systems which apparently do not obey Beer's law. Although this can sometimes cause problems in concentration determination, study of the deviations themselves can often be used to provide valuable structural or other information. The great majority of apparent deviations from Beer's law arise because concentration variation causes changes in the distribution of chromophores in solution. In these cases, if the absorption contribution from each species is considered, Beer's law is obeyed by each species and it is the variation in species which produces the apparent deviation from Beer's law. An example of such a system is shown in Fig. 2.24, which illustrates the spectrum of coproferrihaem in aqueous phosphate buffer over a wide concentration range. The pH and ionic strength are constant and these spectral changes are produced solely by variation in the ferrihaem concentration. Since these spectra (plotted in terms of molar absorption coefficient) are not coincident, it is clear that Beer's law is not obeyed. Figure 2.25(a) shows a typical Beer's law plot for part of these data over a relatively narrow concentration range, and Fig. 2.25(b) shows the variation of ε with concentration plotted logarithmically over a much wider concentration range. (The value of ε would be constant if Beer's law was obeyed.) It may be seen that the usual type of Beer's law plot over a narrow concentration range is not sufficiently sensitive to detect major deviations. This example serves to illustrate the importance of

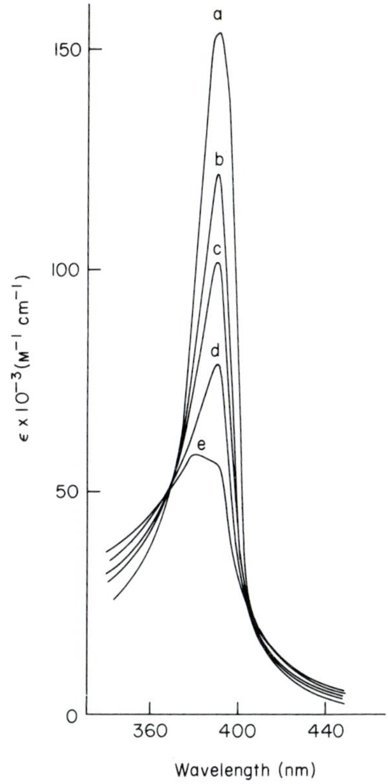

FIG. 2.24 Spectra of coproferrihaem in phosphate buffer (pH 6·97) showing variation due to aggregation. Concentrations were (a) $3{\cdot}97\times10^{-8}$M; (b) $6{\cdot}36\times10^{-6}$M; (c) $2{\cdot}78\times10^{-5}$M; (d) $9{\cdot}93\times10^{-5}$M; (e) $3{\cdot}18\times10^{-4}$M.

carrying out measurements over a sufficiently wide concentration range before assuming that Beer's law is obeyed.

It may be seen in Fig. 2.24 that there are two wavelengths at which the ε values for all the spectra is the same. These points are known as *isosbestic points*. At these wavelengths the ε values of the two species contributing to these spectra are identical and all mixtures must therefore display the same ε value. Since it is very unlikely that a third absorbing species would also have identical ε values at these wavelengths, the presence of isosbestic points is generally taken to be diagnostic for the presence of two absorbing species only. In this case, these and other data have shown that the two species are monomeric and dimeric ferrihaems. The dimers are linked via an oxo-bridge between the iron atoms which leads to distinct spectral differences between monomeric and dimeric species. Spectra such as those in Fig. 2.24 may be

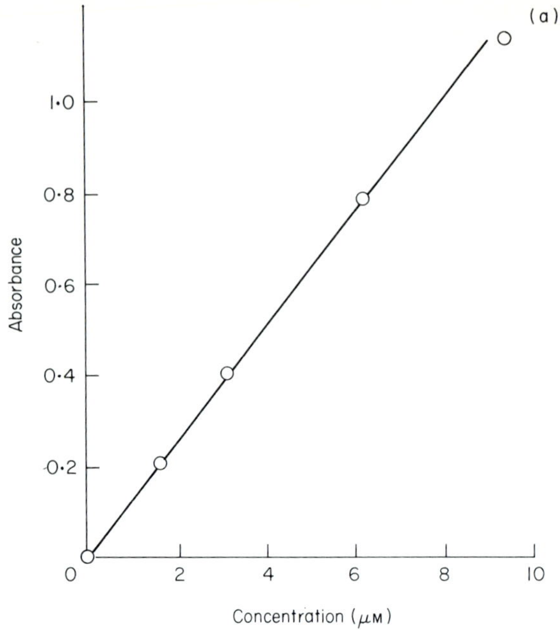

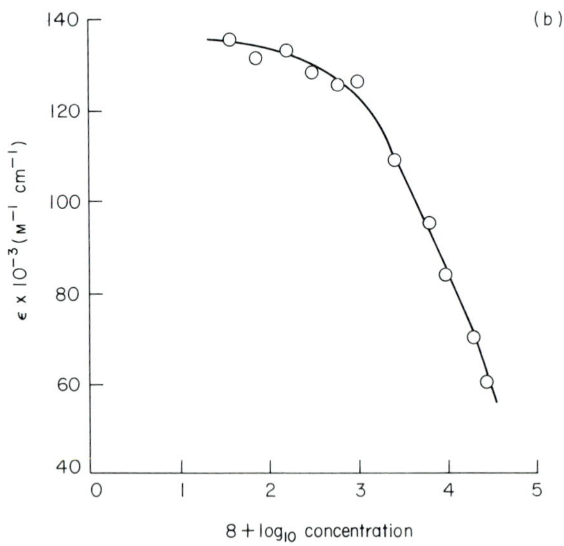

FIG. 2.25 Deviations from Beer's law for coproferrihaem. (a) conventional Beer's law plot for coproferrihaem; (b) logarithmic plot over a wide concentration range.

readily used to determine the equilibrium constant for dimerization and kinetic data for dimerization using the temperature jump method (see p. 52).

Difference Spectroscopy and its Applications

We have seen (p. 26) that the λ_{max} for a given chromophore always occurs in the same region of the spectrum but that there is often a slight dependence upon the chromophore environment. In some ways this is a nuisance, since the details of the solvent, pH, etc. must always be reported with the spectral data. However, for macromolecules such as proteins, the slight dependence of spectra upon environment offers the possibility of determining structural information regarding the conformation of the macromolecule.

Because the changes produced by different environments are relatively small, it is more accurate and convenient to measure directly, the difference between the chromophore in the two environments rather than make an absolute measurement of each one separately. This can be achieved very readily in a double-beam spectrophotometer by placing one of the solutions in the reference compartment and the other in the sample compartment. The spectrophotometer will then respond to the difference in absorption and the measurement of the variation of this difference with wavelength produces a *difference spectrum*. It is possible to define a *difference molar absorption coefficient* ($\Delta\varepsilon$) at any wavelength by the relation

$$\Delta A = \Delta\varepsilon \, c \, l \tag{2.5}$$

where ΔA is the difference in absorption between the chromophore in the two environments. It follows that $\Delta\varepsilon$ can have both positive and negative values (see Fig. 2.26).

It is possible to change the environment in a number of ways, but change in solvent composition and pH have given the most information. A great deal of work has been done, particularly on proteins, but it is only possible to select a few examples in this brief discussion.

Solvent perturbation

One way in which the environment of a chromophore can be varied significantly is to alter the solvent composition. From the point of view of the protein the solvent is represented by all of the non-protein molecules it experiences around it. It would not be sensible to change the solvent completely from water to a non-aqueous system since denaturation would result. However, significant solvent effects can be induced by use of mixtures of water

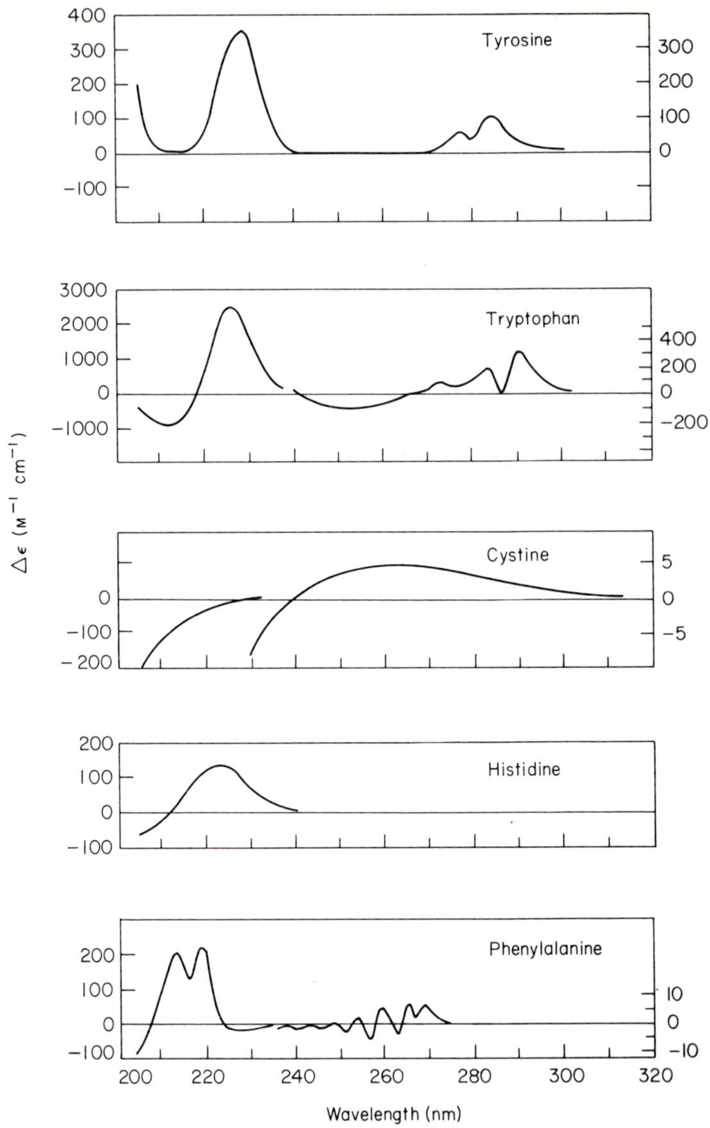

FIG. 2.26 Solvent difference spectra of amino acids produced by 20% (v/v) ethylene glycol/water. (Spectra taken from J. W. Donovan, (1969). *J. Biol. Chem.* **244**, 1961)

with polyhydroxy compounds such as ethylene glycol or glycerol without loss of the native protein structure. Difference spectra for solvent perturbation of the chromophores commonly found in proteins are shown in Fig 2.26, which shows that the difference molar absorption coefficient for tryptophan in the ethylene glycol system is considerably larger than that for other amino acids.

Perhaps the most important application of the solvent perturbation technique is in calculating the "degree of exposure" of certain chromophores. This is based on the premise that residues on the exterior of the protein will be exposed to solvent and will contribute to a difference spectrum. Conversely residues buried in the interior of the protein can make no contribution. In this respect, UV spectroscopy can give valuable structural information. Two examples are now considered.

Solvent perturbation difference spectra of rabbit muscle aldolase are shown in Fig. 2.27 at pH 6 and pH 1. Inspection of Fig. 2.27 suggests that these difference spectra are due primarily to tyrosine and tryptophan. The difference absorption coefficients around 230 nm are clearly greater at pH 1 than at pH 6. This is consistent with the idea that in the native conformation (pH 6) some residues are not exposed to solvent, whereas at pH 1, denaturation has led to considerably more exposure. It is possible to quantify these differences in exposure by a curve-fit procedure using the known difference spectra of tryptophan and tyrosine (Fig. 2.26). The best fit for the difference spectra in Fig. 2.27 is obtained by assuming that 18 tyrosine and two tryptophan residues are exposed at pH 6 and 46 tyrosine and 12 tryptophan residues are exposed at pH 1. This suggests that, in native rabbit muscle aldolase, 28 tyrosine and 10 tryptophan residues are buried in the interior of the protein. It is, of course, essential to show that the solvent perturbation itself does not alter the protein conformation. This is usually done by checking the linearity of the dependence of the perturbation ($\Delta\varepsilon$) with increasing concentration of the perturbant.

One of the very first studies using the solvent perturbation technique was carried out by Herskovits and Laskowski on ribonuclease. Measurements were made at about 287 nm where the tyrosine difference spectrum shows a peak but the tryptophan difference absorption coefficient is zero (Fig. 2.26). Thus measurements made at this wavelength reflect only changes in the tyrosine residues, of which ribonuclease contains six. Some of the results of this work are shown in Fig. 2.28 for two perturbants, glycerol and sucrose. The results are shown in terms of the ratio of $\Delta\varepsilon$ at various pH values to $\Delta\varepsilon$ for the native protein at pH 6. By using such ratios, results using different perturbants can be directly compared. Figure 2.28 shows that $\Delta\varepsilon_{obs}/\Delta\varepsilon_{pH5-6}$ increases as the pH decreases, due to loss of the tertiary structure as the protein denatures, and the corresponding exposure of additional tyrosine groups. It is clear that the value of $\Delta\varepsilon_{obs}/\Delta\varepsilon_{pH6}$ approaches 2 as the pH

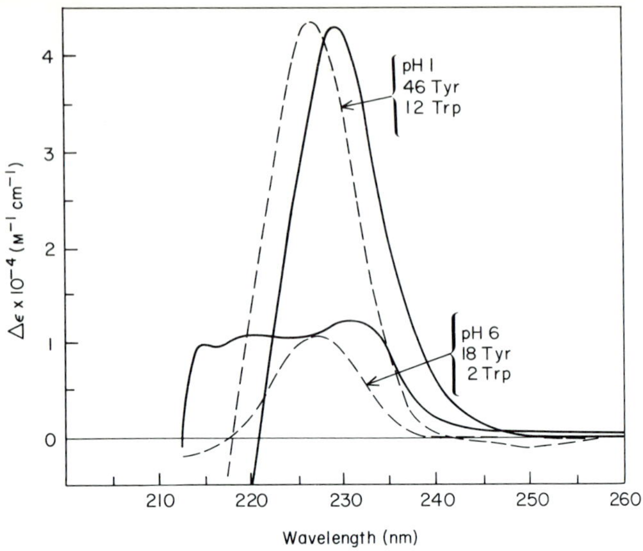

FIG. 2.27 Solvent perturbation spectra of rabbit muscle aldolase. The observed difference spectra (~) are compared with calculated perturbation difference spectra (- - -) for 46 mol of tyrosine plus 12 mol of tryptophan (at pH 1) and for 18 mol of tyrosine plus 2 mol of tryptophan (at pH 6). (Data taken from J. W. Donovan, (1969). *J. Biol. Chem.* **244,** 1961)

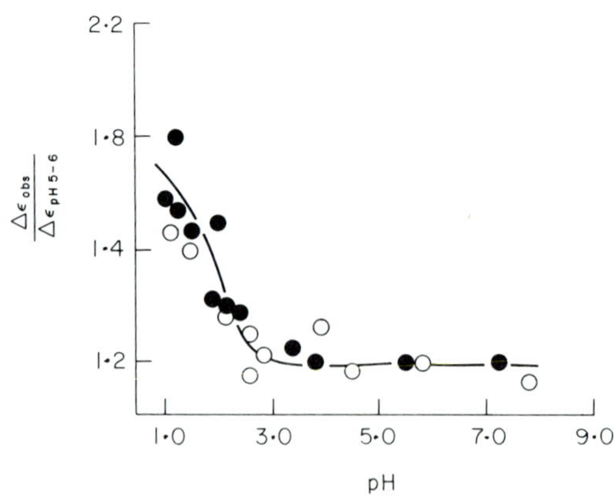

FIG. 2.28 Solvent perturbation spectral data for ribonuclease produced by 20% sucrose (●) and 20% ethylene glycol (○). (Data taken from O. B. Wetlaufer, (1962). *In* "Advances in Protein Chemistry", Vol. 17, (Eds C. B. Anfinsen, M. L. Anson, K. Bailey and J. T. Edsall), 304–405. Academic Press, London and New York.)

decreases. Since ribonuclease contains six tyrosine residues, this suggests that, in the native protein, three are buried and three on the surface of the protein. These results are in good agreement with an earlier study using the spectrophotometric titration method (see pp. 61–65) which showed that only three of the six tyrosine residues of ribonuclease could be reversibly titrated. These conclusions have since been confirmed by the X-ray structure.

Thermal perturbation

In this brief account, it has not been possible to consider the effect of temperature upon an absorption band. For a molecule such as a protein, there are clearly several factors which contribute to the absorption and which are temperature dependent. These include intrinsic solvent effects, change in solvent concentration due to expansion of the solvent, effects on ionizable groups and changes in conformation. Although thermal perturbation spectra are easily measured, the precise explanation of the difference spectrum may be difficult. However, the technique may sometimes be used in an empirical fashion for following denaturation.

pH Perturbation

A great deal of information can be obtained by direct measurement of difference spectra of proteins at different pH values. An example of this approach is considered on p. 63.

Spectrophotometric Titrations

Principle

An acid–base titration involves addition of, for example, alkali to a weak acid solution until neutralization is complete. By measurement of the increasing volume of alkali added and the incremental pH changes, the concentration of the weak acid and its pK value can be determined. Similar measurements can often be carried out by following the change in absorbance with pH as alkali (or acid) is added. This can only be done, of course, when dissociation of an acid group involves a spectroscopic change in the chromophore. Such measurements are called *spectrophotometric titrations*. For proteins, this usually amounts to a titration of the phenolic groups of tyrosine, by measurement of the change in absorption at 295 nm (see p. 29 and Fig. 2.4), or sometimes observation of histidine dissociation by measurement at 210–230 nm. However, in principle it may also be possible to obtain information from dissociation of acid–base functions in prosthetic groups of proteins e.g. haemoproteins, or by the use of extrinsic probes (chromophores which bind at specific positions in a protein and effectively "report back" on the environment they find, see Chapters 3, 4, 6, 7). Since changes in absorption of a complete protein produced by pH changes are often relatively small, a reference solution of similar absorption properties is generally used and the measured absorption changes then produce a difference spectrum (see p. 57).

It must be borne in mind that a protein is a complex assembly containing many acid–base functions and the p*K* values of amino acid residues in proteins are rarely identical to those of the free amino acid. A spectrophotometric titration can therefore hardly ever be regarded in terms of the dissociation of isolated groups in a protein. Often chromophores (e.g. tyrosine) have their spectra altered indirectly by the dissociation of other neighbouring, but non-absorbing, groups. In such cases, the non-absorbing groups will also be titrated. For this reason, in a spectrophotometric titration, the actual consumption of acid or alkali is not usually recorded and the situation is simplified by measurement only of pH and absorbance. In this way the p*K* values of chromophores only are determined. This selectivity can be used in conjunction with potentiometric titration, to distinguish between the acid–base behaviour of groups which ionize in the same p*K* range, for example the ε-amino group of lysine and the phenolic group of tyrosine.

From these considerations, it may be seen that the chief uses of spectrophotometric titrations of proteins are in determining the p*K* of ionizable chromophore groups and in obtaining conformational information about the protein. An example of each will now be considered.

Determination of p*K*

Before determination of p*K* values, it is necessary to check that the ionization is reversible. This may be carried out by measurement of the titration curve by gradual addition of alkali and then repeating this measurement after restoration of the original pH by a single addition of acid. However, a better procedure is to carry out a reverse titration by incremental addition of acid, looking for superimposition of the two curves. Lack of a "hysteresis loop" indicates true reversibility.

Under reversible conditions, from the Henderson–Hasselbalch equation

$$pK' = pH - \log\left(\frac{\alpha}{1-\alpha}\right) \tag{2.6}$$

where p*K*′ is the apparent p*K* value and α is the degree of dissociation. From eqn (2.6), it may be seen that the p*K*′ value is equal to the pH when the acid is half dissociated and this is usually the simplest way to estimate p*K*′. This is illustrated in Fig. 2.29 which shows schematically, a typical spectrophotometric titration curve for a protein and the determination of pK'.

The value thus obtained is not the thermodynamic or *intrinsic* pK because of the effect of the total charge on the protein. Using a simple model, eqn (2.6) can be modified, to take account of this and becomes

$$\mathrm{p}K = \mathrm{pH} - \log\left(\frac{\alpha}{1-\alpha}\right) = \mathrm{p}K_{\mathrm{int}} - kZ \tag{2.7}$$

where pK_{int} is the intrinsic pK, Z is the charge on the protein and k is a constant. Figure 2.30 shows data obtained from spectrophotometric titrations for fibrinogen. The intrinsic pK value is the intercept on the ordinate.

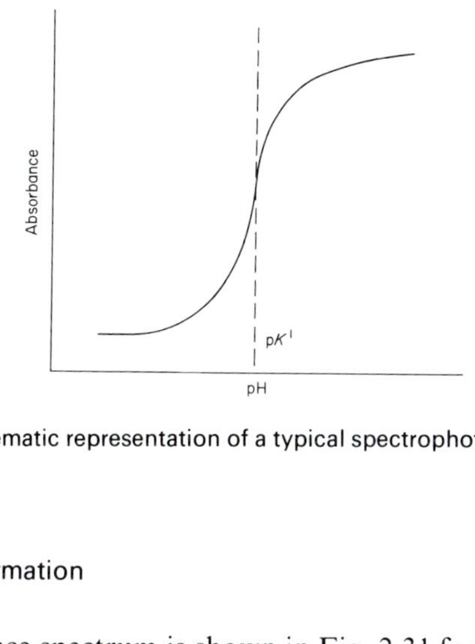

FIG. 2.29 Schematic representation of a typical spectrophotometric titration.

Conformational information

A typical difference spectrum is shown in Fig. 2.31 for the dimeric protein ovomacroglobulin (molec. wt $6{\cdot}6 \times 10^5$). Information regarding conformation may be obtained in the following ways.

1. The sign of the difference spectrum is positive. This usually signifies that a larger number of chromophores are exposed to solvent in the reference than in the sample, i.e. that the chromophores on ovomacroglobulin become more exposed to solvent as the pH is decreased from 6·5.

2. The types of chromophore involved and the number of them may be determined. The perturbation of phenolic chromophores produces difference spectrum peaks at 280 and 288 nm ($\Delta\varepsilon^{288} = 700$) and perturbation of the tryptophan chromophores produces a difference spectrum peak at 295 nm

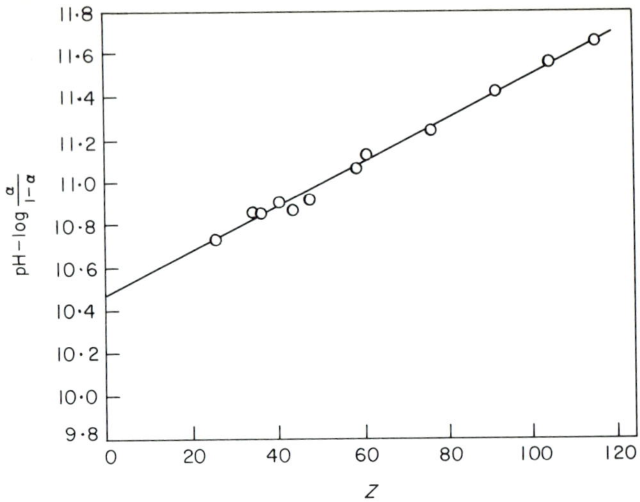

FIG. 2.30 Data from spectrophotometric titrations for fibrinogen, obtained at 25°C in 0·25 M KCl. The value of Z was calculated from the potentiometric titration curve for a molecular weight of 10^5. (Data taken from E. Mihalyi. (1968). *Biochemistry* **7**, 208)

($\Delta\varepsilon^{295} = 1600$). Bearing in mind the relative values of the molar difference absorption coefficients, it is immediately clear from Fig. 2.31 that there is considerably more perturbation of the phenolic chromophores than the tryptophan chromophores. An approximate idea of the number of tyrosine and tryptophan residues involved can be determined by dividing the height of the 288 nm difference peak in Fig. 2.31 by 700 and correspondingly, the height of the shoulder at 295 by 1600, which yields the result: 26 tyrosine residues and four tryptophan residues. These approximate values can then be used as starting values in a curve-fit exercise in an attempt to generate the observed spectrum in Fig. 2.31. Subsequent manipulation shows that the best fit is obtained by assuming that 24 tyrosine residues and two tryptophan residues are perturbed. This combination generates the broken line in Fig. 2.31, which is in good agreement with the observed spectra. (The wavelength displacement occurs because the curve fit uses data for free amino acids rather than peptide residues.)

3. Confirmation of this interpretation comes from complementary solvent perturbation experiments with ethylene glycol. These indicated exposure of 32 tyrosine residues and eight tryptophan residues per subunit at neutral pH, but 56 tyrosine residues and 10 tryptophan residues per subunit at acid pH. Solvent perturbation therefore suggests the exposure of 24 tyrosine residues

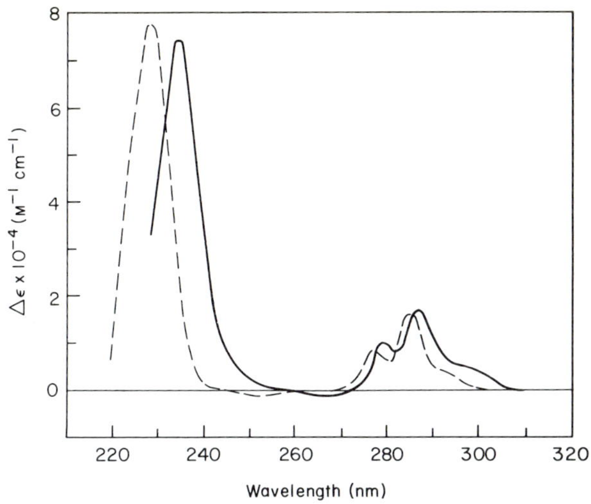

FIG. 2.31 Acid difference spectrum of ovomacroglobulin expressed in terms of molar difference absorption coefficient of the subunit. Solid line:absorption difference observed when a pH 6.5 solution is measured against a pH 2.8 solution as reference. Dashed line: difference spectrum calculated for exposure of 24 phenolic and 2 indole chromophores per subunit. The subunit contains approximately 100 phenolic and 30 indole chromophores, of which 32 and 8 respectively, are exposed to solvent at pH 6.5. (Data from J. W. Donovan, C. J. Mapes, J. G. Davis and R. D. Hamburg, (1969). *Biochemistry* **8,** 4190.)

and two tryptophan residues when the pH is decreased from neutrality, in complete agreement with the acid difference spectrum.

Nucleic Acids

As we have seen (p. 32), the absorbance of a native DNA molecule is significantly less than the additive absorbance of its component nucleotides. This *hypochromism* is due to the interactions between bases in the native structure. The spectrum of native DNA therefore reflects its conformation and it is potentially possible to obtain structural information from spectroscopic studies. Nucleic acids can be denatured in various ways such as exposure to heat and extremes of pH. During the denaturation process, the double helix of the native DNA unwinds and the two strands separate as random coils. This process and its converse (renaturation) can be followed by study of the absorbance change at 260 nm.

The hyperchromic effect

Figure 2.32 shows spectra of *Escherichia coli* DNA at different tempera-

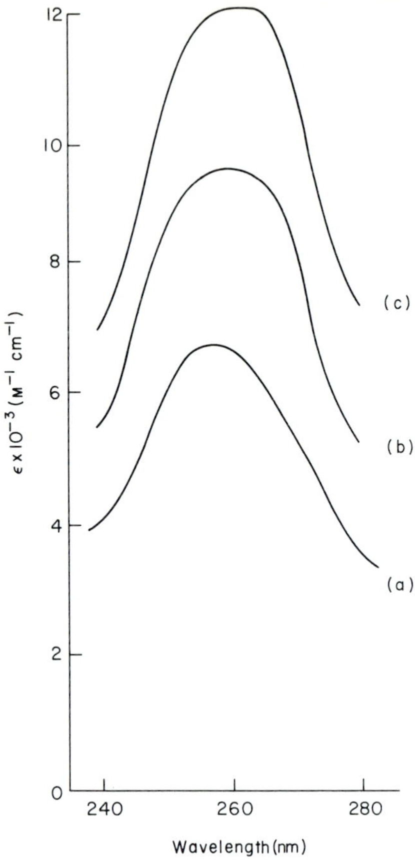

FIG. 2.32 Hyperchromism of DNA from *E. coli* in 51% methanol. (a) Spectrum obtained at room temperature; (b) spectrum obtained at 95°C; (c) reconstructed spectrum of the component deoxyribonucleotides of *E. coli*. (Data taken from H. R. Mahler, B. Kline and B. D. Mehrotra, (1964). *J. Molec. Biol.* **9**, 801.)

tures. In Fig. 2.32(a), the DNA is present in the native, double-helical form, whilst Fig. 2.32(b) shows the same sample heated to about 95°C. At this temperature, the DNA exists almost entirely in single stranded coils. The increase in absorption or *hyperchromism* accompanying this denaturation is evident. Figure 2.32(c) shows a reconstructed spectrum of the component deoxyribonucleotides of *E. coli*. Clearly a further hyperchromic effect occurs upon hydrolysis of denatured DNA to its component nucleotides.

The hyperchromic effect is itself related to the number of A–T (adenine + thymine) pairs in a DNA molecule. This is shown in Fig. 2.33 in which the increase in absorption observed on heating various DNA samples to 10°C above the melting temperature (see below) is plotted as a function of the percent of A–T pairs. An approximate estimate of the A–T content of a DNA sample can therefore be readily obtained in this way.

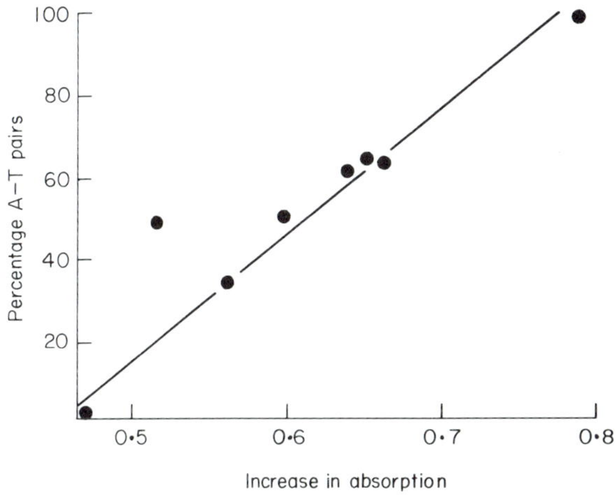

FIG. 2.33 Correlation of DNA hyperchromism with number of A–T pairs. (Data taken from H. R. Mahler, B. Kline and B. D. Mehrotra, (1964). *J. Molec. Biol.* **9**, 801.)

Melting of DNA

Measurement of the extent of the hyperchromic effect allows the possibility of following the progress of denaturation as a function of temperature. Typical graphs obtained in this way are shown in Fig. 2.34. Because the observed transitions are so sharp, this phenomenon has been termed the *melting* of DNA. The melting point, T_m, is defined as the temperature at the midpoint of the transition. The observed value of T_m is dependent upon the base composition of the DNA. Since G–C pairs in the double helix have more hydrogen bonding stability than A–T pairs, the melting point increases with G–C content. In practice, the variation is almost linear as shown in Fig. 2.35 and reasonably accurate estimates of base composition may be made from melting point measurements.

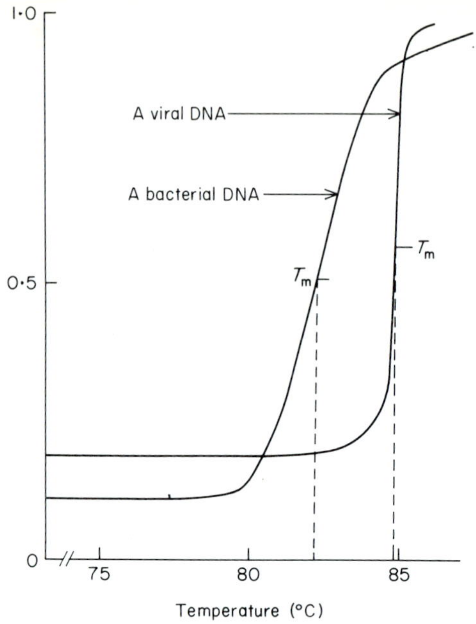

FIG. 2.34 Melting of DNA, Characteristic melting curves for a viral DNA and a bacterial DNA. The former, being smaller, shows a sharper transition. (Taken from A. Lehninger, (1975). "Biochemistry", 875. Worth Publishers, New York.)

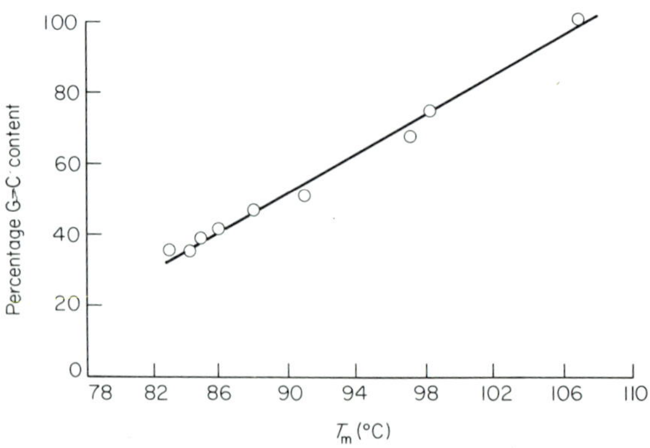

FIG. 2.35 Variation of T_m with G–C content for various species of DNA. (Data taken from H. R. Mahler, B. Kline and B. D. Mehrotra. (1964). *J. Molec. Biol.* **9,** 801.)

Bibliography

There are very many books and review articles, relevant to this chapter. The following list is a small selection from these.

Books

Hague, D. N. (1971). "Fast Reactions" Wiley-Interscience, New York.

Rao, C. N. R. (1974). "Ultraviolet and Visible Spectroscopy", 3rd edn, Butterworth, London.

Sten, E. S. and Timmons C. J. (1970). "Electronic Absorption Spectroscopy in Organic Chemistry", 3rd edn. Edward Arnold, London.

Chapters and reviews

Beavan, G. H., Holiday E. R. and Johnson E. A. (1955). *In* "The Nucleic Acids", Vol I. (Eds E. Chargaff and J. N. Davidson), 493–545. Academic Press, London and New York. (Useful article on nucleic acid spectra.)

Donovan, J. W. (1973). *In* "Methods in Enzymology", Vol XXVII. (Eds C. H. W. Hirs and S. N. Timasheff), 497–547. Academic Press, London and New York. (Two detailed reviews on the difference spectroscopy of proteins and spectrophotometric titrations of proteins.)

Kirtley M. E. and Koshland D. E. Jr. (1972). *In* "Methods in Enzymology", Vol XXVI. (Eds C. H. W. Hirs and S. N. Timasheff), 578–601. Academic Press, London and New York. (A discussion of the use of environmentally sensitive groups attached to proteins.)

Wetlaufer, O. B. (1962). *In* "Advances in Protein Chemistry", Vol. 17. (Eds C. B. Anfinsen, M. L. Anson, K. Bailey and J. T. Edsall), 304–405. Academic Press, London and New York. (An early, but excellent review of the ultraviolet spectroscopy of proteins.)

3 Molecular Emission Spectroscopy (Fluorescence and Phosphorescence)

G. R. Penzer

Introduction

Fluorescence is emitted light. It is generated when a molecule relaxes from one electronic state to another of lower energy. It is distinguished from other types of luminescence (e.g. bio-, thermo-, chemi-, tribo-) by the way the initial excited state is generated. In fluorescence this process is photochemical. Excited states are generated by absorption of photons—exactly the same as the process which is studied by absorption spectroscopy. A natural question, therefore, is "what information do fluorescence spectra provide which cannot be discovered from absorption spectra?" The detailed answer to this question should become clear in the rest of this chapter, but it helps to indicate at the outset why absorption and emission measurements can be expected to signal different messages.

Absorption of a photon by a molecule is a single almost instantaneous interaction. Fluorescence, however, involves two interactions and two photons, absorption and subsequent emission. Each process is practically instantaneous, but there is a time lag (typically $\sim 10^{-8}$ s) between them during which the molecule exists in an electronically excited state. The duration of this lag is variable because it depends on non-radiative processes which deactivate the excited state in competition with its deactivation by fluorescence. The efficiencies of these non-radiative quenching mechanisms depend on the environment of the molecule and hence so does the intensity of emission. Thus the fluorescence of an emitter is a probe (an indicator) of its immediate surroundings in a way that an absorption spectrum is not.

These properties are illustrated in Table 3.1 for riboflavin (vitamin B_2), a typical aromatic fluorophore, and its derivatives FMN and FAD. The molar absorption coefficient is much less solvent dependent than the fluorescence quantum yield (an approximate analogue of absorbance, defined as the ratio of the number of photons emitted to the number of photons absorbed).

It is this *environmental sensitivity*, coupled with *high absolute sensitivity* (you do not need many molecules to generate an observable emission) and the fact that *fluorescence can be observed from all kinds of sample* (solid and liquid, clear and opaque) that have chiefly been exploited by biochemists during the explosion of interest in fluorescence spectroscopy since 1965.

This chapter is broadly divided into four sections. The first part (pp. 72–93)

TABLE 3.1 Environmental sensitivity of the flavin spectrum in aqueous solution (pH 7)[a]

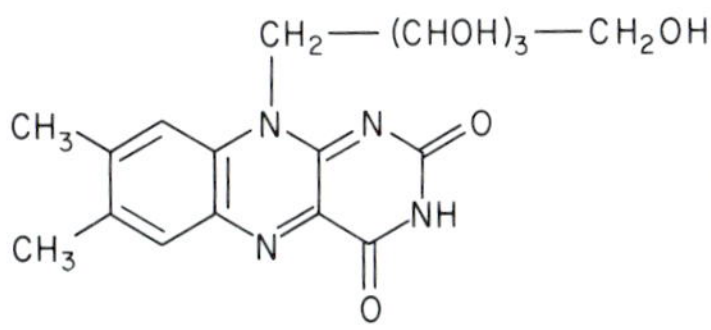

Riboflavin

Environment	Absorption maximum (nm)	Molar absorption coefficient (M^{-1} cm^{-1})	Fluorescence maximum (nm)	Quantum yield of fluorescence
Riboflavin	445	12 000	525	0·25
FMN	445	12 000	525	0·25
FAD	445	11 000	525	0·05
FAD in lipoic acid dehydrogenase	455	11 000	520	0·11
FAD in thioredoxin reductase	455	11 000	520	0·04
FAD in L-amino acid oxidase	460	11 000	not observable	
FMN in old yellow enzyme	465	11 000	530	0·01
FMN in lactate oxidase	460	11 000	not observable	

[a] Spectral maxima and molar absorption coefficients do not vary widely in non-aqueous solvents (e.g. the absorption maximum in dioxan is unchanged while emission is at 520 nm) but quantum yield is more sensitive (e.g. riboflavin in dioxan—0·71; FMN in 50% aqueous DMSO—0·20; FAD in 50% aqueous DMSO—0·12).

contains a discussion of matters of principle, i.e. the nature of fluorescence and its molecular basis. This is followed (pp. 94–102) by a consideration of the practical ways to measure fluorescence and an appraisal of the strengths and weaknesses of fluorescence methods as ways to study biochemical problems (pp. 102–107). Finally, some promising strategies are considered in more detail (pp. 107–113).

Fluorophores

Not all molecules display easily observed fluorescence. Those which do are called fluorophores. Typically such molecules possess low-lying excited states (such that excitation is possible with radiation wavelengths greater than 220 nm, i.e. energies less than 550 kJ mol^{-1}). Molecules which only absorb at shorter wavelengths may easily disintegrate on excitation because of the high energy of the photon involved. Indeed, photochemical reactivity at wavelengths *greater* than 220 nm is commonplace and can cause major problems for the fluorescence spectroscopist.

As we have seen in Chapter 2, the organic molecules which absorb with wavelengths greater than 220 nm are chiefly those with extended. π-electron orbitals—aromatic compounds like riboflavin and the amino acid tryptophan, or molecules with conjugated double bonds like vitamin A (retinol, (I)).

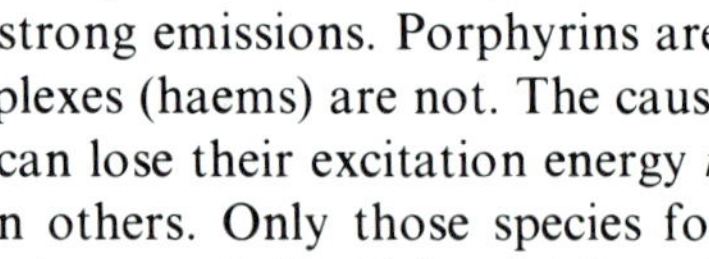

(I)

However, not all such molecules fluoresce efficiently. The nucleic acid bases, for example, are practically non-fluorescent, even though a variety of similar heterocyclics display strong emissions. Porphyrins are intensely fluorescent, whilst their iron complexes (haems) are not. The cause of these variations is that some molecules can lose their excitation energy *non*-radiatively a great deal more easily than others. Only those species for which non-radiative processes are slow, emit powerfully. Table 3.2 lists the common biological fluorophores, the biological chromophores whose fluorescence is too weak easily to be measured, and a number of non-biological fluorophores which find use as fluorescent indicators in biochemical systems.

Transition metal ions and their complexes have low-lying excited states, but are not usually fluorescent. However, a number of the lanthanides escape rapid non-radiative quenching and generate observable emissions. Such ions can be used as analogues for some biologically important ions which do not

TABLE 3.2 Common fluorophores in biochemistry

(a) Natural fluorophores	(c) Fluorescent indicators
Aromatic amino acids	Dansyl chloride (dimethylaminonaphthalene sulphonyl chloride)
Pyridine nucleotides (reduced)	ANS and derivatives (anilinonaphthalene sulphonate)
Pyridoxal	1,N^6-ethenoadenine and 3,N^4-ethenocytosine
Flavins	Pyrene-3-sulphonate
Vitamin A	Ethidium bromide
Porphyrins	Formycin
Chlorophyll	Anthroylstearate and similar derivatives
Steroids	Chlorotetracycline
"Y"-bases (in *t*RNA)	Lanthanide ions
	Fluorescein isothiocyanate
(b) Non-fluorophores	
Nucleotides and nucleic acids	
Lipids	
Carbohydrates	
Carboxylic acids and esters	
Non-aromatic amino acids	
Common metal ions	

have an easily observed spectroscopic property like fluorescence by which their behaviour can be followed. For example, terbium (III) interacts with high-affinity calcium sites on membrane proteins, and europium (III) interacts at magnesium binding sites on *t*RNA.

The Emission Process

Energy

Most organic molecules in their ground states are singlets (all electrons are paired, see Chapter 2). On absorbing radiation one electron is excited to a higher energy state but without a change of spin. Such states are thus excited singlet states and the excited state with lowest energy is the first excited singlet. Fluorescence emissions occur when first excited singlets relax to ground states. If the initial absorption generates a higher excited state this will usually relax quickly and non-radiatively to the first excited singlet which may then fluoresce. For example riboflavin has four intense absorption bands in the visible and ultraviolet regions. Only the longest wavelength band (centred at 445 nm) corresponds to the ground-state→first excited singlet transition: the others are caused by transitions to higher excited states. However, a single emission spectrum ($\lambda_{max}^{emission} \sim$ 525 nm) is observed which is independent of excitation wavelength (Fig 3.1).

An *emission* spectrum is observed when a sample is excited at a single wavelength, and the intensity of emission is scanned as a function of emission wavelength (as in Fig 3.1). An *excitation* spectrum is measured when a single emission wavelength is chosen (somewhere within the band covered by the emission spectrum) and the emission intensity is recorded as a function of excitation wavelength. An excitation spectrum usually looks very similar to the fluorophore's absorption spectrum because the efficiency with which higher excited states relax to the first excited singlet state is very high. Thus, whatever the wavelength of absorption, the first excited singlet is eventually generated at a rate proportional to the absorbance at the excitation wavelength. Therefore fluorescence emission intensity is proportional to the absorbance at the excitation wavelength.*

It might be expected that the fluorescence emission spectrum of a molecule would overlap its longest wavelength absorption because both involve the transition between ground and first excited singlet states. The spectra in Fig. 3.1(b) illustrate a general observation that this prediction is incorrect. Emission

* This statement must be modified when inner filter effects are important (see p. 99).

(a)

(b)

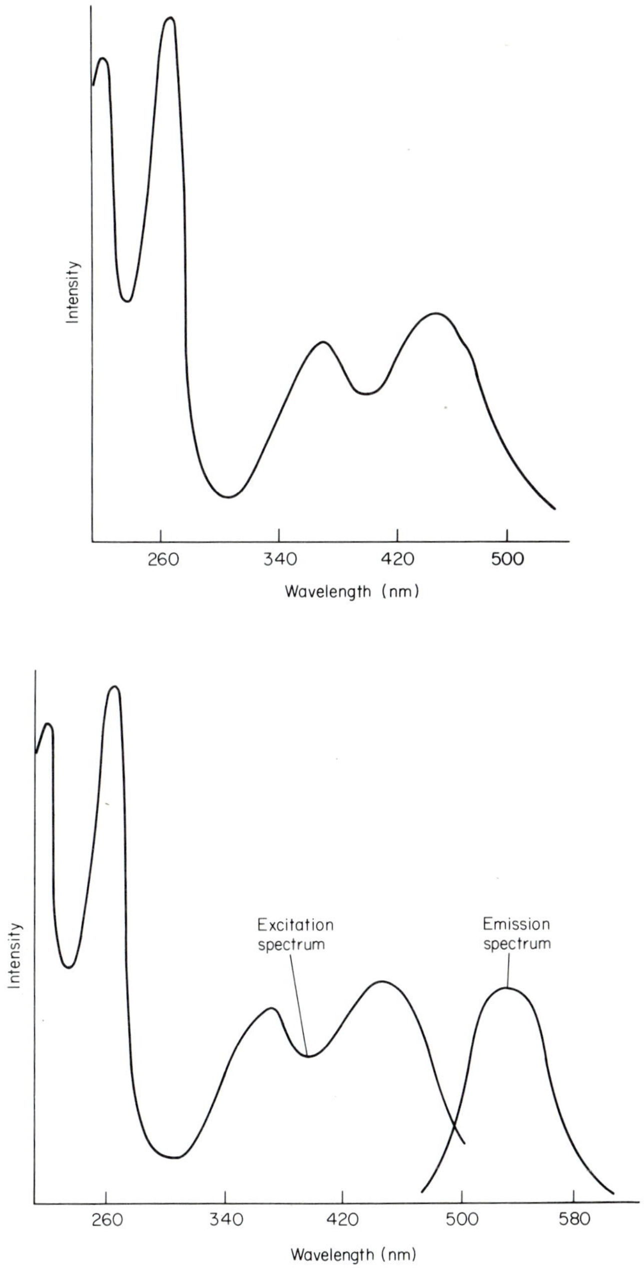

FIG. 3.1 (a) Absorption spectrum and (b) fluorescence excitation and emission spectra of aqueous FMN (pH 7).

spectra are at longer wavelengths than (red-shifted from) the corresponding absorptions. There are two main reasons for this.

1. At room temperature almost all molecules are in the lowest vibrational energy level of the particular electronic state they occupy. This applies equally to ground and excited singlet states. The band of wavelengths corresponding to normal absorption has the transition between zero vibrational levels at the *lowest* energy. In contrast the band comprising the emission spectrum has the zero–zero transition at its *high* energy edge (Fig. 3.2). Thus if this effect alone applies, emission should approximately mirror absorption, and overlap only at the edge with the zero–zero transition. The spectra in Fig. 3.1 however indicate that the actual shift is greater than predicted as there is no significant overlap at all.

2. A molecule takes about 10^{-15} s to absorb a photon. During so short a period, atomic nuclei move negligible distances (typical vibration frequencies are 10^{13} s^{-1}). Thus the atomic positions in an excited-state molecule and its surrounding solvation shell are initially identical to the positions in the

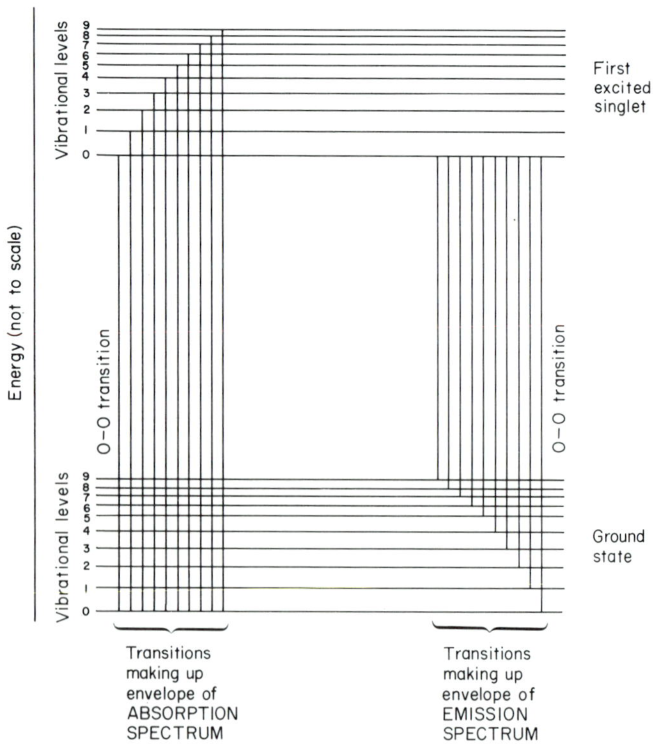

FIG. 3.2 Energy level diagram for ground state ↔ first excited singlet transitions.

ground-state from which it has been generated. (This is called the Franck–Condon principle.) But it is unlikely that the most stable atomic configuration will be identical in ground and excited states. An electronic transition naturally alters the electron distribution within a molecule (its polarity) and hence it is probable that a slightly changed configuration of atoms and surrounding solvent molecules becomes the configuration of lowest energy. During the lifetime of a first excited singlet (typically 10^{-8} s) such rearrangements, minimizing excited state energy, occur. On emission there is a similar effect in reverse. The ground-state is initially formed with non-optimal atomic positions and solvation. These effects reinforce each other in causing emission to involve a smaller energy change than absorption (Fig. 3.3). The magnitudes of

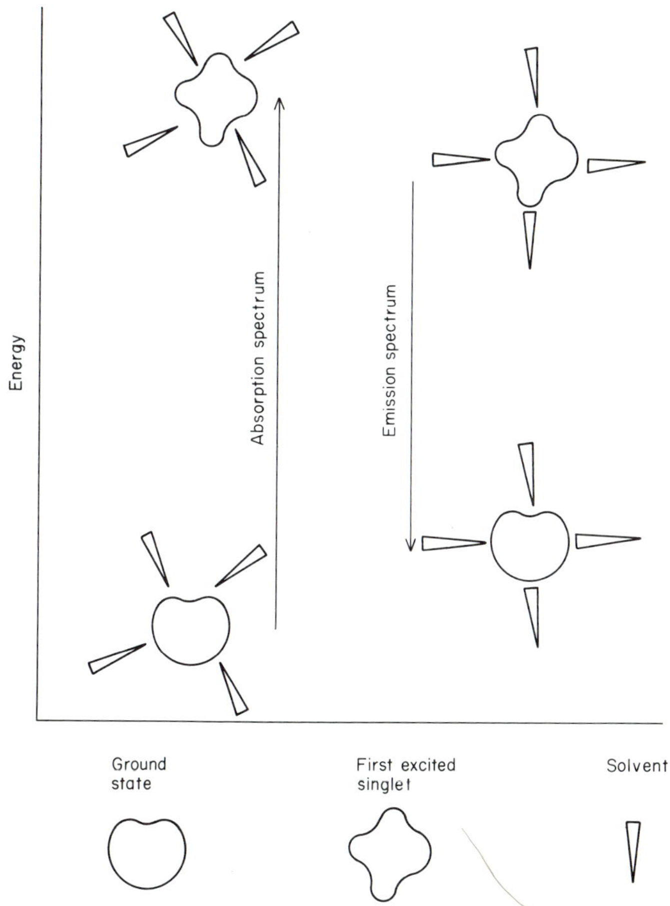

FIG. 3.3 The Franck–Condon principle and effects of solvation differences between states on the energies of electronic transitions.

such red shifts depend on solvent. A polar solvent with strong solvating interactions (like water) tends to cause larger shifts than a weak non-polar solvent. The data for riboflavin in Table 3.1 illustrate this trend (the emission maximum in dioxan is at shorter wavelengths than that in water) though the effect in this instance is small.

Intensity

The intensity of light absorption by a chromophore (expressed in Beer's law—see p. 16) depends on the product of the amount of sample being studied and its molar absorption coefficient, which in turn depends on the overlap between electron orbitals of ground and excited states. The same factors govern fluorescence intensity. However the quantity of sample under observation is no longer a simple product of total concentration and path length because the relevant species is not the ground-state, but the first excited singlet.

Normally, fluorescence measurements are made with a small but constant (steady-state) fraction of total fluorophore molecules in the first excited singlet state by illumination of the sample with a constant intensity of exciting radiation. The situation is represented by:

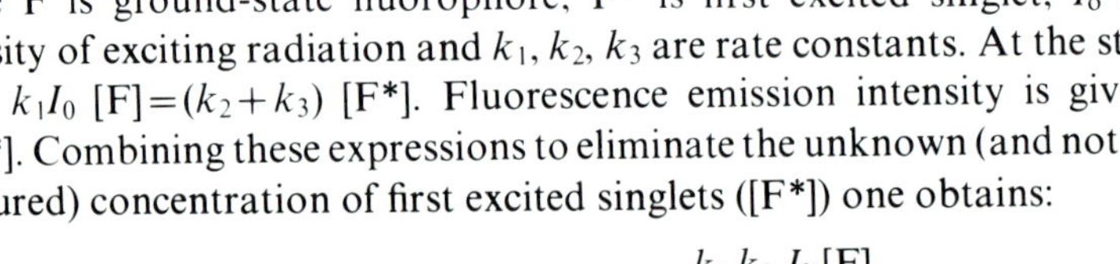

where F is ground-state fluorophore, F* is first excited singlet, I_o is the intensity of exciting radiation and k_1, k_2, k_3 are rate constants. At the steady-state, $k_1 I_0$ [F]$=(k_2+k_3)$ [F*]. Fluorescence emission intensity is given by k_2[F*]. Combining these expressions to eliminate the unknown (and not easily measured) concentration of first excited singlets ([F*]) one obtains:

$$\text{Fluorescence intensity} = \frac{k_1 k_2 I_0 [\mathrm{F}]}{k_2 + k_3} \tag{3.1}$$

Except under abnormally intense illumination, the concentration of ground-state fluorophore molecules, [F], approximates to total fluorophore concentration, because $< 1\%$ of the molecules are excited at any instant. k_1 is related to the intrinsic probability of absorption of a photon by a ground-state molecule (and hence to the absorption coefficient): k_2 is similarly related to the intrinsic probability of emission by the first excited singlet. Both increase when overlap between the appropriate orbitals is improved.

Thus the total emission intensity from a sample is proportional to each of:

(i) k_1
(ii) I_0
(iii) [F]
(iv) $k_2/(k_2+k_3)$.

The first three of these are either constant (k_1) or independently variable (I_0 and [F]). The fourth quotient, however, is less quantitatively predictable, primarily because k_3 is a composite rate constant covering several non-radiative deactivation processes, some of which are sensitive to minor changes in solvent composition or temperature.

Quenching of Fluorescence

The decay of first excited singlets to ground states by fluorescence is a first order process (just like radioactive decay, for example). Typical rate constants (k_2) for the process are 10^6–10^7 s^{-1}. The competing non-radiative mechanisms all convert the excess energy of the first excited singlet into thermal energy, often initially to excited vibrational levels of the fluorophore itself, but ultimately (after collisions) to the thermal energy of the solvent. Under circumstances where such processes "win" the competition with radiative energy loss, *quenching* of fluorescence is said to occur.

Radiationless transitions

The restrictions for non-radiative transitions between electronic states are even more limiting than those specified by the Franck–Condon principle for radiative ones. Not only must the atomic positions in the two states be identical, but also the *energies* must be equal. This can be represented diagrammatically for diatomic molecules by Morse curves which display the energy of a molecule (on the ordinate) as a function of the distance separating its two atoms (Fig. 3.4). Only at the point where Morse curves for two electronic states cross can non-radiative transitions between the states occur. Similar diagrams for polyatomic molecules are too complicated to represent on paper because of the large number of atomic positions involved. Qualitatively, however, the situations in polyatomic and diatomic molecules are similar, and the notions derived from simple Morse curves are applicable to more complex species. The name given to energy release by non-radiative transitions between singlet states is *internal conversion*.

When considering the details of non-radiative deactivation processes it is necessary to consider triplet as well as singlet energy levels. A triplet state

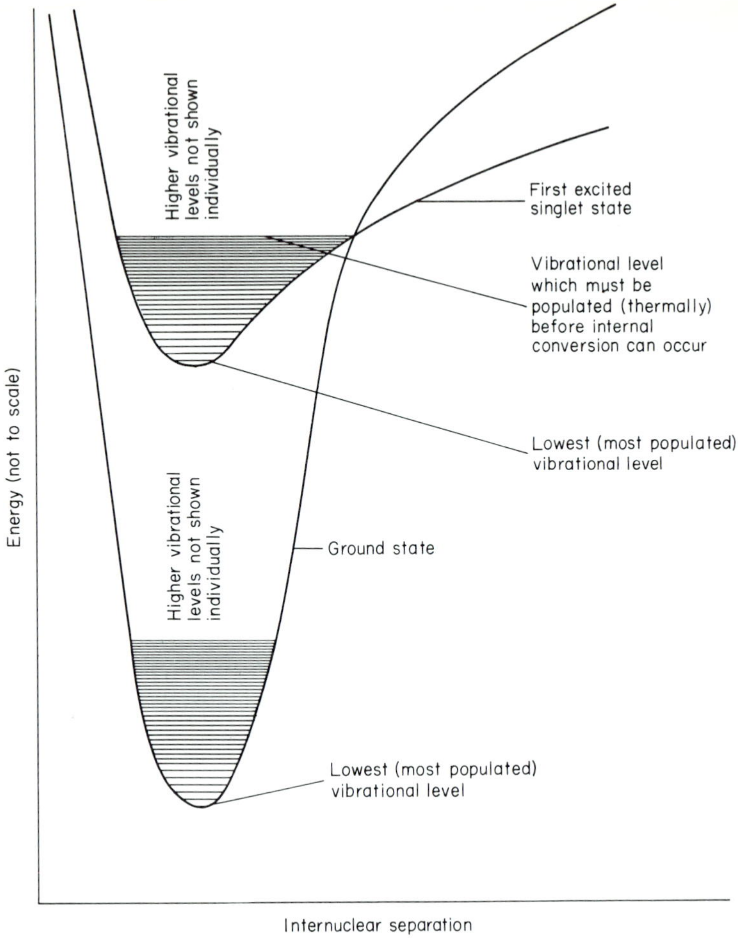

FIG. 3.4 Morse curves for ground and excited states illustrating the conditions for internal conversion.

contains two unpaired electrons and is formed from a singlet state when one electron changes its spin orientation. This is impossible for a ground-state because of the rule that no two electrons with parallel spins can share the same orbital. However, once an excited state is generated the electron involved no longer shares an orbital and is therefore free to adopt either spin orientation. Thus every *excited* singlet state has a corresponding triplet level. It is slightly more stable for unpaired electrons in separate orbitals to have parallel spins than paired ones (Hund's rule) so triplet energy levels are always slightly lower than the singlet states with which they correspond.

Although triplet states are more stable than the corresponding singlets they are not so easily generated. This is because the probability of an electron changing spin (which must occur when a triplet is formed from a singlet) is very low. Thus direct excitation from a ground-state singlet to a triplet level by absorption of a photon of suitable energy is so improbable an event that such absorptions are usually too weak to detect. Triplet states can often be generated quite easily, however, from nearby excited singlets. The process, singlet→triplet or the reverse, is called *inter-system crossing*. It requires that the potential energy surfaces of the two states cross at an easily accessible energy (just as with radiationless transitions between singlets) and that there are microscopic magnetic perturbations to which the electron spin can couple in order to change its direction. Such perturbations are produced by the random motion of charged, polar, polarizable and paramagnetic species in fluid solutions. The situation resembles longitudinal (spin–lattice) relaxation of nuclear spins in nuclear magnetic resonance (see p. 248).

When considering radiationless transitions there are three important factors, two of which affect internal conversion and inter-system crossing in different ways.

1. How much activation energy is required for the molecule to achieve a configuration where its potential energy equals that of another electronic state? There is not much experimental data on this point, and activation energies will vary widely for different fluorophores. Possibly the usual situation is that activation energies for inter-system crossing (excited singlet→triplet and triplet→ground state) and for internal conversion between excited singlets are both significantly smaller than that for internal conversion between excited singlets and the ground state. This is the case for riboflavin where measurements of some of the activation energies have been made (Fig. 3.5).

2. Are there adequate magnetic fluctuations in the system to permit inter-system crossing when the potential energy surfaces of singlet and triplet states cross? Generally the answer for fluid solutions is yes. The random movements of charged, polar and polarizable molecules inevitably generate transient magnetic effects. Such effects are enhanced by paramagnetic species (like metal ions with unpaired electrons, e.g. Mn(II)) and molecules or ions containing heavy atoms with large loosely held electron clouds (e.g. I^- or CH_3I). Lowering the temperature reduces perturbations because molecular motion is slower. In viscous and solid solutions, particularly those from which all paramagnetic species (including dissolved O_2) have been removed, adequate perturbations can be virtually excluded so that inter-system crossing is greatly reduced.

3. Can the newly generated state (after internal conversion or inter-system crossing) lose energy rapidly thereby preventing reversal to regenerate the

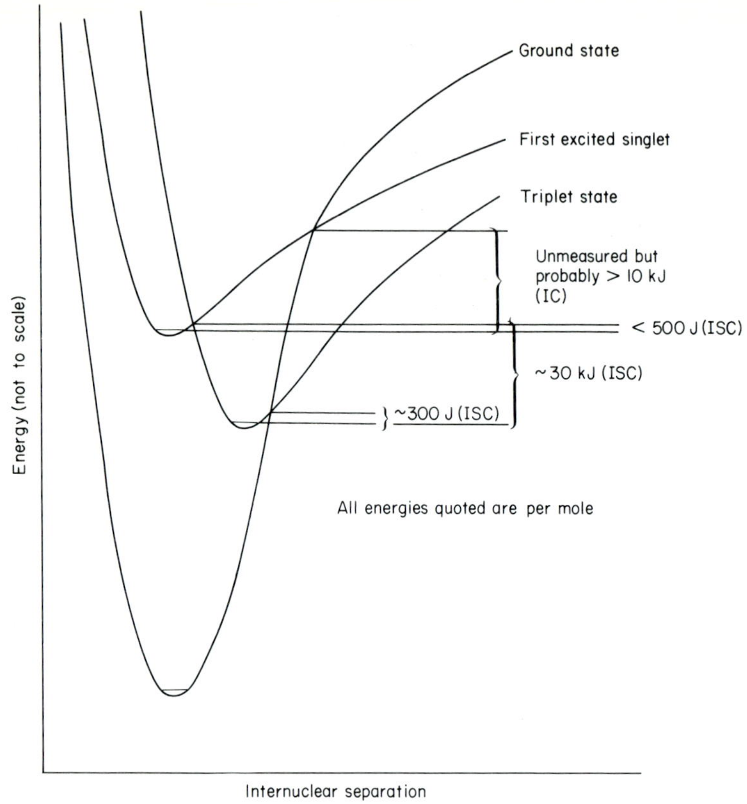

FIG. 3.5 Activation energies of internal conversion (IC) and inter-system crossing (ISC) for riboflavin.

initial state? In fluid solutions excess energy is usually lost readily to vibrational and translational modes in the solvent molecules. In highly viscous and solid solutions where collision rates are reduced energy loss is still possible, though slower. In the vapour phase, however, particularly at low pressures where collisions are infrequent, the rate at which excess energy is lost can be drastically cut.

None of these general considerations suggest immediate reasons why radiationless quenching totally dominates excited state deactivation processes for some molecules (e.g. the nucleotides) but not others (e.g. riboflavin). Weak fluorophores can have either slow rates of emission (low k_2-values as a result of small overlap between excited state and ground-state orbitals), fast rates of non-radiative quenching (high k_3-values) or both. Often high k_3-values are apparently the crucial factor.

A common substituent which can sometimes deactivate excited states effectively is $—NR_2$, presumably because the many vibrational modes associated with the amino function and its ready protonation to $—NHR_2^+$ in both ground and excited states provide good routes for the degradation of excited state energies. Thus you can rationalize (for example) the low fluorescence of cytosine and the much higher fluorescence of its derivative 3,N^4-ethenocytosine(II) (itself an approximate analogue of non-fluorescent adenine). In the etheno-derivative, the amino group of cytosine is tied into a ring. This reduces its vibrational freedom and also its capacity to cause rapid excited state deactivation. A similar effect occurs when the free amino group of adenine is incorporated into the third ring of highly fluorescent 1,N^6-ethenoadenine (III).

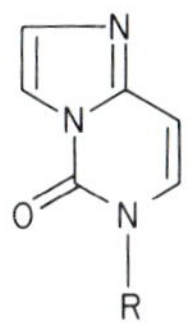

(II)

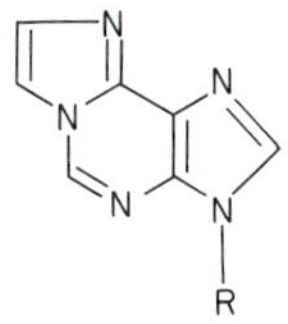

(III)

Added quenchers

Internal conversion depends little on a molecule's environment at constant temperature beyond the extent to which solvation changes alter the relative energies of ground and excited states. The dependence of inter-system crossing on magnetic perturbations in the environment makes its situation different. Compounds which enhance such perturbations increase inter-system crossing when added to a solution of fluorophore and hence quench fluorescence. Often the dependence of fluorescence intensity on quencher concentration follows the kinetic law derived by assuming that quenching requires a bimolecular collision between quencher and fluorophore. Hence it is called *collisional quenching*. The law is derived making the usual steady-state assumption for the concentration of first excited singlets (eqn 3.1):

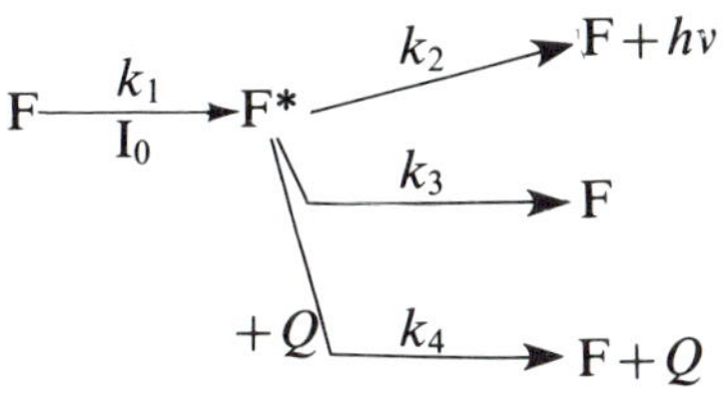

$$\text{Fluorescence intensity} = \frac{k_1 k_2 [F] I_0}{k_2 + k_3 + k_4 [Q]} \tag{3.2}$$

Defining E_Q as the fluorescence intensity with quencher concentration $[Q]$ (and hence E_0 as the intensity in the absence of quencher) it therefore follows that:

$$\frac{E_0}{E_Q} = 1 + \frac{k_4 \quad [Q]}{k_2 + k_3} \qquad (3.3)$$

The quotient $k_4/(k_2+k_3)$ is a constant. It is often known as the Stern–Volmer quenching constant (K_{SV}). Typical values of K_{SV} for a good quencher (like the iodide ion) in aqueous solution at 300 K are 5 to 50 M^{-1}. A value of $K_{SV}=50$ M^{-1} means that the bimolecular quenching rate constant ($k_4 = K_{SV}(k_2+k_3)$), is in the range 5×10^8 to 5×10^9 $\text{M}^{-1}\text{s}^{-1}$ for typical fluorophores with lifetimes 10–100 ns (i.e. $k_2+k_3 = 10^7–10^8 \text{s}^{-1}$). This is close to the diffusion controlled rate constant.

The susceptibility to quenching by fluorophores in a macromolecule has been used to assess degrees of exposure to solvent in much the same way as solvent perturbation difference spectroscopy (see p. 57). Some of the most interesting results have come from observing the sensitivity of tryptophan emission in chymotrypsin and some other proteins to quenching by oxygen. Even those residues known to be buried in the centre of the macromolecule (from X-ray diffraction) are readily accessible to oxygen because their fluorescence emissions are easily quenched. This suggests that in fluid solution the protein is a loose vibrating structure through which small uncharged molecules (like oxygen) can diffuse quite easily.

There is a second major class of quenchers, *static quenchers*. Such molecules do not follow the Stern–Volmer law but exert their effect by forming non-fluorescent complexes. Typical static quenchers for aromatic fluorophores are themselves aromatic. They form sandwich complexes with the planes of fluorophore and quencher rings parallel. Often formation of such complexes scarcely perturbs the absorption spectrum. Presumably the reduced fluorescence efficiency is caused by enhanced non-radiative deactivation of the first excited singlet. The closely held complexing molecule probably enhances inter-system crossing through the magnetic effects of its electrons.

There are several experimental distinctions between collisional and static quenching as well as adherence, or otherwise, to the Stern–Volmer law. One is temperature dependence of quenching. This is always positive for collisional quenching but frequently the reverse for static quenching (fewer molecules of complex existing as temperature increases). A more definitive criterion is excited state lifetime. Collisional quenching shortens the measured lifetime (by increasing the overall efficiency of deactivation processes). Static quenching, however, does not affect excited state lifetime at all. It functions by reducing the number of molecules capable of forming fluorescent excited singlets (i.e. it reduces the concentration of effective fluorophore to the concentration which is *not* complexed).

Collisional and static quenching together explain why the fluorescence of a macromolecule is usually different (sometimes less and sometimes more) from the combined emissions of its component fluorophores. For example, the three aromatic amino acids (Trp, Tyr and Phe) are the only residues in proteins which fluoresce. Yet total protein fluorescence is seldom the numerical sum of the emissions of these three amino acids combined in the proportions in which they occur in the polymer. Sometimes the protein fluorescence is higher (probably in such cases several residues are protected inside the folded macromolecule from the collisional quenching which occurs with free amino acids). Sometimes it is lower (because increased static quenching is caused by interactions between fluorophore residues and other parts of the folded protein chain). The balance struck between conflicting quenching and other effects (e.g. energy transfer, see below) depends on the exact shape to which the polymer chain folds. Hence fluorescence is sensitive to molecular conformation and can be used to detect changes in protein (and other polymer) shapes.

A variety of other quenching mechanisms are known in some systems. The formation of both excimers (a dimer between first excited singlet and ground state of a fluorophore) and exciplexes (a complex between first excited singlet and a ground state molecule of a different compound) leads to modified emission properties. Normal fluorescence is quenched but often emission from the excimer (or exciplex) itself is observed. Another effect deactivating excited states is resonance energy transfer.

Resonance Energy Transfer

If a solution of fluorophore contains molecules of a second compound whose absorption spectrum overlaps fluorophore emission some emitted photons may be reabsorbed to produce excited states of the acceptor species. A similar result is achieved by the non-radiative process called energy transfer. As different electronic states have different electron distributions there is a change in dipole during a transition between two states. If two molecules can undergo transitions of equal energy but opposite sign their transition dipoles can couple together. The excited state energy of one molecule (the donor fluorophore) can thus be transferred to the other (the acceptor) which may then itself fluoresce (Fig. 3.6). This process does *not* involve emission and reabsorption of a photon.

The probability of energy transfer between two molecules depends on several factors. Amongst these are the overlap between donor emission and acceptor absorption spectra, the distance separating the molecules and their

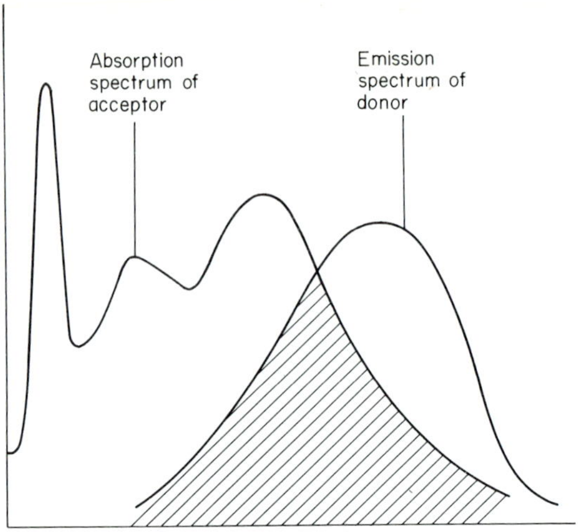

FIG. 3.6 Overlap of donor emission spectrum and acceptor absorption spectrum—a necessary condition for resonance energy transfer. The hatched area is a measure of overlap. The greater this area as a proportion of the total donor emission spectrum, the more efficient is energy transfer.

relative orientation. The efficiency of energy transfer by a donor molecule (T) is defined by

$$T = 1 - (\phi_D/\phi_D{}^0) \tag{3.4}$$

where ϕ_D is the quantum yield of donor fluorescence in the presence of acceptor and $\phi_D{}^0$ is the quantum yield with no acceptor. The value of T can thus be measured by experiment. It is related to R (the distance separating donor and acceptor) and R_0 (the distance at which energy transfer and direct fluorescence by the donor are equally probable) by

$$\mathrm{T} = \frac{(R_0/R)^6}{1 + (R_0/R)^6} \tag{3.5}$$

R_0 is given by

$$R_0^6 = \frac{9000(\ln 10)\, \phi_D \kappa^2}{128\, \pi^5\, n^4\, N} \int_0^\infty \frac{f_D\, \varepsilon_A\, d\bar{\nu}}{\bar{\nu}^4} \tag{3.6}$$

where N is Avogadro's number, n is the refractive index of the solvent, κ^2 depends on the mutual orientations of donor and acceptor transition dipoles and is ~0·67 for an assembly of molecules in rapid random motion, and the integral is the overlap integral for the fluorescence spectrum of the donor normalized to unity (f_D) and the absorption spectrum of the acceptor (ε_A) divided by the fourth power of the wavenumber ($\bar{\nu}$). It is therefore possible to calculate R_0 from the spectra of donor and acceptor and other available data providing reasonable values of the orientation factor κ can be assumed. Combining the calculated R_0 with the measured transfer efficiency (T), values for R, the actual average separation of donor and acceptor molecules, can be calculated from eqn (3.5). Typical R_0 values are in the range 1·0–5·0 nm, and this is the range over which energy transfer measurements provide an approximate "spectroscopic ruler".

Resonance energy transfer plays an important role in determining the spectra of macromolecules containing several fluorophores. For example the emissions of most proteins are dominated by tryptophan in spite of the presence of tyrosine and phenylalanine. This is because tryoptophan absorption and emission are at the lowest energy of the three aromatic residues, and in most proteins the other two transfer much of their own excitation energy to tryptophan more rapidly than they can emit it as fluorescence. Thus a tryptophan residue (or residues) acts as an energy sink for all excitation energy in the macromolecule. A similar effect occurs in chloroplasts. The accessory pigments (like carotene) which absorb short wavelengths of the visible spectrum pass their excitation energy to chlorophyll by energy transfer. Indeed even within the assembly of chlorophyll molecules some (as a result of the environment) absorb at lower energies than others. It is these molecules (P 700) which act as the ultimate acceptors for the excitation energy.

Effects of Physical Conditions

Temperature

In the previous sections the dependence of a fluorophore's emission on its environment has been described. Variations occur because changed electron distributions in both ground and excited states affect transition probabilities (intensity) and energies (wavelength), and also because of changes in the efficiency of non-radiative quenching. Normally temperature changes do not affect radiative processes, but the fact that non-radiative deactivation requires activation energy means its efficiency increases as temperature rises. Thus fluorescence intensities generally fall when temperature is increased (Fig. 3.7).

Solvent

Gross solvent changes (e.g. from H_2O to $CHCl_3$) have their parallel in

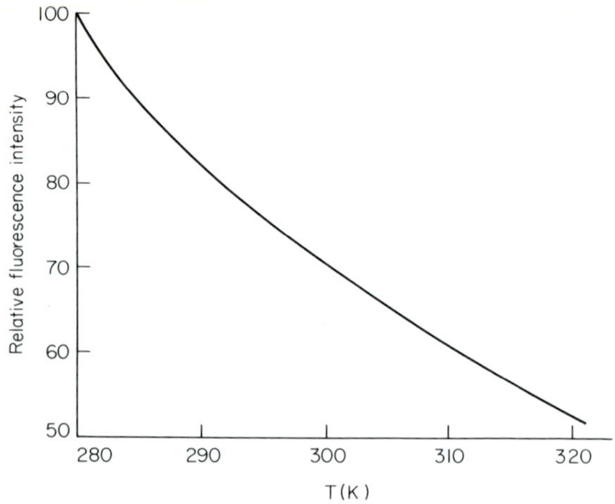

FIG. 3.7 Fluorescence intensity of a typical fluorophore (ethenoAMP in aqueous solution) as a function of temperature.

biological systems. A molecule may be partitioned between an aqueous environment (like serum) and something approaching a hydrocarbon solvent (like the centre of a protein or membrane) and this can have profound effects on its emission. The changed quantum yields of FMN and FAD when bound to proteins (see Table 3.1) illustrate this point.

pH

Apparently smaller environment changes can also have major effects on fluorescence. A prime example is variation of pH. Often the spectra of acid and base forms of a molecule are grossly different. Sometimes they are shifted one from the other: sometimes only one of the two forms emits significantly at any wavelength. This is no surprise. Protonation can easily affect the π-electron system of a molecule. Consider the dimethylaminonaphthalene fluorophore which is used in protein end group analysis (dansyl chloride). When not protonated the lone pair from the N can be delocalized over the whole ring system. But when the molecule is protonated these electrons form a σ-bond to H and are hence removed from the π-electron system.

Excited states of molecules participate in acid-base equilibria too, but often display pK_a values shifted by several units from those of the ground states. Thus an observed pH-dependence of emission may reflect an excited state pK_a, a ground state pK_a, or a combination of both. Rapid proton exchange between fluorophore and solvent occurs significantly even at pH-values well removed from a pK_a. This can provide an effective quenching mechanism for excited states (presumably by offering good routes for energy loss to the solvent).

Lifetimes

There is usually a difference between the observed lifetime and the intrinsic radiative lifetime of an excited state. The former is shorter than the latter. It is defined as the time taken for the emission of a fluorophore to fall to $1/e$ of its initial value in the absence of further excitation. The situation must therefore be imagined where illumination is sharply stopped and the subsequent fluorescence decay monitored (Fig. 3.8). The observed lifetime (τ) is related to the rate constants for deactivation (both radiative and non-radiative—see p. 83) by

$$\tau = \frac{1}{(k_2 + k_3 + k_4[Q])} \qquad (3.7)$$

Typical values for solutions at 300 K are listed in Table 3.3

The intrinsic lifetime of a fluorophore (τ_o) is related only to the rate constant for fluorescence emission by

$$\tau_0 = \frac{1}{k_2}$$

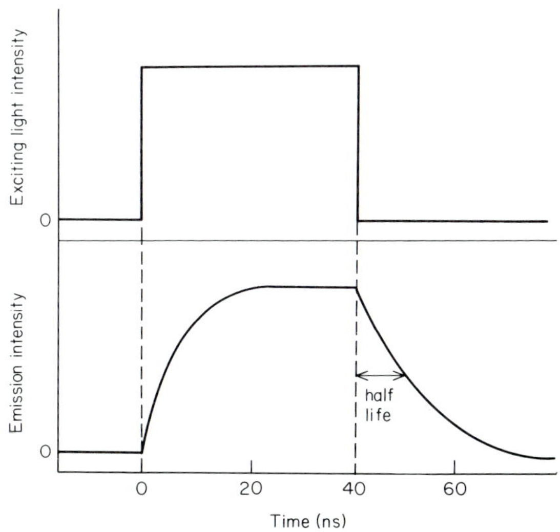

FIG. 3.8 Excitation and emission pulse shapes used in measuring fluorescence lifetimes. A perfect rectangular excitation pulse (as shown) is ideal, but in practice no source has instantaneous rise and decay times.

TABLE 3.3 Fluorescence lifetimes

Fluorophore	Measured lifetime (τ) (ns)[a]
NADH	0·4
NADH bound to lactate dehydrogenase	1·5
FAD	2·5
Fluorescein	4·4
FMN	4·7
Retinol	5·0
Chlorophyll *b*	5·9
Chlorophyll *a*	6·9
EthenoAMP	20
Quinine sulphate	20
Dansyl-lysine	23
Pyrenbutyric acid	193

[a] Sample intrinsic fluorescence lifetimes (τ_0) for some of these fluorophores ($\tau_0 = \tau/(\text{quantum yield})$) are: FMN, 19; retinol, 250; chlorophyll *a*, 25; chlorophyll *b*, 74 (all in ns).

It therefore equals the observed lifetime only in the absence of all non-radiative deactivation. It is a close relative of molar absorption coefficient (ε) as both depend on the overlap between ground and excited state orbitals. An approximate rule of thumb is that $\tau_0 \approx 10^{-4}/\varepsilon_{max}$ s. Thus the strongest absorbers ($\varepsilon_{max} \approx 10^4/\text{M}^{-1}\ \text{cm}^{-1}$) tend to have the shortest intrinsic lifetimes ($\tau_0 \approx 10$ ns).

The measurement of fluorescence lifetimes requires more expensive equipment than constant illumination fluorescence spectrometers. Both the nearly instantaneous extinction of the exciting light and the rapid decay of emission which follow it make for experimental problems (Fig. 3.8). However, they can be solved, and lifetime studies contain, in principle, more information than steady state fluorescence measurements. For example human serum albumin has a typical protein fluorescence spectrum dominated by tryptophan emission. However, the fluorescence lifetime measured depends on the wavelength at which the instrument is set within the broad emission band. This indicates that the band comprises overlapping spectra from residues in different environments and with different lifetimes. This kind of information is harder, if not impossible, to obtain from steady state fluorescence studies.

Anisotropy

When a solution of fluorophore is excited with plane polarized light only those molecules with a component of the excitation transition dipole parallel to the plane of polarization will be excited. If emission occurs rapidly (before significant molecular motion) the fluorescence will also be polarized. Conversely if many random molecular collisions and rotations happen before emission the ordered orientation of the excited molecules will be lost and the observed emission will be unpolarized. Thus studies of fluorescence anisotropy (fluorescence polarization) can provide information about rates of molecular motion. It is fortunate (from this point of view) that the timescale of fluorescence (1–1000 ns) overlaps well the rates of rotation of typical biological macromolecules in aqueous solution at room temperature.

Steady state polarization measurements are easily made by inserting polarizing filters in the excitation and emission light paths of a standard fluorometer. The intensities found with the emission polarizer orientated parallel ($E_{\parallel}$) and perpendicular ($E_{\perp}$) to the excitation polarizer are combined to give the steady state polarization (p)

$$p = \frac{E_{\parallel} - E_{\perp}}{E_{\parallel} + E_{\perp}}$$

Under normal conditions with emission observed at 90° to excitation the maximum possible value of p is 0·5. This is found for a rigid solution in which the transition dipoles for absorption and emission of the fluorophore are parallel (normally the case when excitation is in the long wavelength absorption band direct to the first excited singlet). When the transition dipoles are not parallel, p is less than 0·5 even for a rigid solution, and at the limit (transition dipoles perpendicular), p is $-0{\cdot}3$. Whatever the relative orientations of transition dipoles, p tends to zero as rotation rates increase relative to the fluorescence lifetime (Table 3.4).

Molecular rotational relaxation times (ρ) can be calculated from polarization measurements provided that the limiting polarization (polarization in a rigid solution, p_0) and fluorescence lifetime are known, using the relation

$$\frac{1}{p} - \frac{1}{3} = \left(\frac{1}{p_0} - \frac{1}{3}\right)\left(1 + \frac{3\tau}{\rho}\right)$$

The value ρ is defined as the time after which the average cosine of the net angle rotated by a molecule is e^{-1}. In qualitative terms this is the time it takes an average molecule to rotate about 90°. A second rotational time called the rotational correlation time, θ, is sometimes used instead of the relaxation time. The values of θ and ρ are related by $\rho = 3\theta$.

Non-spherical molecules have more than one rotational relaxation time (for rotation about different axes) but this kind of polarization measurement only gives an average value. The average does not itself necessarily possess great significance, but monitoring alterations in polarization is often a good way to detect molecular changes. Variations in p may be caused by alterations in τ or

TABLE 3.4 Polarization of FMN fluorescence in glycerol/water mixtures of several viscosities

Viscosity	p_{445}[a]	p_{375}[b]
($mN\ m^{-2}\ s$)		
∞ (by extrapolation)	0·50	0·35
1 500	0·44	0·31
1 000	0·39	0·28
500	0·30	0·21
1	0·01	<0·01

[a] The value p_{445} is the polarization measured for excitation at 445 nm. At infinite viscosity this reaches the theoretical maximum (showing that the transition dipole of this absorption is parallel to the transition dipole for emission).
[b] The value p_{375} is the polarization measured for excitation at 375 nm. The maximum this reaches is 0·35 showing the transition dipole is not parallel to the transition dipole for emission (in fact the angle between the two is about 30°).

ρ or both. Whichever, it can often be taken to signal a conformation or aggregation change in a macromolecule. For example the change in rotational mobility of lysine residues in poly-L-lysine when it undergoes a helix–coil transition has been studied by labelling the ε-NH_2 groups with dimethylaminonaphthalene fluorophores by reaction with dansyl chloride (see p. 104) and observing the change in polarization with conformation. Predictably polarization values are higher for the α-helix than for the more flexible random coil.

More precise molecular information comes from analysis of fluorescence anisotropy decay curves (that is from fluorescence lifetime studies using polarized light). Anisotropy, A, is closely related to polarization, p, and is defined by

$$A = \frac{E_{\parallel} - E_{\perp}}{E_{\parallel} + 2E_{\perp}}$$

Both A and p fall as fluorescence decay progresses because molecules emitting 100 ns after excitation have had more opportunity to move at random than those emitting after 10 ns or 1 ns.

The advantage of using anisotropy rather than polarization in lifetime studies is that $A(t)$, the anisotropy, time t after a decay curve starts, can be directly related to the rotational relaxation time (ρ) and is independent of fluorescence lifetime (τ).

For a rigid spherical molecule

$$A(t) = A_{(t=0)} \times e^{-3t/\rho} \text{ (or } A(t) = A_{(t=o)} \times e^{-t/\theta})$$

Anisotropy decay curves for non-spherical molecules are not simple exponentials (see below). Fluorescence lifetimes can be obtained from anisotropy decay curves using the related expression

$$S(t) = S_{(t=0)} \times e^{-t/\tau}$$

where S is defined by

$$S = E \ + 2E_{\perp}$$

A general theoretical expression for molecules of any shape describes anisotropy decay curves in terms (ultimately) of three rotational correlation times (for rotation about three axes). It can be fitted to the experimental results, and if the data are good enough the three correlation times can be calculated. However, although anistropy decay studies are theoretically capable of yielding this kind of molecular information the data are usually too inaccurate and they are often applied like steady state polarization, to detect changes rather than to yield structural details.

Phosphorescence

Phosphorescence is the radiative process by which a triplet state decays to a ground-state singlet. It requires a change of electron spin and so is a "forbidden" transition. This makes it an improbable event and phosphorescence is rarely observed in fluid solutions. Intrinsic phosphorescence lifetimes are in the range 1–1000 ms and longer (compare τ-values for fluorescence) so non-radiative deactivation processes easily dominate in fluids.

However, at low temperatures (e.g. 77 K) and in solid solutions the quenching processes are much less efficient and phosphorescence spectra can be studied. The lowest triplet level has less energy than the first excited singlet (see p. 80) and so phosphorescence spectra are observed at longer wavelengths than fluorescence.

The extreme experimental conditions necessary to observe phosphorescence limit the value of the effect in studying biological samples. When it can be used, however, (and it is sometimes possible to prepare samples of pure protein at 77K with conformations little different from those of biological importance) it provides a useful complement to fluorescence. It provides a different timescale (ms not ns) and a different (longer) range for energy transfer distance measurements. Its greatest value, however, is probably in analytical work where one deals with dead samples anyway. There are situations where measuring concentrations by phosphorescence avoids interfering emissions (from impurities in the sample) which prevent simple fluorescence measurements.

Fluorescence Spectrometers

Basic components

The basic components of a fluorescence spectrometer (fluorometer) are:

(i) a light source for excitation;
(ii) an optical filter or monochromator with which to select the excitation wavelength;
(iii) a sample compartment;
(iv) a filter or monochromator to select the emission wavelength;
(v) a detector with which to measure the intensity of emission;
(vi) a meter or recorder on which to display the detector's response

To maximize sensitivity the exciting light must be intense, the optics must avoid light losses by scattering and absorption and must effectively collect as much of the emitted radiation as possible, and the detector must respond to very low levels of light. To maximize spectral resolution (thus enabling separation of closely spaced spectral bands) high performance monochromators are necessary. These inevitably reduce sensitivity by rejecting all but a narrow bandwidth of radiation. Thus the demands of sensitivity and selectivity conflict. The ideal compromise varies according to the application involved.

Experimental arrangements

A typical fluorometer is shown in Fig. 3.9. The light source is usually a high pressure xenon arc lamp (originally developed for use in lighthouses) which provides a continuum of radiation from 220 to 700 nm. The bandwidth of the grating monochromators is probably 5–10 nm (if it is fixed) though it may be adjustable by varying an appropriate slit width. The spectral range is about 200–800 nm. Emission is observed at 90° to the direction of the exciting beam to minimize the direct detection of scattered incident light. The detector is a photomultiplier. Many instruments have two interchangeable detectors with different characteristics—one sensitive to short wavelength radiation (below ~550 nm) and the other designed for the longer wavelengths. All lenses, mirrors, windows, etc. are, of course, quartz because glass absorbs ultraviolet radiation strongly.

The sensitivity of fluorometers is usually quoted in terms of the lowest concentration of a standard fluorophore (commonly quinine sulphate) detectable. Typical values are a few parts per billion (10^9). Converted to molarities, this means that 10^{-8} M solutions can be studied. A sample volume of 1 cm^3 is ample so only 10^{-11} mol of material is required. In practice it is not very

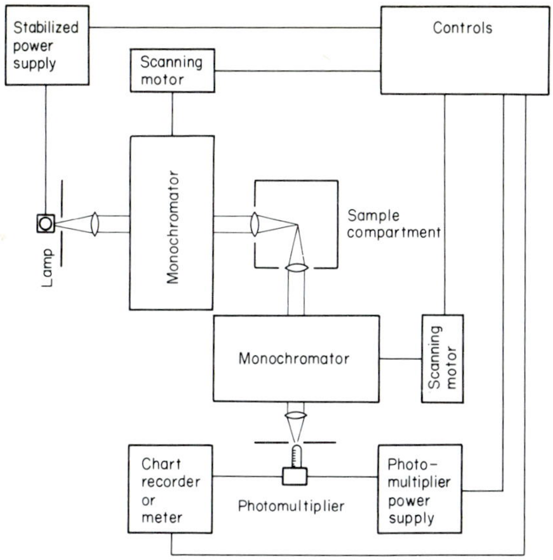

FIG. 3.9 Block diagram of a typical spectrofluorometer.

difficult to increase even this high sensitivity by several orders of magnitude if required. Naturally the sensitivity with which an instrument detects different fluorophores varies according to their different quantum yields and the instrument's varying sensitivity at different wavelengths.

Fluorometers usually have only a single beam (compare spectrophotometers, pp. 23–25). Solvent blanks, when necessary, are performed separately. The light source of a double beam instrument has to be split into two equal components, halving the intensity incident on the sample and hence the sensitivity. However, in fluorometers the (single) beam is often split unequally, with a small fraction (perhaps 5%) shining directly onto a reference photomultiplier. This is because a major cause of irreproducibility in fluorometers is instability of the light source. The efficiency of discharge lamps falls with age, they are sensitive to minor fluctuations in voltage, and the arc itself sometimes moves physically within the lamp causing the intensity of light entering the excitation monochromator to vary. Fluctuations in the signal from the reference detector reflect changes in the light source and can be used to correct for them.

Simpler instruments than that outlined in Fig. 3.9 are often adequate for particular purposes. In analytical work, for example, an instrument may be required routinely to measure the emission of a single fluorophore. In these circumstances suitable filters can probably replace monochromators and a simple low pressure light source with a convenient emission line can replace a high pressure source emitting continuously across the spectrum. Depending on the sensitivity needed, a small photomultiplier or even a photocell can replace the high performance detector and, if the instrument uses the visible region of the spectrum, glass optical components are good enough.

Further sophistication

Instruments can be made more sophisticated in a host of ways. Sensitivity can be improved by using a more powerful light source and a better detection system (cooled photomultiplier or phase sensitive detection or photon counting). Other measurements can be incorporated into a single instrument (e.g. the absorbance, optical rotation and fluorescence of a sample might be monitored simultaneously). The range of light emission measurements possible can be increased to include polarization of fluorescence, phosphorescence (this requires a built-in delay of about 1 ms between illumination and the observation of emission to exclude fluorescence from what is detected), or fluorescence microscopy.

Automatic corrections can be incorporated. Neither the emission intensity of the source nor the sensitivity of the detector are constant across the spectrum. Thus both excitation and emission spectra recorded by instruments in common use are not the true spectra of the fluorophore. Excitation spectra combine the true spectrum with the emission spectrum of the source and emission spectra combine the true spectrum with the detector's response spectrum. They are easy (if tedious) to correct manually once the source and detector characteristics are known, but it is certainly more convenient to have an instrument which makes the corrections itself and prints out absolute spectra. Fluorescence detectors can be linked to many other instruments including stopped flow and temperature jump (see p. 48), liquid chromatographs, autoanalysers, centrifugal fast analysers, etc. In all these applications automatic signal storage and processing through a direct link to a suitably programmed computer can save much time.

Experimental Considerations

The sample

The first fluorometer a person uses is likely to be a standard one (Fig. 3.9). Samples are normally studied as dilute solutions contained in 10 mm clear-sided cuvettes. Any solvent can be used providing it does not absorb radiation in regions overlapping the fluorophore's excitation and emission wavelengths. In many instruments the actual volume of solution observed (i.e. in the light path) is not much more than 1 mm^3, so much smaller sample cuvettes can be used providing they are properly aligned in the instrument.

However clear a solution may be, it will always scatter some of the incident light, but if it contains suspended particles, or is solid, the increase in scattering may cause a serious problem. Often it is best to observe only the front face of the sample in these circumstances, arranging it at an angle to the exciting light (but not 45° to avoid direct reflection of incident light into the detection optics) as shown in Fig. 3.10. Many biological samples contain occasional

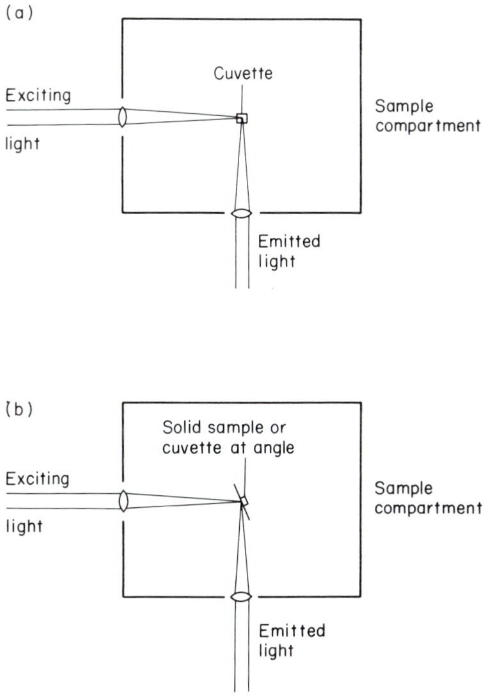

FIG. 3.10 Sample compartment geometry for (a) normal and (b) front face fluorescence measurements.

solid particles (perhaps of denatured protein, etc.) even when the intention was to prepare a clear solution. Rigorous exclusion of such debris is often difficult. It usually helps to obtain stable fluorescence signals when such solutions are stirred constantly (by a suitable paddle in the cuvette arranged to avoid interference with the light paths). This keeps all particles in rapid motion and prevents them from settling or causing other anomalies by floating slowly through the light path. Stirring also helps to reduce photochemical effects by ensuring that all the molecules of fluorophore present in the cuvette contribute to the signal rather than those in a small static patch of illuminated solution alone. Any molecules which react are thus diluted into the whole volume of solution (Fig. 3.11).

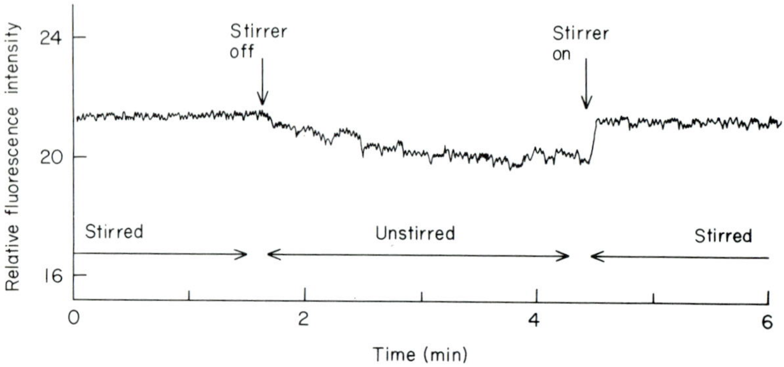

FIG. 3.11 Effect of stirring on the emission signal from a typical photosensitive fluorophore (FMN). The falling and variable signal whilst the sample is unstirred is caused by photochemical destruction of FMN in the fluorometer light path and random diffusion of unreacted molecules into and out of this small volume of observed solution.

Choice of wavelength

Careful choice of the wavelengths used for excitation and emission pays dividends. If the solution under investigation contains fluorescent impurities it is possible to minimize their contribution by choosing an excitation wavelength at which their absorption is small and emission wavelengths at which they do not emit (Fig. 3.12). As well as scattering incident light directly (Rayleigh scatter) most solutions display less intense Raman scattering from the solvent at somewhat longer wavelengths. This is caused by solvent molecules absorbing a quantum of vibrational energy during the scattering process, hence producing scattered light of lower energy than the excitation beam (see Chapter 4). These scatter peaks may overlap emission spectra and distort them (Fig. 3.13). It helps to choose excitation wavelengths well removed from emission wavelengths. However, choosing to excite at wavelengths less than 300 nm often reduces sensitivity because the intensities of most excitation sources fall steeply in this region. Thus a compromise between sensitivity and freedom from interference is often required.

The high sensitivity of fluorescence spectroscopy makes it particularly vulnerable to interference from impurities in the solvent. Indeed these often limit the detection level for an added fluorophore more than absolute instrument sensitivity. For example normal double distilled deionized water has an

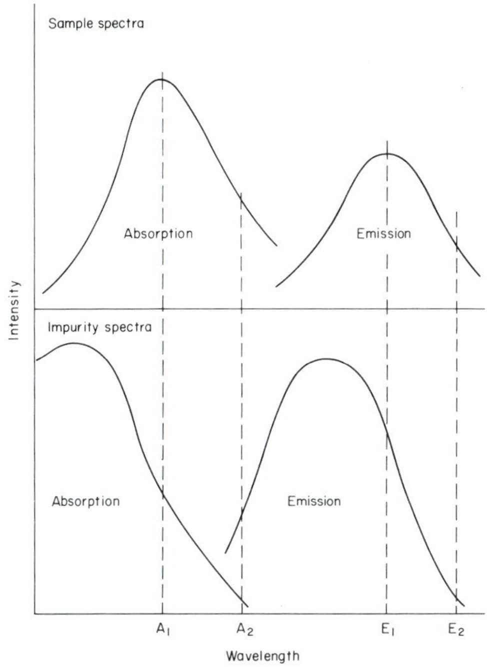

FIG. 3.12 Minimization of interference from impurity fluorescence by choice of wavelength. Although A_1 and E_1 are the absorption and emission maxima of the sample, better analytical results are obtained using A_2 and E_2 which reduce impurity emission approximately 100-fold (10×10) whilst reducing sample emission four-fold (2×2).

easily detected emission. It can be further purified by distillation from permanganate through a heated collar (to prevent diffusion of impurities along the surface of the still from boiler to receiver). Emission from this water is negligible, but even five minutes in a polythene washbottle allows it to dissolve easily detectable levels of fluorophore (Fig. 3.14).

Inner filter effect

Fluorescence emission intensity and fluorophore concentration are not proportional (Fig. 3.15). Their precise relationship varies from one instrument to another because it depends on the fluorometer's optical geometry. The cause of the non-linearity is the *inner filter effect*.

In most instruments both excitation and emission optics focus on the centre of the sample (assuming this is a solution contained in a 10 mm cuvette) and to a first approximation it is only luminescence generated at this focus which is

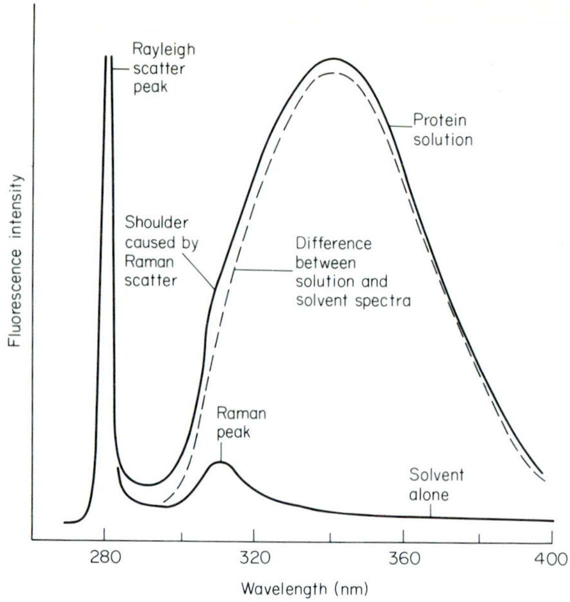

FIG. 3.13 Distortion caused to a true emission spectrum by Raman scattering from the solvent. The spectra are those of a typical protein containing Trp in neutral aqueous solution (10 μg ml^{-1}).

detected. Changing the concentration of fluorophore (e.g. doubling it) has two effects.

(i) There are now twice the number of potential emitters in the observed volume of solution. This alone would double observed fluorescence intensity.

(ii) There are also double the number of absorbing species in the volume of solution through which the exciting beam passes before reaching the focus. These reduce the intensity of light incident on the observed volume of solution thereby tending to lower the observed fluorescence.

These conflicting effects produce results like those in Fig. 3.15. At low absorbances ($\leqslant$0·1) the first effect dominates and emission intensity and fluorophore concentration are almost proportional. However at higher values fluorescence fails to keep pace with fluorophore concentration and above a certain absorbance (for many instruments between 0·7 and 1·0) emission intensity actually *falls* as concentration is increased. Thus when fluorescence is used analytically it is necessary to plot a standard calibration curve and not simply to assume that fluorescence and concentration are proportional. The calibration must use exactly the same solvent as that used for the true measurements because of the different quenching effects of solvents and other dissolved solutes (see p. 83).

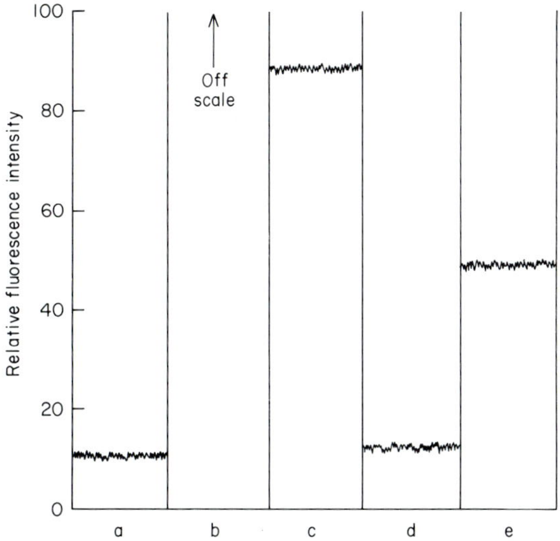

FIG. 3.14 Problems with fluorescent impurities in water. Samples were: (a) empty (air-filled) cuvette; (b) singly distilled water; (c) doubly distilled deionized water; (d) doubly distilled deionized water redistilled from permanganate through a heated collar; (e) as (d) but after storage for 5 min in a carefully cleaned polythene bottle. All were excited at 320 nm with emission observed at 400 nm.

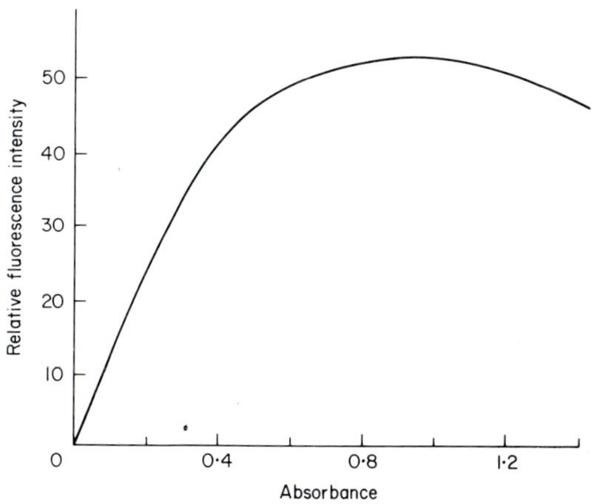

FIG. 3.15 Typical dependence of fluorescence intensity on sample absorbance demonstrating the inner filter effect.

Measurement of Fluorescence Lifetimes

Only the most common technique for measuring fluorescence lifetimes will be outlined here. There are others which will easily be found in the references at the end of the chapter.

The excitation source is the heart of a lifetime fluorometer. It is usually a gas-filled discharge lamp which produces a rapid stream of identical pulses of light. Typically a pulse lasts a few nanoseconds and has very rapid rise and decay times (i.e. it produces an approximately square pulse). The pulse is repeated perhaps every millisecond. Each pulse excites the fluorescence of the sample but emission lags behind the excitation because of the lifetime of the excited state (Fig. 3.8). The fluorescence is detected with a fast-responding photomultiplier (often set up for photon counting). The emission from a single pulse is very weak, but a large number of pulses can be summed by suitably synchronizing the source and the detection system. Nonetheless lifetime fluorometers are generally less sensitive than continuous-illumination instruments.

The decay curves obtained must be carefully analysed to check that they correspond to a single exponential decay and not the sum of two or more (indicating more than one type of emitting species). Lifetimes can be deduced by the normal methods for first-order kinetics (e.g. measuring the half-life of the decay and using the relation $\tau = \tau_{\frac{1}{2}}/\ln 2$).

Emission decay anisotropy (see p. 92) can be measured with a lifetime fluorometer fitted with suitable polarizing filters in excitation and emission beams. This further reduces sensitivity because the polarizing filters each absorb well over half the light incident on them.

Fluorescence Spectroscopy in Biological Systems

Advantages

How then can fluorescence measurements best contribute to the solution of biological and biochemical problems? The following list gives some of the advantages of the approach.

(i) Fluorometry has high absolute sensitivity which, in favourable cases, is unequalled by other common spectroscopic techniques. Careful choice of wavelengths can make it highly selective in the study of mixtures.

(ii) It is normally non-destructive and can be used to study intact biological samples. Linked with microscopy it can be used to investigate very small systems (single cells).

(iii) Common commercial fluorometers are very simple to operate. Measuring a fluorescence spectrum is no more difficult than measuring an absorption spectrum.
(iv) The limit on speed of fluorescence measurements is usually human. With suitable display or signal storage facilities, rapid fluorescence changes can accurately be monitored (down to 1 ns and below). Fluorescence detection thus fits easily on to fast reaction equipment. Rapidly scanning monochromating devices are being developed (e.g. tunable optical filters) so that complete spectra (over several hundred nm) can be obtained in less than 1 ms.
(v) Fluorometry is very flexible—accepting samples of almost any shape, size and physical state. By using light pipes for excitation and emission beams, samples too awkward to fit into an instrument can be observed.

Weaknesses

The following lists some of the problems in applying the technique.

(i) Not all molecules one might wish to study emit observable fluorescence.
(ii) Fluorescence excitation can lead to photochemical reaction modifying, perhaps unacceptably, the sample.
(iii) Much of the theory of fluorescence emission and quenching is only understood qualitatively. Thus an observed fluorescence change cannot often be related quantitatively to its molecular cause. Detailed interpretations of many observations are therefore impossible.
(iv) Fluorescence spectra are generated by excited state molecules whereas what is usually of biological interest is the parent ground state. For example the pH-dependence of a fluorophore's emission may reflect either an excited stake p*K* or a ground state p*K*. It is therefore essential carefully to relate what is observed to the biologically relevant context.
(v) Few instruments measure, and few reports in the literature quote, absolute fluorescence parameters (e.g. quantum yield, corrected spectra, etc.). More often observations are only internally related (e.g. it is stated that on adding X to Y the emission intensity at wavelength λ increased by $Z\%$). This makes correlation between work in different laboratories difficult.
(vi) The non-linearity of fluorescence with fluorophore concentration means that instruments used analytically must be individually calibrated for each new application. When it is sensitive and specific enough, absorption spectroscopy is better than fluorescence for simple analytical applications.

Some of these weaknesses can be overcome. The last two simply require time and effort, and there are already signs that more care is being given to reporting results in a way which makes fluorescence spectroscopy easily repeatable. Photochemical reactivity interference can be limited by stirring, selecting instrument conditions which minimize exciting light intensity, and by adding triplet quenchers (usually it is the triplet state which reacts photochemically).

Probes

The major strategy which offers ways to circumvent the main weaknesses of fluorometry in biochemistry is the use of fluorescent indicator molecules (fluorescent probes). These are fluorophores which are added to the system of interest. Their emission is then monitored and interpreted in terms of the properties of the system (see Table 3.2 for a list of some common probes). Use of probes gets round some problems but it introduces others.

Potential advantages in the use of probes

Fluorescent probes have the advantage that they can be used to study systems with little or no intrinsic fluorescence (e.g. nucleic acids) In addition, they can be chosen with spectroscopic properties which do not overlap those of any other molecules in the system hence avoiding problems arising from interference. In principle probes permit the experimenter to choose the exact site he wishes to study in a biological system. For example, if a protein is being investigated, only Trp, Tyr and Phe environments can be observed by intrinsic fluorescence. There are probes which react specifically with other side chains

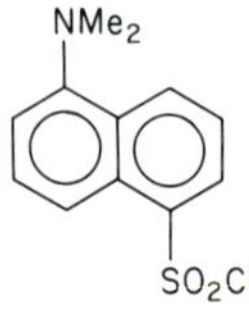

IV

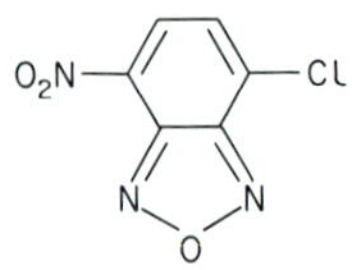

V

(e.g. dansyl chloride, 1-dimethylaminonaphthalene-5-sulphonyl chloride (IV), labels the ε-NH_2 groups of Lys and NBD-chloride, 7-chloro-4-nitrobenzo-2-oxa-l, 3-diazole (V), reacts with the —SH of Cys and, in some cases, the —OH of Tyr) thus extending the range of what can be studied. Similarly, probes linked to substrate analogues are directed to a specific binding site on a protein or membrane (e.g. 1, N^6-ethenoATP binds at the ATP site on many enzymes).

Probes can be designed to possess emission properties sensitive to particular types of environment change. For example a conjugated polyene (like all *trans* retinol) undergoes significant extension when excited to the first excited singlet. If such modification is hindered by a microenvironment of high viscosity the observed absorption and emission maxima are closer than at low viscosity. Thus retinol can be used to measure microviscosity. Excimer fluorescence can be used to study diffusional mobility. Pyrene-3-sulphonate forms excimers which emit at a different wavelength from pyrene monomers. Excimer formation requires collision between a first excited singlet and a ground state molecule during the lifetime of the excited state. When diffusion is slow this is reduced. Thus the mobility of probe molecules is reflected in the relative intensities of the monomer and excimer emissions.

Finally if the probe molecule alone is electronically excited the biological system itself remains in the ground state.

Possible problems in the use of probes

There are factors on the debit side too. Fluorescent probe emission properties cannot be interpreted to give biological information unless the effects governing the probe's behaviour are well understood. Thus extensive investigation of the spectroscopic properties of the molecule in free solution and simple model systems should be undertaken before the probe molecule is used to study complex biological molecules.

Adding a probe inevitably perturbs the system to which it is added. It is therefore essential to check that such perturbations do not cause major changes in biological behaviour. This applies equally to covalent probes (sometimes called labels) and non-covalent ones which interact only through weaker binding forces (like electrostatic interactions, hydrogen bonding, etc.). Thus any work involving probes must involve the controls necessary to discover whether or not the system retains its normal biochemical activity.

It is also necessary to ensure that one can distinguish between the spectroscopic properties of bound (reacted) and unbound (unreacted) molecules. The best probes are those displaying a large change in fluorescence (preferably an increase of intensity) on binding, so that wavelengths can be chosen where unreacted probe has negligible emission even if it is present. Examples are:

(i) NBD-chloride (V) is itself not fluorescent. It forms fluorophores by reacting with side-chain groups on a protein (usually —SH) (VI).

(ii) Fluorescamine (4-phenylspirofuran-2- (3H), 1-phthalan-3, 3′-dione

(VII)) reacts rapidly with NH_2 groups and slightly more slowly with water. As a reagent it is stored in a non-aqueous solvent. On addition to an aqueous solution it reacts with NH_2 groups to produce a fluorescent product (VIII). Excess unreacted reagent subsequently reacts with water to produce a non-fluorescent compound.

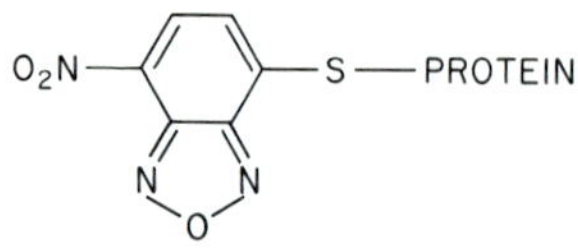

(VI)

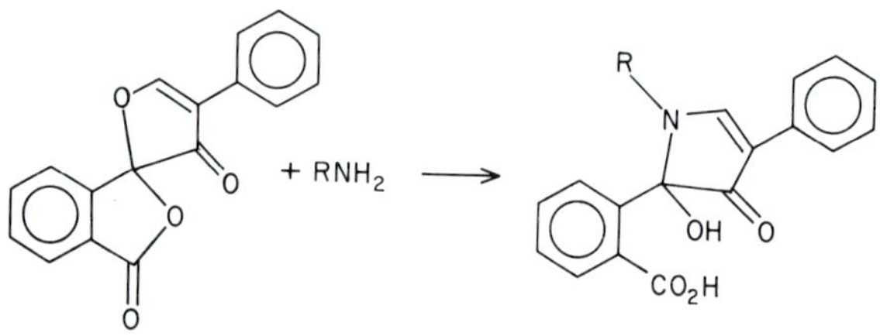

(VII) (VIII)

(iii) Anilinonaphthalene sulphonates (e.g. 1-anilinonaphthalene-8-sulphonate, 1,8ANS) have very weak emission in aqueous solution because of effective quenching. Rapid exchange of protons on the nitrogen (and hence rapid vibration) probably causes this (see p. 83). In non-aqueous solvents where rapid proton exchange is reduced the emission intensity increases many-fold and the emission maximum shifts to shorter wavelengths (Table 3. 5). Often, when 1, 8 ANS binds to a macromolecule, a similar loss of vibrational freedom ensues and large fluorescence enhancements result. These can be observed at

TABLE 3.5 Fluorescence properties of 1,8 ANS in different solvents

Solvent	Quantum yield	Emission maximum (nm)	Measured lifetime (ns)	Intrinsic lifetime (ns)
Water	0·004	555	0·55	138
Methanol	0·21	493	6·05	29
Ethanol	0·36	480	8·85	25
Dioxan	0·50	473	11·8	24

wavelengths where the fluorescence of unbound ANS contributes negligibly.

The nucleotide analogues 1,N^6-ethenoadenine and 3,N^4-ethenocytosine are examples of probes which lack this favourable characteristic. Often when using them (e.g. to study nucleotide-binding enzymes) one observes relatively small emission changes caused by the low proportion of enzyme-bound molecules against a large background emission of unbound molecules.

Applications

The consequences of these strategic considerations are that fluorescence measurements are mainly used for the following purposes.

1. To measure low concentrations (both in research work and routine analytical biochemistry. The ease with which the technique can be automated and applied to small samples is of great value.

2. To detect changes (even though molecular causes may be obscure). For example adding butan-2-ol to the enzymes papain causes a change in protein fluorescence. Presumably this is caused by an altered conformation. The nature of the new state is unclear from the fluorescence, but that it exists is certain. If necessary the kinetics of the change can be studied by monitoring fluorescence and so on. This kind of work, though perfectly possible with purified compounds (like papain), is probably most valuable for studying complex systems comprising several components where other techniques (which can be used with pure compounds) are much harder to apply. For example, fluorescence has been used to monitor the kinetics of energization of electron-transporting particles from beef heart mitochondria.

3. To deduce something about the molecular properties of a biological system (by applying a knowledge of the molecular causes of the various emission parameters). This often involves considerable speculation and is best attempted by supplementing evidence from fluorescence with the results of other techniques. It is rare for the nature of a molecular event or structure to be *proved* by fluorescence spectroscopy, but perhaps less so for hypotheses to be proposed on the basis of emission studies which can subsequently be upheld or disproved by more rigorous techniques. For example the emission spectrum of a protein may suggest that tyrosine residues are completely absent. But it would be a brave and foolish person who stated dogmatically that there was certainly no tyrosine present, because it is possible that a residue might be subject to very efficient quenching and hence fail to emit significantly. The implication of the fluorescence result must be substantiated by another tech-

nique (in this case amino acid analysis or even simply absorption spectroscopy).

The remainder of this chapter contains only a very few examples, in the above three categories, chosen to hint at the wide range of fluorescence experiments open to biochemists. It should be treated as a starting point for further thought, not a comprehensive catalogue.

Analysis

Anyone who doubts the wide range of applications of fluorescence in analysis should glance at Udenfriend's book (see references). Fluorophores can be detected directly, but molecules without intrinsic emissions must first undergo some fluorogenic reaction. In both cases, most biological samples require "cleaning up" before the analysis to remove as much of the irrelevant material as possible. Details of the clean-up procedure vary according to the source of the sample and the compound being analysed, but typically some or all of the following steps will be used: removal of solid debris; precipitation and removal of dissolved proteins and nucleic acids; chromatography; solvent extraction; dialysis, etc. The minimum "clean-up" consistent with an accurate, reproducible assay is the goal.

Use of fluorophores in qualitative analysis, often as tracers, is also very valuable. It is commonplace to study thin layer and paper chromatograms under light to detect compounds either by their intrinsic fluorescence, or by their quenching of the background fluorescence of the chromatographic supporting material. Similarly one can use fluorescence to detect fluorophores eluted during column chromatography or after electrophoresis. Often it is worth labelling non-fluorescent species with a fluorophore in order to increase the ease with which they can be traced at low concentration. The classic example of this is immunofluorescence. One of the most selective forms of biological assay known involves the specific interaction of an antibody with its antigen. The methods used to study the interaction vary depending on circumstances, but the basic technique is to separate antibody–antigen complex from unreacted antibody and measure how much antibody it contains. In immunofluorescence the antibodies are labelled with powerful fluorophores (e.g. fluorescein, by reaction with fluoresceinisothiocyanate, or the dimethylaminonaphthalene group, by reaction with dansyl chloride), and the antibody–antigen complex is detected by its fluorescence emission. Antigens can be detected within fixed single cells in this way by using fluorescence microscopy. The technique can also be applied as a quantitative assay.

Sometimes the kinetics of a reaction can be studied by using fluorescence to monitor concentration changes. A common application is study of those enzymes which react with pyridine nucleotide coenzymes (e.g. many glycolytic enzymes, dehydrogenases etc.). The compounds NADH and NADPH are strongly fluorescent ($\lambda_{max}^{emission} \sim 435$ nm) but NAD^+ (and $NADP^+$) are not. Thus rates of oxidation or reduction of these coenzymes can be monitored by fluorometry (reduction more easily because an increase of fluorescence from zero can be detected more sensitively than a decrease from a high value). This can be applied in enzyme and substrate assay methods as well as in research into kinetics and reaction mechanisms. Fast reaction techniques can be applied to supplement monitoring steady state kinetics (see pp. 48–54). Clearly, any of the fluorophores listed in Table 3.2, may be determined fluorometrically.

Monitoring molecular changes

Fluorescence can be used to investigate binding between molecules (typically, in biochemistry, at least one species is a macromolecule or supramolecular structure, like a ribosome or a membrane). The emission properties of either or both of the interacting molecules can be monitored. The interactions of coenzymes (pyridine nucleotides, pyridoxal derivatives, flavins) and proteins have been widely investigated. For example a simple titration demonstrates that when riboflavin binds to *Azotobacter* flavodoxin its fluorescence is quenched. The strength, but not the rate, of binding is less at high flavodoxin concentrations than at low ones. This indicates that the protein aggregates, and that the polymeric form binds flavin less strongly than monomers because the rate of dissociation of coenzyme from polymer is greater than its rate of dissociation from monomer.

One difficulty when studying interactions of fluorescent ligands is that the spectra of bound ligand may well be very weak in comparison with the fluorescence of excess unbound ligand (see probes, p. 105). One can try to overcome problems of this kind by studying polarization of fluorescence changes (if a small ligand binds to a macromolecule its polarization normally increases because its rate of rotation is reduced). But sometimes (e.g. when ethenoATP binds to pyruvate kinase) even this tactic has limited success because the ligand retains some rotational mobility within its binding site.

There are many situations where the interacting molecules are not fluorescent or, if they are, the emissions are not sensitive to the association of interest. For example isoleucine:*t*RNA ligase displays normal protein fluorescence, but this does not alter in the presence of the enzyme's various substrates (Ile, ATP, PPi). However it interacts with TNS (a close analogue of ANS, 2-*p*-toluidinylnaphthalene-6-sulphonate) to enhance TNS fluorescence. This emission is now sensitive to binding interactions, and dissociation constants for the substrates have been determined. Control experiments have also been performed to show that TNS does not significantly alter the enzyme's kinetic reactivity.

The value of monitoring two emissions when studying an interaction is illustrated by work on the binding of the tetrapyrrole bilirubin (see Fig. 9.15, p. 364) to serum albumins. Three rates are observed on mixing bilirubin with albumin (Fig. 3.16):

(i) a rapid fall of intrinsic protein fluorescence ($\tau_{\frac{1}{2}} < 10$ ms);
(ii) a slower increase in bilirubin emission ($\tau_{\frac{1}{2}} \approx 50$ms);
(iii) a further fall in protein fluorescence at an even slower rate of ($\tau_{\frac{1}{2}} \approx 300$ms).

The three discrete events may be:

(i) rapid binding of bilirubin and protein;
(ii) conformation change of the bilirubin within the protein binding site;
(iii) a slow migration of the bilirubin molecule within the binding site.

A second class of effects often monitored by fluorescence are macromolecule conformation and aggregation changes (including denaturation). For example the conformation of ferricytochrome *c* is often monitored by following changes in the absorption spectrum of the haem group. However, the

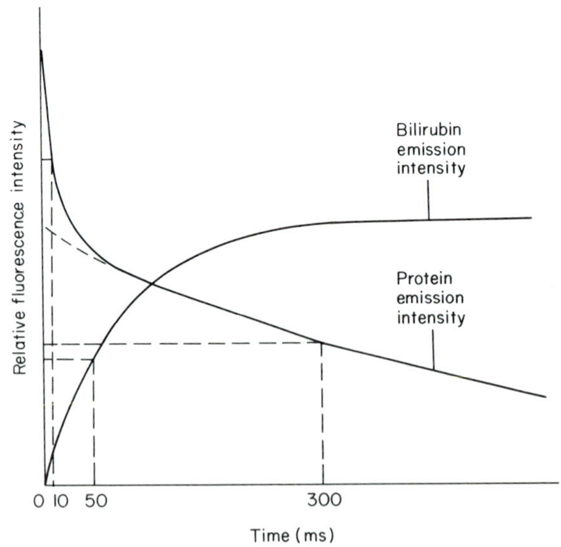

FIG. 3.16 Rates of change of bilirubin and serum albumin fluorescence intensities during binding.

tryptophan fluorescence of the molecule is also a sensitive measure and can indicate events not reflected in the behaviour of the haem spectrum. The fluorescence increases 30-fold on unfolding the protein in guanidine hydrochloride. The shape of the curve (Fig. 3.17) indicates that the major co-operative transition (which is also reflected by haem absorbance changes) is followed by "residual unfolding" as more denaturant is added.

Beef liver L-glutamate dehydrogenase monomers aggregate to form large multimacromolecular complexes. An attempt has been made to discover whether these aggregates are more nearly spherical or linear assemblies by studying fluorescence polarization (the polarization of a sphere is lower than

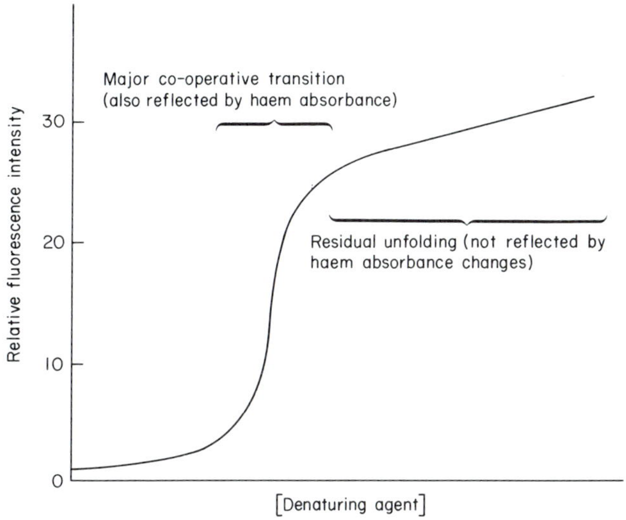

FIG. 3.17 Fluorescence intensity changes of ferricytochrome *c* during denaturation.

that of a linear molecule with the same molecular weight). Such large aggregates rotate slowly (relaxation time ~ 1000 ns), so a fluorescent label with an unusually long fluorescence lifetime was chosen, pyrenebutyric acid (lifetime ~ 200 ns). As expected, fluorescence polarization increases with protein concentration. The concentration dependence of this increase is closer to what one predicts for linear aggregates than spherical ones, though the fit with theory was imperfect. This is not surprising because gross simplifying assumptions (e.g. perfect linear or spherical aggregation as the two situations considered) were made in the theoretical treatment.

The quantity of experiments using many different fluorophores to detect and monitor events in biological membranes is much too great to review here. Great ingenuity has been exercized to extract the last drop of molecular insight from the experimental results. The work using 1,8 ANS to investigate submitochondrial particles is an example. When particles and probe interact there is a fluorescence enhancement. Emission is further increased by addition of substrates (e.g. succinate, NADH, ATP), but this response is abolished in the presence of uncouplers. The kinetics of the uncoupling effect are biphasic, indicating that ANS "feels" two distinct molecular events (probably an initial fast membrane conformation change followed by a slow dissociation of ANS from the membrane because the conformation change alters the dissociation constant).

Information about molecular structure

Transfer RNA molecules typically contain a number of modified nucleic acid bases, some of which are fluorescent. Some (the "Y" bases) are based on guanine and resemble the etheno derivatives (of A and C) discussed above. Another example is 4-thiouridine (IX). The distances separating some of these

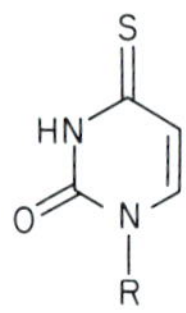

(IX)

bases and other fluorophores linked specifically to different regions of the *t*RNA molecules have been estimated by measuring the efficiencies of energy transfers between them. One can compare the results with the distances calculated from the known crystal structure of yeast *t*RNA^{Phe} (Table 3.6). The agreement, though imperfect, suggests that the conformations detected by the two sets of experiments are similar. This provides support for the contention that the crystal and solution structures of the molecule are almost the same.

An example of the use of fluorescence anisotropy in studying molecular

TABLE 3.6 Comparison of *t*RNA dimensions by fluorescence energy transfer and X-ray crystallography

Distance (nm) Fluorescence	X-ray diffraction
2·4	1·6
3·8	4·2
5·5	5·2
3·6	2·5
>6·5	6·7

structure which again links solution fluorescence measurements with crystallographic work is provided by measurements on *Staphylococcus aureus* endonuclease. The emission anisotropy decay from this protein is described by a single exponential, suggesting the molecule is approximately spherical. The measured relaxation time (ρ) corresponds to a molecular volume of about 40 nm^3 (calculated by applying the relationship $\rho = 3\eta V/kT$ where η is viscosity and V molecular volume). The "dry" volume calculated for the molecule from its molecular weight and partial specific volume is 20 nm^3. The difference is explained if the protein is surrounded by an approximately unimolecular layer of strongly held water molecules in solution. Something similar is presumably the case in crystals because the molecular volume calculated from crystal unit cell dimensions is also about 40 nm^3. This kind of discrepancy between dry volume and volume measured by fluorescence is observed similarly for most globular proteins.

Fluorescence properties can be used roughly to "fingerprint" a structure. This has been tried with different lysozymes. The crystal structure of hen egg white lysozyme is well known, and its emission properties have also been thoroughly catalogued. They have been compared with those of the enzyme isolated from other sources (e.g. human, rat and mouse lysozymes). On comparing both spectra and their sensitivities to perturbations like pH changes and substrate binding, one finds, for example, that the human and rat enzymes are very similar. This suggests there must be a high degree of homology in the distribution of fluorescent (aromatic) residues throughout the structures of the two molecules, in spite of a 20% difference in the amino acid sequences.

Further Reading

General introductions to fluorometry

Cundall R. B. and Gilbert A. (1970). "Photochemistry", Nelson, London.
Jones D. N. (Ed.), (1976). "Introduction to the Spectra of Biological Polymers", Academic Press, New York and London.
Radda, G. K. and Dodd, G. H. (1968). *In* "Luminescence in Chemistry", (Ed. E. J. Bowen), Van Nostrand, London.

More advanced general texts

Chen R. F. and Edelhoch H. (1974). "Concepts in Biochemical Fluorescence", Marcel Dekker, New York.

Guilbault G. G. (1973). "Practical Fluorescence, Theory, Methods and Techniques", Marcel Dekker, New York.

Articles on particular topics

Fluorescence lifetimes

Rigler, R. and Ehrenberg, M. (1973). *Q. Rev. Biophys.* **6,** 139.
Yguerabide, J. (1972). *In* "Methods in Enzymology", Vol. XXVIC, (Eds C. H. W. Hirs and S. N. Timasheff), 498. Academic Press, London and New York.

Analysis

Udenfriend, S. (1962). "Fluorescence Assay in Biology and Medicine", Academic Press, New York and London.

Membranes

Radda, G. K. and Vanderkooi, J. (1972). *Biochim. Biophys. Acta* **265,** 509.

Probes

Brand, L. and Gohlke, J. R. (1972). *Ann. Rev. Biochem.* **41,** 843.

Proteins

Penzer, G. R. (1974, 1975, 1976). *In* "Amino Acids, Peptides and Proteins", Vols 5, 6 and 7, Chemical Society, London.

Enzymes

Dalziel, K. (1962). *Biochem. J.* **84,** 244; *J. Biol. Chem.* (1963), **238,** 2850.
Holbrook, J. J. and Gutfreund, H. (1973). *FEBS Lett.* **31,** 157.

4 Vibrational Spectroscopy

P. Gans

Introduction

Vibrational spectroscopy has been used very extensively in biochemistry and there is an enormous amount of published information on the subject. In this chapter, therefore, we cannot attempt to summarize the results obtained so far. Rather, we must attempt to outline the principles of the techniques and give examples of the types of application that have been reported, with the aim of indicating the scope and validity of vibrational spectroscopy applied to biochemical problems.

Ten years ago this chapter would have been entitled "Infrared Spectroscopy". From the mid-1960s Raman spectroscopy underwent a renaissance due to the use of lasers. More recently, resonance Raman spectroscopy has emerged as a powerful tool for biochemists. At the same time techniques and instrumentation in IR spectroscopy have undergone a minor revolution. We shall therefore allot more space to the description of experimental techniques than would have been thought desirable but a few years ago, giving particular emphasis to those features which are valuable in biochemical applications. In many instances the power of these new techniques has yet to be harnessed for the solution of biochemical problems, so that we must look forward to future developments, as well as discuss those of the past.

To understand the interpretation of IR and Raman spectra we need first to examine the nature of molecular vibrations, for it is these that give rise to the spectra. Even though IR and Raman spectroscopies are experimentally totally different, a molecular vibration may give rise to features in either or both of them. The features may be due to an isolated structure in the molecule, such as a >C=O or >C=C< group, a specific type of molecule such as a steroid or carbohydrate, or to a characteristic of a macromolecule, such as a protein chain conformation. An understanding at the molecular level of the origins of spectral features is of fundamental importance if biochemically meaningful conclusions are to be drawn from them.

The Diatomic Molecule

Most of the basic aspects of molecular vibrations are displayed by the simplest molecule, a diatomic with formula AB (e.g. CO, N_2). To begin with, let us assume that the molecule is distorted from its equilibrium state by an increase in the length of the bond, δr, and that the force tending to restore the molecule to its equilibrium state is proportional to δr. This is a statement of Hooke's law, which applies to any spring if the displacement δr is sufficiently small. It follows from quantum mechanical considerations, that the energy of the molecule resulting from this distortion, the vibrational energy E_{vib}, is given by eqn (4.1),

$$E_{\text{vib}} = (v + \tfrac{1}{2})h\nu \qquad (4.1)$$

where h is Planck's constant, v is a quantum number which can take any integral value including zero, and ν is the frequency of vibration (s^{-1}). The displacement δr itself varies periodically in time with frequency ν as shown in eqn (4.2), where δr_0 is the maximum

$$\delta r = \delta r_0 \sin(2\pi\nu t + \phi) \qquad (4.2)$$

value of δr, and ϕ is an arbitrary phase angle. This form of motion is called simple harmonic motion, and can easily be visualized. The atoms A and B separated by a distance r_e move apart, come together, and move apart until the interatomic distance is r_e again ν times per second. The vibration frequency is determined by the masses of the atoms A and B, and by the Hooke's law proportionality constant k, as shown in equation (4.3),

$$\nu = \frac{1}{2\pi}\left[\left(\frac{1}{m_A} + \frac{1}{m_B}\right)k\right]^{\frac{1}{2}} \qquad (4.3)$$

where m_A and m_B are the masses of atoms A and B. We may note that if the mass of either atom A or atom B is changed by isotopic substitution the vibration frequency will change. An increase in isotopic mass will cause the frequency to decrease. For diatomic molecules the frequency shift can be calculated directly from eqn (4.3).

If the molecule is made to gain or lose a quantum of energy ΔE equal to $h\nu$, the vibrational quantum number v will change by one unit. The energy quantum may be directly absorbed (or emitted) as IR radiation of frequency ν giving rise to an IR spectrum, or it may be attached to a quantum of light of frequency much greater than ν, giving rise to a Raman spectrum. In the latter case, in practice the molecule is irradiated with high-frequency radiation such as visible light, and the light scattered from the molecule is analysed. The scattered light contains radiation at the incident frequency with additional radiation at that frequency plus or, more frequently, minus the vibration frequency. In Raman spectroscopy the vibrational energy is absorbed from a radiation source of much larger energy, with a consequent small decrease in the latter.

The transition $v=0 \rightarrow v=1$ is called the fundamental transition and is the only one that can be observed in the simple harmonic approximation. (Any transition with $\Delta v=1$ is in principle allowed, but for all vibrations of frequency above about 400 cm^{-1}, only the $v=0$ state is appreciably occupied at room temperature. A transition in which $v \neq 0$ is called a hot transition). However, in real molecules, Hooke's law is only valid for very small vibrations, and consequently transitions such as $v=0 \rightarrow v=2$ and $v=0 \rightarrow v=3$ can sometimes be observed; the former is called the first overtone, the latter the second overtone and so forth. Overtones are readily observed in the resonance Raman effect, and may also be seen in IR absorption. They are rarely found in Raman spectra. Because of anharmonicity the frequencies of overtones are not exact multiples of the fundamental frequencies, as implied by eqn (4.1). Rather, they are always less than the value calculated in the harmonic approximation.

Fundamental vibrations have frequencies in the approximate range $1{\cdot}0 \times 10^{12}$ to $1{\cdot}2 \times 10^{14}\ s^{-1}$, corresponding to wavelengths of 300 000–2500 nm; thus the wavelength is longer than that of visible light ($\sim$400–700 nm) and the frequency is smaller. It is customary to quote vibration frequencies in the unit of the number of waves per centimetre (cm^{-1}) (see Chapter 1). In this unit, vibration frequencies range from 30 to 4000 cm^{-1}. It is clear that many hundreds or thousands of vibrations will take place during the time required for even the fastest chemical reaction, so it is very unlikely that any kinetic effects will be observed. Put another way, the lifetime of a vibrational excited state is much shorter than the lifetime of a chemical entity during a reaction such as, for example, chemical exchange. The vibrational spectrum will therefore always show the sum of effects from individual species, whereas an NMR spectrum, operating on a much longer time-scale, can show individual entities separately or as an average, depending on the reaction rate (see Chapter 6).

Polyatomic Molecules

A non-linear molecule with N atoms has a total of $3N$-6 vibrations. Not all of these vibrations will have different frequencies, but nevertheless a large molecule will have a large number of distinct vibration frequencies. The actual modes of vibration of the molecule are termed the *normal modes*, and for each normal mode it is possible in principle to write an equation similar to eqn (4.3). Now, it may be shown that

$$4\pi^2\nu^2 = (\text{mass factor}) \times (\text{force-constant factor}) \tag{4.4}$$

The mass factor may involve the masses of many atoms, just as the force-constant factor may involve many bonds. For large molecules it is, in general, possible to write out eqn (4.4) only approximately for any normal mode.

It is useful to classify the vibrations of molecular fragments as follows. A *stretching* vibration is just as in a diatomic molecule AB. A *bending* vibration is most simply defined as involving a change in the angle between two bonds. Other modes called wagging, rocking, or twisting are particular forms of the bending vibration.

Experience has shown that many chemical groupings exhibit vibrational frequencies in fairly well-defined zones. An example, which we shall discuss in greater detail later, is the amide >C=O group in peptides, whose stretching frequency is found in the range 1630–1660 cm^{-1}. The C—O stretching mode is coupled, or mixed, with other modes to a small extent. In contrast, calculations have shown that the amide II band in peptide spectra, which is found in the range 1515–1550 cm^{-1}, consists of 40% C—N stretching and 60% N—H deformation modes. We cannot therefore describe the amide II vibration in terms of one single grouping, but must be content to adopt the empirical approach—all peptide amides have a band in the region indicated. The precise nature of the vibration responsible for the band must remain unknown.

The use of characteristic frequencies to identify chemical groupings is very well established, and many books will be found giving spectra-structure correlations. We cannot possibly summarize all that work here, but Table 4.1 illustrates how C—O stretching frequencies can be used to differentiate between various simple compounds that contain carbonyl groups. Note, for example, how an amino acid that contains a free—NH_2 group may be distinguished from one that contains an—NH_3^+ group.

Of the various phenomena that can occur in polyatomic molecules, but not in diatomics, two are worth a special mention. Firstly, *combination bands* may be observed as a result of the simultaneous excitation of two distinct normal modes. Combination bands can sometimes be useful in structure determina-

TABLE 4.1 Selected C—O stretching frequencies

	cm^{-1}
Acyclic ketones	1725–1700
Aryl ketones	1700–1680
Diaryl ketones	1670–1660
Saturated aliphatic aldehydes	1740–1720
Aryl aldehydes	1715–1695
Saturated aliphatic acids	1725–1700
Saturated aliphatic esters	1750–1720
Primary amides	approx. 1690
Secondary amides	1680–1630
Tertiary amides	1670–1630
Amino acids	1665–1585
α-Amino acid hydrochlorides	1755–1730

tion; for example, aromatic substitution types can be identified by the pattern of their combination bands in the region 1700–2000 cm^{-1}. More usually, combination bands simply increase the complexity of a spectrum. Secondly, *Fermi resonance* can result in the shift of a fundamental frequency. For this to happen an overtone or combination must have a similar frequency to that of the fundamental which is to be affected. At the same time the intensity of the overtone or combination in resonance is increased, so it may appear that a fundamental band has split into two components. Isotopic substitution causes frequency shifts, though their magnitude is less easy to calculate theoretically in polyatomic molecules.

Intensities

Although the intensity of an IR or Raman band depends on the nature of the normal mode, certain generalizations are possible which help in the interpretation of the spectra.

Infrared absorption

The intensity of an IR absorption is proportional to the amount by which the dipole moment of the molecule changes during the vibration. Therefore, changes involving highly polar groupings tend to give intense IR absorptions. Most doubly bonded groups are highly polar, so >C=O, >C=N—, >P=O and similar groups give rise to high-intensity IR bands. However, the

>C=C< group may have a low or even zero intensity in the IR. If the substituents are placed symmetrically about the centre of the double bond, the IR intensity must be zero because there is no net dipole moment change when the >C=C< bond is stretched. As we shall see, this vibration can be observed in the Raman spectrum, pointing to the complementary nature of this other form of vibrational spectroscopy. Other groupings which may be difficult to see in the IR spectra include the dioxygen group in respiratory proteins, and the disulphur link in cystine bridges.

An important point to realize is that different vibrations of one group may give rise to strong bands in either the IR or Raman spectra respectively. This is particularly true of species with a porphyrin structure, because of the high symmetry of the ring system.

Raman scattering

The intensity of a Raman band is proportional to the change in *molecular polarizability*. The polarizability gives a measure of the dipole moment *induced* in a molecule by an external electric field. More precisely, it is the proportionality constant relating magnitude of the induced moment to the strength of the external field. The polarizability change is intuitively much harder to visualize than a dipole moment change. There are instances of vibrations which have zero Raman intensity, though non-zero IR intensity. Large polarizability changes occur with vibrations involving polarizable atoms, typically a second- or third-row element such as S, Cl, Br, or I. With the first-row elements C, N and O of which the bulk of biologically important molecules are largely constructed, the atomic polarizability is small, but high Raman intensities can result from systems involving double bonds or delocalized electrons (aromatic compounds in particular).

When plane polarized light (see Chapter 5) is used to excite the Raman spectrum of a solution (as is now universally the case), the Raman radiation may retain that polarization or not. In the latter case the band is said to be *depolarized.* The depolarization ratio, ρ, defined as the ratio of intensities of the perpendicularly and parallel polarized Raman radiation, is 3/4 for a depolarized band. If the molecular vibration does not alter the symmetry of the molecule the Raman band will be polarized, with $\rho < 3/4$, and in certain cases, e.g. tetrahedral molecules, ρ may approach zero. This is a very useful aid in determining the nature of the normal mode.

Resonance Raman scattering

When a molecule is irradiated with light corresponding to an electronic transition, typically at the peak of a visible absorption band, the intensity of certain Raman bands can be greatly enhanced. The enhancement is governed by the mixing of vibrational and electronic wave functions. Two cases can at present be distinguished, though the theory is far from complete. If a single electronic excited state is involved, only totally symmetric vibrations can couple. These are those vibrations, discussed above, which do not alter the symmetry of the molecule and have $\rho < 3/4$ in the normal Raman effect. The result of the coupling is an enormous enhancement of the intensity of the coupled vibration. The amount of resonant enhancement depends on the strength of the electronic transition. For a typical transition with molar absorption coefficient of approximately 10 000 $\text{M}^{-1}\,\text{cm}^{-1}$, the enhanced factor may be 100–1000. Thus, there is a highly selective enhancement of the intensity of certain vibrations. The depolarization ratio of the enhancement band may be anything between 0 and 3/4 depending, *inter alia*, on the shape of the molecule. This enhancement is called A-term enhancement.

A second type of resonant enhancement occurs when one vibration is coupled to two electronic excited states. Any vibration of the molecule can in principle couple in this mode. The depolarization ratio may vary from 3/4 upwards, to infinity in theory. This "anomalous" polarization helps to characterize these B-term enhancements and to distinguish them from the A-term enhancements involving totally symmetric vibrations.

The resonant enhancement process is even more selective than has been indicated so far. In practice, the electronic excitation will be accompanied by some distortion of the molecule. It is only those vibrations that tend to promote this distortion and have the correct symmetry that are resonance enhanced. This opens up the possibility of using resonant labels (as spin labels are used in EPR spectroscopy). Indeed, the label may be part of the normal molecules; selective excitation of the 4-thiouracil residue in certain transfer RNAs is described below.

One further aspect of A-term resonance Raman enhancement, of little direct value to biochemists, is that the overtones of the totally symmetric vibrations may also have enhanced intensities. Since the overtones are always less intense than the fundamentals, little new information results. However, analysis of an overtone progression will give an estimate of the vibration's anharmonicity.

Infrared Spectrometers

With the description of the experimental techniques we shall really get to grips with the different forms of vibrational spectroscopy. Throughout, since we have biochemical applications in mind, we shall emphasize those techniques relevant to aqueous solutions.

The basic design of the IR spectrophotometer remained unchanged from the introduction of the first commercially available instrument around 1947 for a period of about 25 years. Improvements were made in performance, in the light source, optics, detection and electronics, but the design remained that of the so-called optical-null instrument. These spectrophotometers are available at remarkably low prices and will be found in most biochemical laboratories. In the last two or three years the scene has altered radically. Interferometers covering the mid-IR region have become available. These instruments require a dedicated computer, are relatively costly, but extremely versatile. In addition, the optical-null design has been replaced by a ratio-recording system of much greater photometric accuracy and precision. When used together with a data-processing computer both these instruments make possible experiments which cannot be performed with an optical-null instrument, particularly those involving aqueous solutions.

Optical-null spectrophotometers

The essential elements of this device are shown in Fig. 4.1. A source, usually a red-hot bar of ceramic material, emits IR radiation at all frequencies. The beam is split into two parts; one part passes through the sample and the other through a reference material. The beams are recombined alternately by the chopper and passed through the monochromator, detector and amplifier systems to provide a signal that moves the recorder pen up or down. As the absorption by the sample increases, the pen moves downward and the attenuator comb moves into the reference beam, so that "absorption" in sample and reference beam are balanced. This is the optical-null, at which there is no difference in the beam intensities, hence no alternating current in the out-of-balance signal.

The photometric accuracy and precision of this instrument are typically no better than 1% transmittance, being limited by the electromechanical system associated with the reference attenuator. Consider what happens when the sample absorbs 90% of the radiation (equivalent to an absorbance of 1). The attenuator comb comes in to reduce the reference beam intensity to 10% of the total. The detector is then dealing with only 10% of the normal energy, and sensitivity decreases considerably. For higher sample absorptions the instrument goes "dead".

These characteristics are particularly relevant to the spectroscopy of aqueous solutions. Water absorbs very strongly except in the regions 2860–1720 and 1540–950 cm^{-1}, known as "windows". The "windows" in D_2O are 2130–1250 and 1140–820 cm^{-1}. Cells with path lengths of 75 μm or

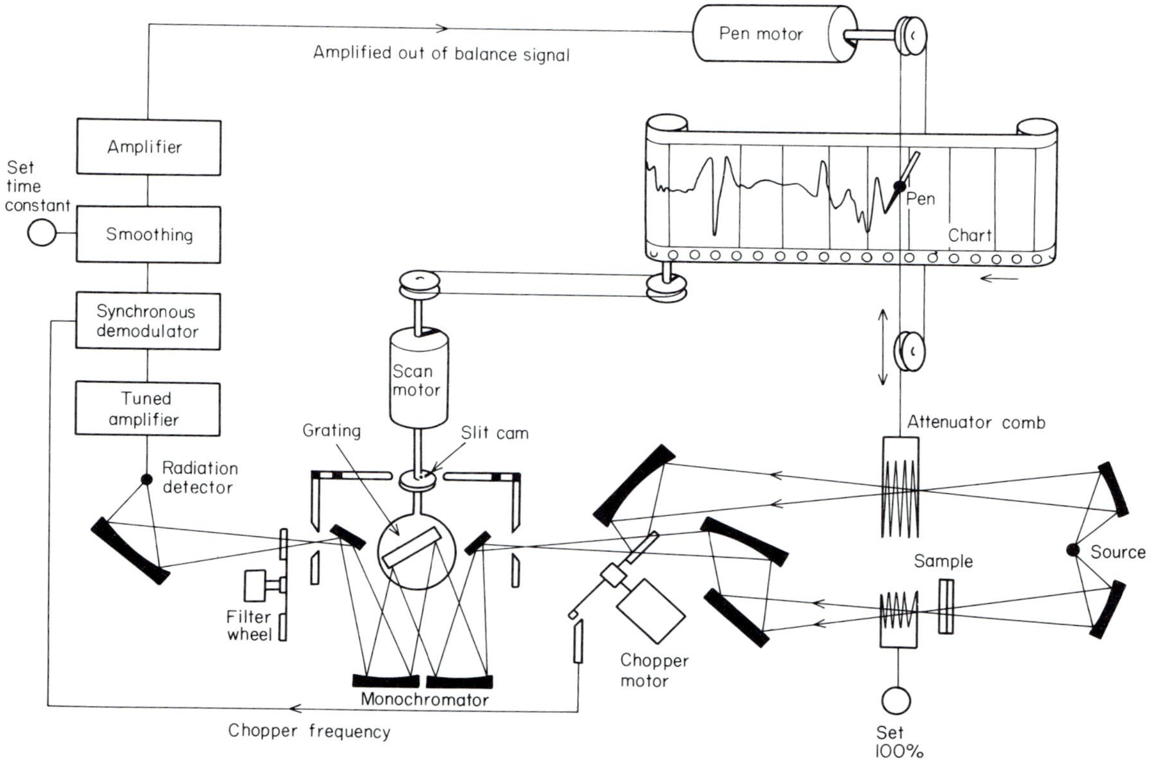

FIG. 4.1 A hypothetical optical-null infrared spectrophotometer. Reproduced with permission from "Laboratory Methods in Infrared Spectroscopy", 2nd edn (Eds R. G. T. Miller and B. C. Stace), (1972). Heyden, London.

less must be used; at 1100 cm^{-1}, 75 μm of H_2O transmit only 2% of the incident radiation. At about 1600 cm^{-1}, a region of interest in proteins, 50 μm of D_2O transmit 50% of the radiation. Clearly, great care must be exercised when using an optical-null instrument at low transmittances, since spurious results can easily be obtained when the instrument is nearly "dead".

A ratio-recording spectrophotometer

The optical layout of such an instrument is very similar to that of the optical-null. However, there is no reference attenuator. Instead the intensity ratio $I_{sample}/I_{reference}$ is recorded directly. On one instrument, the Perkin–Elmer model 580, the photometric accuracy is specified at 0·2% transmittance. The precision is not specified, but repeatability is within 0·05% transmittance. This represents a 20-fold improvement on the precision of an optical-null instrument.

The full potential of the ratio-recording instrument can only be realized if a computer is used to work up the spectrum. There are two facets to this. Firstly, the spectrum can be scanned n times to increase the signal-to-noise ratio by approximately $\sqrt{n}$ (as in NMR spectroscopy). This is advantageous when a dilute biosample is observed against the background of a strong water spectrum. Then, the background can be removed digitally, by subtracting a stored background spectrum. The sort of results that can be achieved are illustrated in Fig. 4.2. The technique is so powerful that one can imagine spectra being obtained from very thin samples of H_2O solutions even in the 1600 cm^{-1} region of intense water absorption.

Interferometers

Interferometers were originally developed for the far-IR region (below 250 cm^{-1}, say). Recently, improved technology has made it possible to use an interferometer to record spectra in the usual range up to about 4000 cm^{-1}. A typical layout is shown in Fig. 4.3. Light from the IR source is collimated and directed at a beam-splitter. Half the beam goes towards a fixed mirror, and half to a moving mirror. When the beams are recombined by the beam-splitter there may be a phase difference between the two beams resulting from the optical path-length difference (opd). The moving mirror travels at constant velocity, causing the opd to change systematically. The light that emerges from the beam-splitter is modulated in the characteristic form known as an interferogram. This interferogram is modified by passage through a sample according to the sample's absorption frequencies. The resultant interferogram must be transformed by computer (using the fast Fourier transform, as in

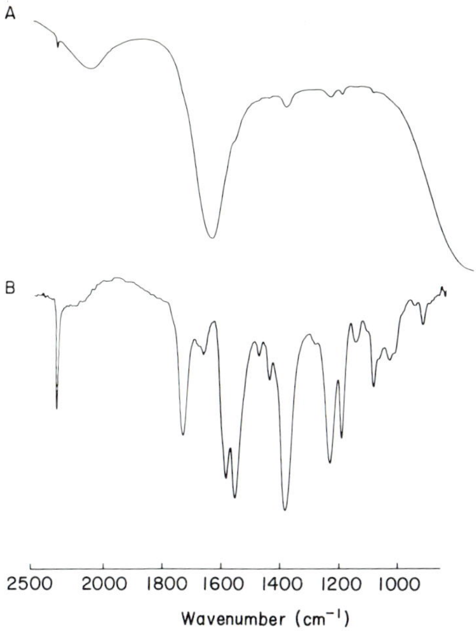

FIG. 4.2. Infrared absorption spectrum of soluble aspirin recorded in 2% solution between BaF_2 plates at 7 μm separation. Spectrum kindly supplied by Perkin-Elmer Limited. A, original spectrum obtained in one scan. B, spectrum A after subtraction of water spectrum and ×15 expansion. Spectrometer model 580B.

NMR spectroscopy, see Chapter 6) into a spectrum. The ratio of sample to reference spectrum is also calculated by computer, and this ratio-record is what the user finally sees.

The interferometer requires a dedicated computer to store data and perform the transform, but this computer can also be used for spectral accumulation and data treatment. The question then naturally arises, is there any clear advantage to using an interferometer as opposed to a ratio recording spectrophotometer?

For biochemical applications the situation at present is finely balanced, with the interferometer having a slight edge in the matter of frequency accuracy. This arises from the fact that the moving mirror position is monitored by a He/Ne laser with very high wavelength accuracy. Hence for any study involving small frequency shifts or requiring extensive multiple scanning, an interferometer is indicated. For most other applications a ratio-recording spectrophotometer and data processing computer is likely to be a cheaper but satisfactory alternative. For aqueous solutions the performances of the two systems should be comparable.

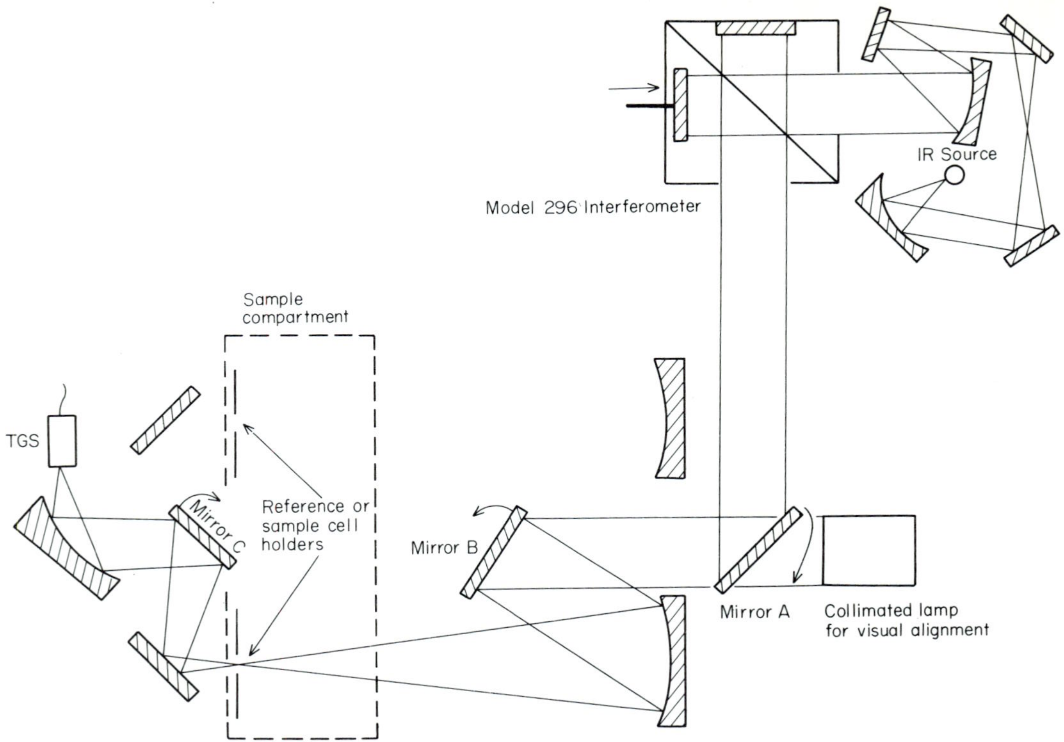

FIG. 4.3 Optical layout of the Digilab FTS-14 interferometer. The mirrors A and B can be moved to redirect the light into the reference cell. Reproduced with permission from "Laboratory Methods in Infrared Spectroscopy", 2nd edn (Eds R. G. J. Miller and B. C. Stace), (1972). Heyden, London.

Sampling Techniques for Infrared Spectroscopy

Specialist sources must be consulted for the many and varied methods of sample preparation for IR spectroscopy. The simplest technique involves grinding the sample in an agate mortar, and suspending the finely ground material in nujol (liquid paraffin); the mull so produced is supported between IR-transparent plates of, for example, sodium chloride. More sophisticated techniques are required for most biopolymers, including proteins, which may be denatured by grinding.

For aqueous solutions there are two alternative techniques. A sample in solution may be examined by transmission, or by (multiple) internal reflection (MIR). For transmission, it is necessary that the cell windows neither dissolve in water nor react with the sample (silver chloride, though insoluble, could attack cysteine and other sulphur-bearing residues). Windows of CaF_2 are the cheapest, but are opaque below about 1050 cm^{-1}. A point worth making is that fused silica is transparent above about 2700 cm^{-1} and can be used to examine solutions in the overtone regions; indeed, much early IR spectroscopy was performed in this way.

In the MIR method the sample is held in a glass or steel container, and is in contact with a suitable prism (e.g. germanium prism) Fig. 4.4. At first sight this seems a more attractive sampling technique than using a transmission cell with its expensive windows and narrow path length. In practice analysts are somewhat divided over the merits of the two systems. The main drawback to MIR seems to be that it requires very concentrated solutions. This requirement might be less demanding when using computer-based data processing; weak features can easily be extracted and expanded against a solvent spectrum if the signal-to-noise ratio is sufficiently large.

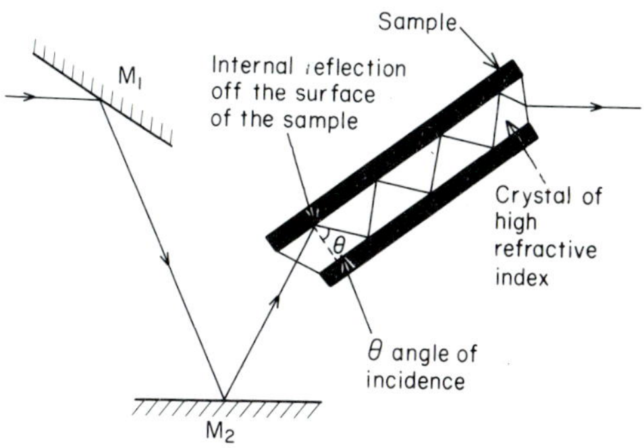

FIG. 4.4 Schematic representation of a multiple internal reflection cell. The angle of incidence θ may be varied by rotating the mirror M_2 and/or the crystal.

These points are illustrated in two sets of aqueous solution spectra. Figure 4.5 shows a spectrum of carbon monoxide bound to sulphmyoglobin (sulph-Mb). The concentration of ferrous sulph-Mb was 7·0 mM. The aqueous solution was held between CaF_2 windows with a spacing of 50 μm. The spectrum was obtained with a Digilab FTS-14 interferometer. Two thousand interferograms were coherently co-added at 2 s per scan.

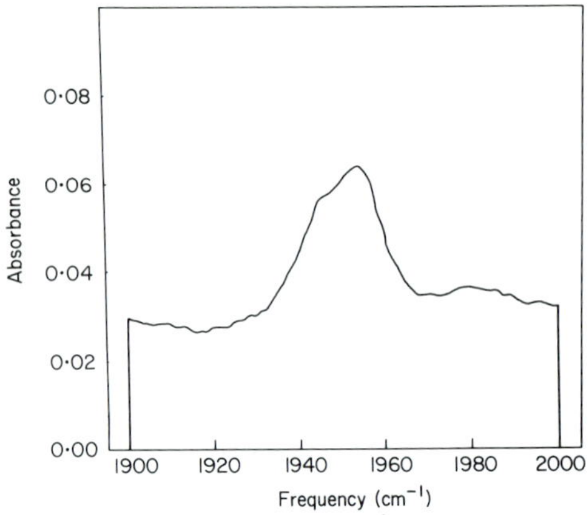

FIG. 4.5. Infrared transmission spectrum of bound carbon monoxide. For details see text. (J. A. Berzofsky, J. Peisach and J. O. Alben (1972). *J. Biol. Chem.* **247**, 3774.).

Figure 4.6 shows spectra of ammonium dihydrogen phosphate taken with a Ge prism MIR device. The effective path was about 10 μm long. Spectra were obtained from an FTS-14 interferometer using 400 scans. These two examples amply illustrate the power of interferometry for the examination of aqueous solutions.

Orientated samples

Many biopolymers, particularly those of a fibrous nature can be placed in a particular orientation in the IR spectrometer. If the IR radiation is polarized, absorption may vary with the angle of polarization. These dichroism studies can yield valuable information.

Raman Spectrometers

The spectrometer set-up is the same for both normal and resonance Raman spectroscopy. A light source (a laser) providing a thin, collimated beam of

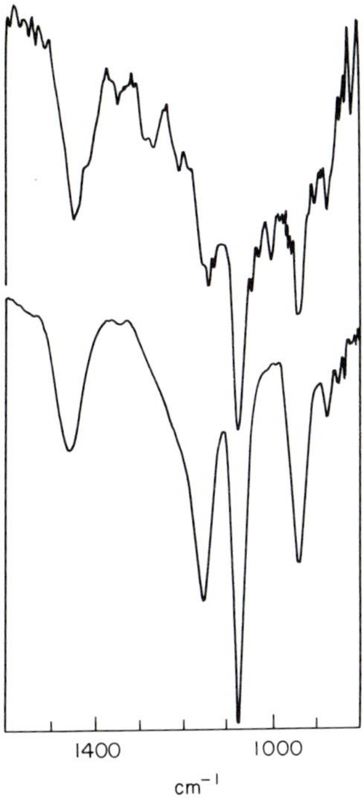

FIG. 4.6 Multiple internal reflection spectra of $NH_4H_2PO_4$ solutions obtained interferometrically. The concentrations were: upper trace 0·05 M; lower trace 0·5 M. (M. J. D. Low and R. T. Young. (1973). *Spectrochim. Acta* **29A**, 1761.)

coherent, monochromatic, plane polarized light is used to illuminate the sample. The scattered radiation is collected at right angles to the input beam, and is passed through a high quality monochromator to a detector. Raman shifts are measured in cm^{-1}, using the laser frequency as zero, and are usually measured on the "red" side of that zero. That is, the Raman radiation has the frequency of the input minus the vibration frequency—a quantum of vibrational energy is lost in the scattered radiation. The corresponding line on the blue side of the input frequency is less intense because it must contain a quantum of vibrational energy collected from a molecule, and usually there are fewer molecules ready to give up vibrational quanta than to receive them. Lines on the red side of zero are called *Stokes lines*; those on the blue side are called *anti-Stokes* lines. A typical optical arrangement for a Raman spectrometer is shown in Fig. 4.7.

The laser which is used for Raman spectroscopy is a device which emits an intense beam of light in the form of a very narrow beam whose diameter is

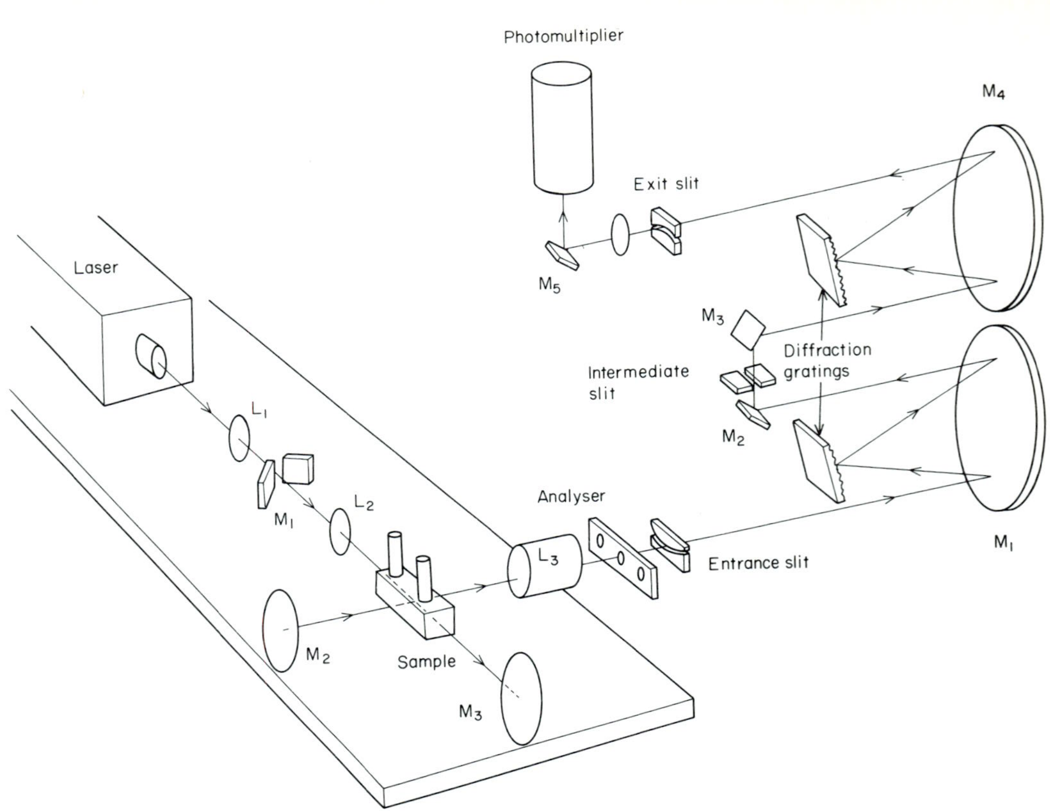

FIG. 4.7 Optical layout of Coderg pHO Raman spectrometer.

usually less than 1 mm. Some typical operating characteristics of useful lasers are given in Table 4.2. All except the dye laser emit radiation at a few precisely defined wavelengths. The dye laser emits a line of precise but tunable wavelength. The tunable range depends on the dye used. For example, rhodamine 6G gives a useful output range of about 570–620 nm with peak power of approximately 800 mW at 580 nm when using a 4 W pump. (The dye laser is pumped by another laser, such as a high power Ar^+ laser.)

The output powers given in Table 4.2 are nominal and vary according to manufacturer and model. The intensity of the Raman radiation is directly proportional to the laser power, so it is advantageous to use the highest power available with three important restrictions—the power should not be so high that thermal decomposition of the sample occurs, the wavelength should not be absorbed by the sample unless the resonance Raman spectrum is being recorded, and the wavelength should be such as to avoid sample fluorescence. A further point to note is that Raman intensity is proportional to ν^4, the fourth power of the Raman frequency, so that the bluer the light, the more intense are the spectra which are obtained.

TABLE 4.2 Operating characteristics of some typical lasers used in Raman spectroscopy

Type	Wavelength (nm.)	Output power (mW)
Kr^+	647·1	1000
He/Ne	632·8	100
Ar^+	514·5	2000
	488	2000
He/Cd	441	50
	325	15
dye	tunable	

Thermal decomposition can be avoided by such strategems as cooling or spinning the sample. The spinning sample technique is essential for resonance Raman spectroscopy.

Fluorescence, which is a major problem with biosamples, can mask the Raman spectrum. Fluorescence may be 1000-fold stronger than the Raman emission in adverse cases. Since fluorescence is dependent upon excitation frequency it can sometimes be reduced by a change, usually towards a redder exciting line. The Kr^+ ion laser is then indicated for fluorescent samples, even though there is a considerable loss in the ν^4 factor and there may be some loss in the detection sensitivity. Mixed Ar^+/Kr^+ lasers are also available, though they give slightly less output power than their single-gas counterparts.

Another approach to the elimination of fluorescence is based on time discrimination. Raman scattering is an essentially instantaneous process, but fluorescence occurs with a delay of more than 1 ns (see Chapter 3); e.g. for tryptophan the delay is approximately 3 ns. Therefore if the laser is pulsed, and the pulses have a width of 0·5 ns or less the fluorescence can be blocked by locking the signal amplifier to the laser pulses. By this technique the fluorescence can be reduced by a factor of around 10. A working system is illustrated in Fig. 4.8. For resonance Raman work it is desirable to have a wide range of laser lines available, so as to match the electronic absorption. A tunable laser is, of course, more convenient.

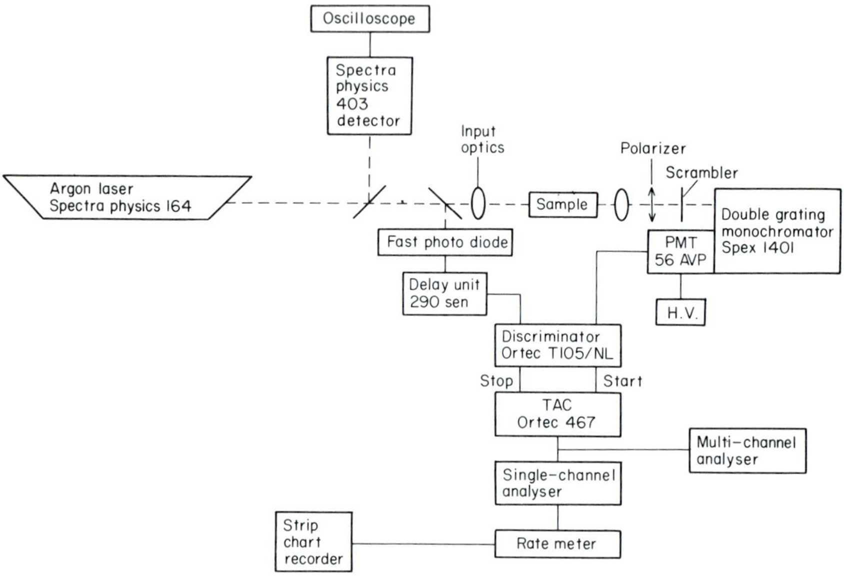

FIG. 4.8. Schematic diagram for equipment operating time-discrimination against fluorescence. (I. W. Shepherd. (1977). *In* "Advances in Infrared and Raman Spectroscopy," Vol. 3, Eds R. J. H. Clark and R. E. Hester, 127. Heyden, London).

Little need be said here about the monochromator and detector. A single-grating monochromator is inadequate for Raman work. In practice two or three gratings are used with additive dispersion, but the "triple" is only needed for work with solid samples at very low frequencies. The recent introduction by Jobin-Yvon of holographically ruled gratings may in fact have rendered the "triple" obsolete.

Sampling Techniques for Raman Spectroscopy

Raman spectroscopy of aqueous solutions is particularly easy. Because the method operates at visible wavelengths glass cells can be used. Secondly, the intensity of the water vibration at 1640 cm^{-1} is low in the Raman effect. Spectra can be recorded in one scan over the whole frequency range, with a lower limit of about 250 cm^{-1} set by Rayleigh scattering. Of course the sample should be free from suspended particles or colloidal material to minimize Tyndall scattering. This requirement is partially incompatible with biosamples, since Tyndall scattering is a well-known technique for determining particle size in, for example, proteins. However, since Tyndall scattering leaves the input frequency unchanged this will only be a problem in the study of low-frequency vibrations.

Raman spectroscopy is essentially a single-beam technique. Difference spectra can be obtained at modest cost by using a rotating or oscillating cell divided into two compartments, one containing the sample and the other containing reference material (usually solvent). Signals reach the detector in alternation and the reference signal can be subtracted from the sample signal electronically. To observe a resonance Raman spectrum, the sample and laser beam must be in relative motion; otherwise absorption of the radiation leads to sample heating and destruction. Usually the motion is achieved by spinning the sample in the laser beam. Apart from this, the only other precaution to be taken is avoidance of self-absorption of the Raman radiation by the sample itself. This is easily achieved by spinning the sample into a fairly thin film, and irradiating into its edge.

Most resonance Raman work has been confined to coloured compounds, that is, compounds with absorption bands in the visible region. The reason for this is that UV lasers are not yet a practical proposition. Light at UV wavelenghts can be obtained by "frequency-doubling" the output of a visible laser, but the output powers are low. One can confidently predict that UV lasers will soon be available commercially, opening up new prospects for resonance Raman spectroscopy.

In the following sections we turn to a discussion of the applications of vibrational spectroscopy to molecules of biochemical interest. Where possible we shall avoid discussion of the spectroscopy of model compounds (e.g. non-naturally occurring homopolynucleotides) although the study of simple model compounds is a necessary preliminary to the study of complex biological materials. The space allotted below to each type of compound is roughly proportional to the amount of published information on it. Applications are discussed by compound type, bringing IR, Raman or resonance Raman evidence to bear wherever appropriate. Because of the technical problems alluded to above, most work on a given class of compound has been performed with one technique only. Hopefully this unfortunate situation will in future be rectified somewhat, because IR and Raman spectra really are complementary, and do not give the same information. Nowhere is this more clearly illustrated than in antiparallel β sheet polypeptides where the symmetry of the system dictates that IR and Raman frequencies shall not coincide.

Proteins

Globular proteins

Historically, attention has been focussed on vibrations associated with the amide group of the peptide linkage. The reason for this is clear. The hydrogen bonding in which the amide group is involved is of great importance in determining the secondary structure of a protein. In homopolypeptides the α-helix and β-pleated sheet secondary structures can be clearly differentiated. From results of studies on these compounds, information on the secondary structure of proteins can be gleaned. But, because proteins have variable amounts of α, β and random coil regions, information tends to be ambiguous. When a protein is denatured, the changes in secondary structure can be followed directly in the amide vibrations.

The so-called amide I band, at approximately 1660 cm^{-1} can be studied by both IR and Raman spectroscopy of aqueous solutions. The solvent background is greatly decreased if D_2O is used instead of H_2O, but exchange can occur with this solvent, complicating the spectra. For α structures the amide I band lies at 1650–1657 cm^{-1}, while for β structures, including both parallel and antiparallel pleated sheets, the range is 1665–1680 cm^{-1}. Random coils vibrate at 1665 cm^{-1}. In a protein, overlap between amide I bands due to different structures usually gives rise to a single band. An important aspect of the β-structure is that Raman and infrared frequencies may not coincide, and the separate observation of both spectra may help to characterize regions of β-structure.

The amide III region between 1230 and 1295 cm^{-1} has hitherto only been studied by Raman methods on aqueous solutions. The β conformation gives an intense line at 1235 ± 5 cm^{-1}, whereas the α conformation scatters only weakly at 1260–1295 cm^{-1}. Random coils give a line near 1245 cm^{-1}. Considerable changes occur in this region when a protein is denatured. The lysozyme system is probably the best-studied one in this respect. Its amide III band has components at 1240, 1262 and 1274 cm^{-1}; on treatment with 6 M LiBr the shoulder at 1274 cm^{-1} almost disappears, the main peak shifts from 1262 to 1245 cm^{-1} and the 1240 cm^{-1} line is no longer observable. The denaturing of lysozyme in LiBr solution is therefore accompanied by increasing irregularity in the secondary structure. One reason why the amide III region is so useful is that it is free of major peaks due to the amino acid side chains. Studies similar to that of lysozyme have been carried out on numerous other proteins, including insulin, ribonuclease, α-chymotrypsin, tropomyosin, α-casein, serum albumin and β-lactoglobulin, with similar results. The insulin case having received the most attention, is shown in Fig. 4.9.

Overall, vibrational spectra of proteins consist mainly of peaks due to the amino acid side chains. In the Raman spectrum, Trp, Phe and Tyr units give intense Raman lines because of the aromatic nucleii in the side chains. Remarkably, the whole Raman spectrum can be reconstructed by superimposition of the spectra of the individual amino acids with a high degree of fidelity. This is illustrated in Fig. 4.10 for lysozyme. Note that the greatest differences are observed in the conformation-sensitive amide I and III bands. However the side chains may sometimes be sensitive to conformational factors. Tyrosine in particular has a doublet at 850 and 830 cm^{-1} whose intensity ratio varies with the tyrosine environment. For tyrosyl groups on the protein surface the lines have an intensity ratio of 10:8, but for "buried" tyrosyl groups the ratio is 5:10 (see Chapter 2 for corresponding evidence from UV spectra).

The S—S stretching vibration in cystine bridges gives a strong Raman line at about 675 cm^{-1}, and the C—S vibration is found near 510 cm^{-1}. The ratio of the intensities of these two lines has been used to deduce the C—S—S angle. In crystalline lysozyme, an angle near to 114° was deduced, but in ribonuclease it was found to be nearer 104°, the same as in crystalline cystine hydrochloride. A combination of the S—S and tryptophan residue information has given rise to some interesting and detailed speculation concerning lysozyme denaturation. Three of the tryptophans (numbers 62,63 and 104) are located in the active site. The closely positioned cystine 64–80 is also near the active site. The speculation is that irreversible thermal denaturation brings about rupture of the S—S bridge and hence the line at 504 cm^{-1} decreases in intensity. Other intensity changes at 1338 and 1363 cm^{-1} indicate changes in the tryptophan environment. These observations lead to the exciting prospect that enzyme mechanisms may be studied directly, using perhaps tryptophan lines to monitor the reaction.

Another feature of protein chemistry that can be studied by vibrational spectroscopy concerns the differences between the material in solution, or as a crystalline or freeze-dried powder. Where the structure in the crystalline state has been determined by X-ray crystallography, it is particularly valuable to know what conformational changes, if any, accompany dissolution. For ribonuclease A the Raman spectra of crystalline and freeze-dried powder were very similar, but the solution shows marked differences at 514 cm^{-1} (S—S stretch) which were interpreted in terms of differences in the cystine bridge angles. The environment of the "buried" tyrosine residues also appeared to change. The spectra are illustrated in Fig. 4.11.

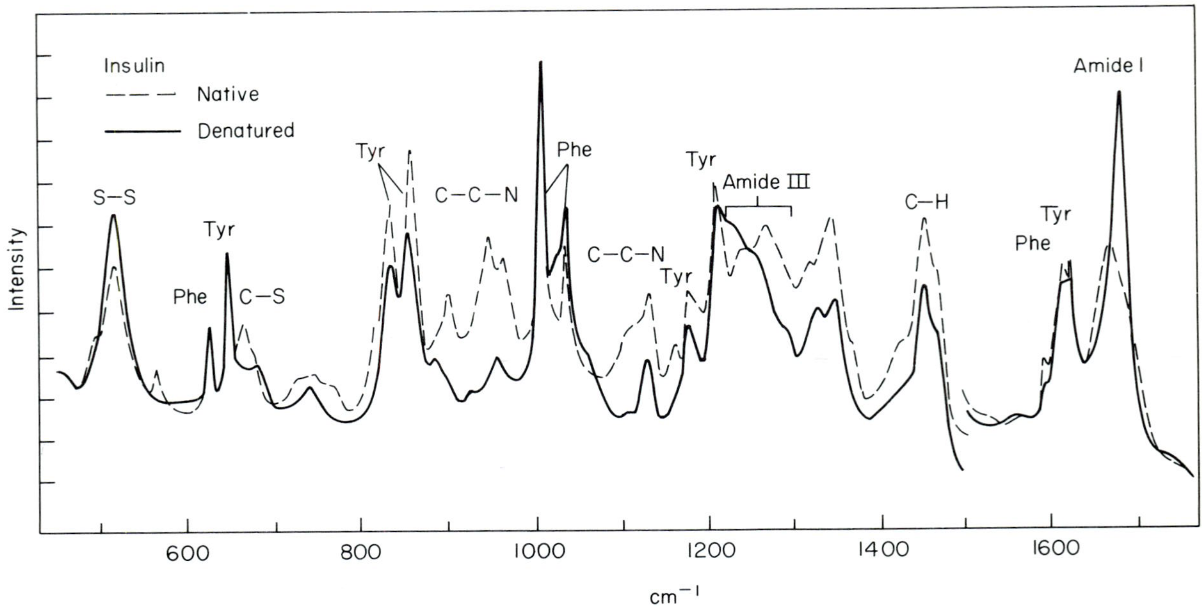

FIG 4.9. Raman spectra of native and denatured insulin in the solid state. (N. T. Yu, C. S. Liu and D. C. O'Shea. (1972). *J. Molec. Biol.* **70,** 117).

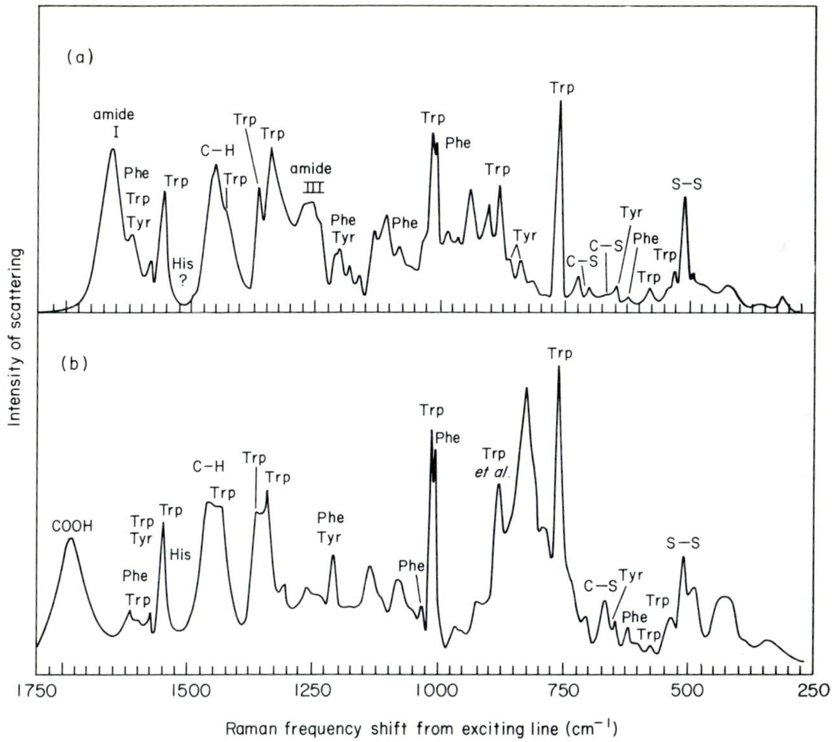

FIG. 4.10 Raman spectra of (a) lysozyme in water pH 5.2 and (b) sum of corresponding amino acid spectra taken at pH 1.0 (R. C. Lord and Nai-Teng Yu. (1970). *J. Molec. Biol.* **50**, 509.)

Fibrous proteins

The fibrous proteins, in contrast to the globular proteins discussed in the previous section, have extended regions of α or β secondary structure. Consequently the vibrational spectra tend to show features similar to those found in the model compounds, i.e. homopolypeptides. Wool, for example, consists of keratin largely in the α form. Most vibrational studies to date have been concerned mainly with the amide I and III bands, and with the determination of proportion of α, β and disordered regions in the structure. Samples studied include silks, hair, feathers, collagen and elastin.

Two areas of study merit particular mention. Stretching of wool is accompanied by a transition from α to β conformations. This can be followed most easily in the amide III region. Treatment with steam, on the other hand causes supercontraction to a cross-β structure, whilst supercontraction caused by

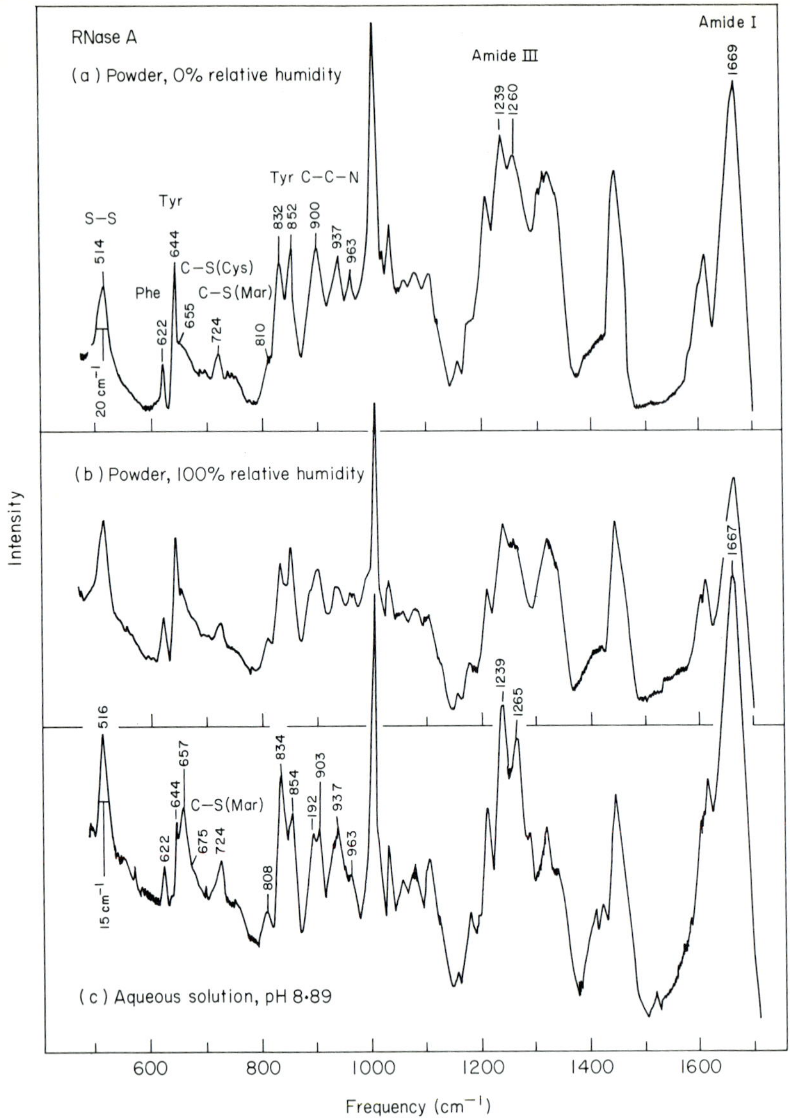

FIG. 4.11 Raman spectra of ribonuclease in powder and solution states. (N. T. Yu, B. H. Jo and C. S. Liu. (1972). *J. Am. Chem. Soc.* **94**, 7572.)

LiBr treatment appears to be accompanied by a chain-disordering similar to that which occurs in lysozyme, given the same treatment.

The conversion of collagen to gelatine is accompanied by very small changes in the Raman-active amide III bands at 1274 and 1248 cm^{-1}, suggesting considerable retention of secondary order. This observation is in conflict with CD results, which indicate only the presence of random coils in gelatine. These contradictions show the value of bringing various techniques to bear on a biochemical problem.

Resonance Raman labels

A promising start has been made in this field. Dye-stuffs with reactive groups can interact with various proteins and the modification of resonance Raman spectra can be used to monitor this interaction. A striking application is in the study of enzyme inhibitors. 4-Sulphonamido-4′-hydroxyazobenzene is a powerful inhibitor of carbonic anhydrase; it was deduced that the inhibitor anions bind directly to the zinc ion at the active site. Other label studies include hapten–antibody interactions and enzyme intermediates.

Metalloproteins

The presence of a transition metal atom in a protein often confers on the material a strong colour, due to charge-transfer transitions, most often of the Cys(S)$\rightarrow M^{n+}$ type (see p. 19). This colour makes possible resonance Raman studies. Since the metal atom is involved both in the charge-transfer process and at the active site of metalloenzymes, we have the opportunity of studying directly the active sites of various enzymes. Those vibrations which distort the molecule in the same way as the electron transition are resonance enhanced, symmetry permitting. Therefore in some cases, notably with Cu(II), the M–ligand stretching vibration is enhanced because the electronic transition is accompanied by an increase in metal–ligand bond length. In others, notably with some octahedral Fe(III) compounds, the charge is transferred to a metal π orbital, with little change in bond length. In these cases the metal–ligand stretching vibration is not enhanced. Lastly, one must note that d–d transitions (see Chapter 2) do not give rise to resonance enhancement.

Oxygen carriers

In Table 4.3. we list the frequencies observed in some oxygenated metalloproteins. The O—O stretching bands have been unambiguously assigned by

TABLE 4.3 Wave numbers for some oxygenated metalloproteins

	cm^{-1}			
Protein	$\nu(^{16}O—^{16}O)$	$\nu(^{18}O—^{18}O)$	$\nu(M—^{16}O)$	$\nu(M—^{18}O)$
Haemocyanin	742	704		
Haemerythrin	844	798	500	478
Oxyhaemoglobin	1385?[a]		567	540

[a] Obtained by low-temperature IR spectroscopy

the isotopic substitution changes. The frequencies observed* for haemocyanin and haemerythrin show that the oxygen has been converted to peroxide, O_2^{2-}, because the vibration frequency lies in the range characteristic of that ion. Both of these proteins contain two metal atoms at the active site, so the overall reaction can be written

$$\text{haemocyanin: } 2Cu(I) + O_2 \rightleftharpoons 2Cu(II)—O_2^-$$
$$\text{haemerythrin: } 2Fe(II) + O_2 \rightleftharpoons 2Fe(III)—O_2^-$$

Thus, direct observation of the vibrations associated with the O_2 entity has given unambiguous evidence concerning electronic structure, which no other spectroscopic technique can provide.

The evidence for oxyhaemoglobin is inconclusive, because of interference from haem bands (see below), and ν(O—O) has not been fully characterized. However, much evidence points to the reduced oxygen being in the superoxide, O_2^-, form, with the reaction being

$$\text{haemoglobin: } Fe(II) + O_2^- \rightleftharpoons Fe(III)—O_2^-$$

Electron carrier and transfer enzymes

Resonance Raman spectra of the simple bacterial protein rubredoxin, which is known to contain one iron atom surrounded somewhat irregularly by four cysteine sulphur atoms, show clear enhancement of the vibrations associated with the FeS_4 group. Stretching modes for Fe—S were identified at 368 and 314 cm^{-1}. Adrenodoxin is known to contain two iron atoms associated with four cysteine ligands and two sulphide ions. Replacement of sulphide by selenide made it possible to identify the Fe—S (Cys) stretching vibration at 350 cm^{-1} and Fe—S stretching at 397 and 297 cm^{-1}. Resonance Raman spectra of ferredoxins have not yet been reported.

* References to original reports quoted in this chapter will be easily found in the review articles given in the Bibliography.

The blue copper proteins also yield information on the metal–ligand vibrations. The resonance Raman spectra are rich in lines, as seen in Fig. 4.12. The assignment, however is controversial. Charge transfer is (Cys)S→Cu(II), but evidently bands other than Cu—S are also enhanced. This spectroscopic problem is at present unresolved.

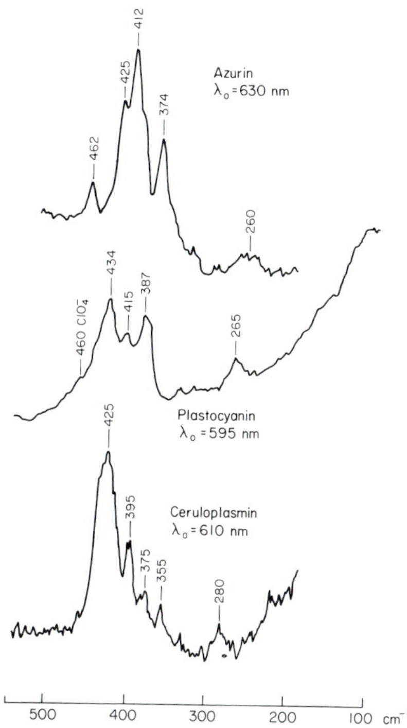

FIG. 4.12 Resonance Raman spectra of three blue copper proteins obtained with excitation from a tunable rhodamine 6G dye laser at the indicated wavelengths. Top: *Pseudomonas aeruginosa* azurin, purity index (A_{280}/A_{600}) 4·1, pH 7·4, 0·6 mM. Middle: Spinach plastocyanin, purity index 1·4, pH 7·4, 1·4 mM. Bottom: Human serum ceruloplasmin, pH 5.8, 0·9 mM. (V. Miskowski, S-P.W. Tang, T. G., Spiro, E. Shapiro, E. and T. H. Moss, (1975). *Biochemistry*, **14**, 1244.)

Iron carriers

Assignment of the resonance Raman spectra of Fe(III) transferrins is also controversial. Chemical modifications and spectroscopic techniques indicate that each Fe(III) is bound to three tyrosine residues and two histidine residues, with one equivalent of bicarbonate being needed for full biological activity. In this case the Fe—O vibrations are not enhanced, and the controversy centres on the nature of the lines that are enhanced—in the region 1150–1610 cm^{-1}; one school favours an assignment of all lines to tyrosine, since charge transfer is Tyr(O)→Fe(III), whereas another assigns one or more bands to histidine. The role of bicarbonate was not clarified by this study. Because of the localized nature of Tyr→Fe(III) charge transfer, no enhancement of bicarbonate bands would have been expected even if the bicarbonate were coordinated to the Fe(III). Transferrins of Cu(II) do show enhancement of the metal–ligand vibrations.

Haem proteins

Compounds containing porphyrin rings give complex resonance Raman spectra exhibiting many interesting features. Excitation within the Soret band (see Chapter 2) at about 400 nm gives rise to A-term enhancement, but excitation within the longer wavelength bands at about 500 nm gives B-term enhancement, with some bands showing inverse polarization, i.e. $1/\rho \approx 0$. An example of the latter is shown in Fig. 4.13. Exact assignment of the porphyrin ring vibrations is very difficult, but changes can be followed empirically, and some useful structural information can be obtained.

Four of the ring modes are sensitive to the electronic nature of the iron

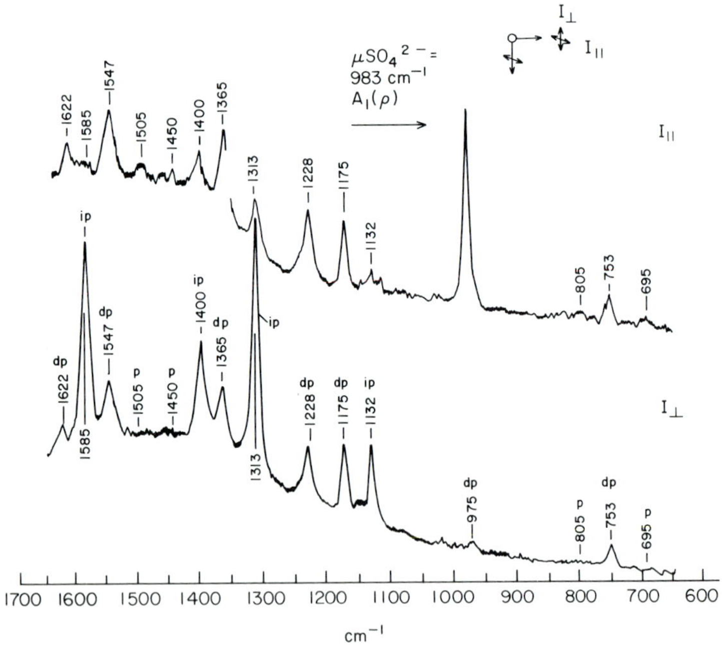

FIG. 4.13 Resonance Raman spectra of ferricytochrome *c* showing inverse polarization of some bands. In a completely inverse polarization the plane of polarization of the incident beam is rotated through 90°. (ip, inverse polarization; dp, depolarized.) The scattering geometry is shown schematically in the diagram, at the top. (Spiro, T. G. and Strekas, T. C. (1973). *Proc. Natl Acad. Sci., U.S.A.* **69,** 2622.)

atom, and characteristic ranges have been established according to whether it is Fe(II) or Fe(III), and whether either is high or low spin. A correlation chart is shown in Fig. 4.14. This chart has a interesting history. Originally it was proposed that ferricytochrome *c* and horseradish peroxidase were anomalous. Recently this proposal was reversed, and now it is proposed that it is the Fe(III) compounds metmyoglobin and methaemoglobin which are anomalous in the same sense as the two fluorides in Fig. 4.14. This "anomaly" is linked to the degree of "doming" of the porphyrin, since doming will tend to reduce π-bonding in the tetrapyrrole structure, thus causing a substantial drop in Raman frequencies. Unfortunately further discussion of this fascinating topic would take us beyond the scope of this chapter.

It is impossible in this brief summary to do justice to the immense amount of work that has been carried out on the various cytochromes involved in electron transport in the mitochondrial chain.

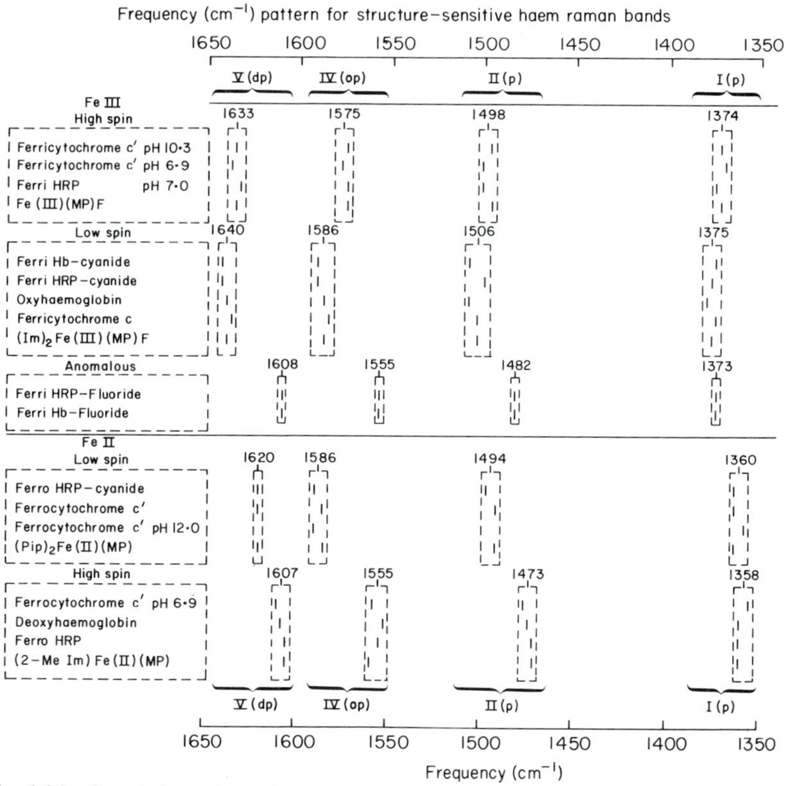

FIG. 4.14 Correlation chart for structure-sensitive Raman lines in haem proteins. HRP = Horseradish peroxidase; MP = Mesoporphyrin 1X; Hb = Haemoglobin; Pip = Piperidine; Im = Imidazole. (p, polarized; dp, depolarized; ap, anomalously polarized.) (Spiro, T. G. and Burke, T. M. (1976). *J. Am. Chem. Soc.* **98**, 5482.)

Nucleic Acids

The vibrational spectroscopy of nucleic acids has not progressed much beyond the stage of study of model compounds. The spectra of the bases and nucleotides are well characterized. The Raman spectra of the latter are similar to those of the bases because of weak scattering by the ribose or deoxyribose component.

The phosphate group gives at least two strong Raman lines at about 1100 and 810 cm^{-1}. These can be used to differentiate between AMP, ADP or ATP. In the nucleic acids themselves both the frequency and intensity of the 1100 cm^{-1} band are independent of main chain conformation. The 810 cm^{-1} band is characteristic of "A-form" main chain conformation, and has been used to estimate the proportion of A-form in various RNAs.

Many of the ring vibrations are lower in intensity in the RNAs than in the nucleotides, a variation that correlates with the decrease in UV absorption wavelength (hypochromism, see p. 32). The fact that the spectrum of an RNA cannot be simulated by superimposition of spectra of synthetic nucleotide-pair polymers in appropriate proportion was demonstrated for a formyl methionine *t*RNA which contains 24 guanilic, 25 cytidylic, 15 adenylic and 8 uridylic acid residues. This suggests that the spectrum contains features due to the macromolecular conformation and leads one to hope that vibrational spectroscopy may ultimately yield some useful structural information on RNAs.

The feasibility of using Raman spectroscopy to study viruses in an aqueous environment has been demonstrated by a study of R17 Virus. The guanine residues in particular showed sensitivity to the nature of protein–RNA aggregation.

The UV absorption of the heterocyclic rings is in the region 260–280 nm, and so, apart from some preliminary work on adenine using frequency-doubled 514·5 nm radiation, little resonance Raman work has been performed. However it has been suggested that Raman intensity changes induced by base-stacking are related to the UV hypochromism by a pre-resonance enhancement process.

The unusual base 4-thiouracil is found in many amino acid *t*RNAs of *Escherichia coli*, and has an absorption band at 330 nm. Resonant enhancement of the 4-thiouracil vibrations has been achieved with 351·1 and 363·8 nm Ar^+ radiation in a rotating cell device. When a stationary sample was used, a photochemically induced bridge-forming reaction was observed in most *t*RNAs, between the 4-thiouracil at position 8 and cytosine at position 13.

These glimpses suggest that once the technical difficulties of operating lasers at UV wavelengths are overcome, the resonance Raman investigation of nucleic acids should be very profitable.

Other Compounds of Biological Interest

Lipids

The interaction of cholesterol with lipids has long been studied and is still far from understood. Raman spectroscopy has been used to study both model systems and membranes. The gel–liquid transition can be followed by monitoring certain Raman intensities and it has been demonstrated by this technique, to quote one example, that cholesterol broadens the gel–liquid transition as a function of temperature, in aqueous solutions of D,L-dipalmitoyl lecithin sonicates. Bands at 1089 and 1128 cm^{-1} show wide intensity variation over the transition.

Chlorophyll

As a porphyrin derivative, chlorophyll gives resonance Raman spectra similar to those of porphyrins. Spectra of spinach chloroplasts and grana were found to be complex and varied with the wavelength of the exciting radiation. Bands due to chlorophyll *b* are more intense than those of chlorophyll *a*. An interesting result of this study was the direct observation of the photosynthetically active carotenoid neoxanthin in very low concentration with excitation at 476·5 nm (Ar^+). The carotenoid's spectrum was much weaker when 466 nm radiation was used.

Polyolefines

The carotenoids give very intense resonance Raman spectra with A-type enhancement of C—C stretching vibrations. These spectra are so intense that they can be obtained directly from a raw carrot or tomato!

The visual pigment rhodopsin consists of the carotenoid retinal bound through a Schiff base linkage to the protein opsin. Raman studies of this and related materials are greatly complicated by their photosensitivity. There is some Raman evidence for photodeprotonation of the Schiff-base nitrogen, based on the value of the C—N stretching frequency, which is consistent with other evidence on the mechanism of photoreception.

Vitamin B_{12}

The Raman spectra of B_{12} derivatives resemble those of the other polyolefines rather than those of porphyrins, with strong resonant enhancement of a C=C band at about 1500 cm^{-1}. As with the porphyrins the Raman spectra are not sensitive to changes in the axial ligands.

Conclusions

In addition to the classes of compound specifically mentioned above many other molecules of biochemical interest have been subjected to vibrational analysis. These include carbohydrates and polysaccharides, steroids and antibiotics. Lack of space alone is responsible for their omission here, as many useful results have been obtained.

In viewing the applications of vibrational spectroscopy to biochemical problems it is clear that the last ten years have seen enormous improvements in technique and range of application. After many years in which it might have seemed that molecules of biochemical interest were too complex for vibrational (mainly at that time IR) spectroscopy, we see that now, direct application of IR and Raman techniques are being made to various biological materials.

One can confidently predict that this revolution will continue to evolve over the next few years. Infrared spectroscopy of aqueous solutions, even at low frequency, should become widespread. The use of computers for data processing should make more quantitative results available. Resonance Raman studies using the label technique or with lasers operating at UV wavelengths will greatly extend the scope of available studies.

The value of vibrational spectroscopy lies in its unique nature, probing the way in which atoms move within a molecule. In this respect it is complementary to the other spectroscopic techniques described in this book and sometimes yields results that seem at variance with them. In this way the advantages and limitations of each technique become known.

Further Reading

The very extensive nature of the literature on vibrational spectroscopy means that this short bibliography is highly selective, concentrating mainly on books and review articles of direct interest to biochemists. These include most of the sources of this chapter.

Books

Barnes, A. J. and Orville-Thomas, W. J. (Eds) (1977). "Vibrational Spectroscopy—Modern Trends", Elsevier, Amsterdam, An excellent brief chapter by Tsuboi on nucleic acids and proteins.

Clark, R. J. H. and Hester, R. E. (Eds) (1975). "Advances in Infrared and Raman Spectroscopy", Vol. 1 Heyden, London, This book has excellent chapters by Frushour and Koenig on Raman spectra of proteins and by Spiro and Loehr on Resonance Raman spectra. Volume 3 of the series has a good chapter by Kiefer on recent techniques in Raman spectroscopy. There are currently six volumes in this series.

Durig, J. R. (Ed) (1976). "Vibrational Spectra and Structure", Vol. 5, Elsevier, Amsterdam. A broad-brush review by Spiro on biochemical applications of resonance Raman spectroscopy. Volume 3 has a chapter on Raman spectroscopy of biopolymers.

Freeman, S. K. (1974). "Application of Laser Raman Spectroscopy". Wiley, New York.
Chapter 10 contains useful summaries on a variety of bio-systems.

Griffiths, P. R. (1975). "Chemical Infrared Fourier Transform Spectroscopy", Wiley, New York, An authoritative account of the construction and use of interferometers.

Jones, D. W. (Ed). (1976) "Introduction to the Spectroscopy of Biological Polymers", Academic Press, London and New York. Separate chapters on IR and Raman work.

Miller, R. G. J. and Stace, B. C. (Eds) (1972) "Laboratory Methods in Infrared Spectroscopy", 2nd edn, Heyden London Good on all practical aspects of IR spectroscopy. Good chapter on MIR techniques.

Reviews

Grossman, W. E. L. (1976). Raman spectroscopy, *Anal. Chem.* **48,** 261R. The companion review to that of McDonald; 31 references to biomolecules.

McDonald, R. S. (1976). Infrared spectrometry, *Anal. Chem.* **48,** 196R. This review is one of a series which appear annually or biennially in *Analytical Chemistry*, covering the whole area of the title. This review has 39 references to medical/biological applications.

Parker, F. S. (1975). Biochemical applications of infrared and Raman spectroscopy, *Appl. Spectrosc.* **29,** 129. 241 references with a very brief discussion of each topic.

5 Circular Dichroism and Optical Rotation

Peter Bayley

Introduction

Circular dichroism (CD) and optical rotatory dispersion (ORD) are two closely related spectroscopic techniques which find particularly important applications in studies of biochemical systems. Both techniques have the specific quality of being sensitive to the shape of molecules. It is well known that a molecule containing a single asymmetric carbon atom is optically active and exhibits the phenomenon of optical rotation. The property of asymmetry is universal in biological systems: amino acids (generally of L configuration), nucleotides (containing the sugars, D-ribose and D-desoxyribose) and carbohydrates (D and L configuration) are the units from which the polypeptides, proteins, polynucleotides and polysaccharides are built up. Since the basic units are asymmetrical the macromolecules derived from them are also asymmetrical and therefore these macromolecules possess the property of optical activity. We use the term "conformation" to designate the three-dimensional structural arrangement of the macromolecule, upon which the distinctive optical properties depend.

It is the purpose of this chapter to illustrate how the techniques of CD and ORD may be used to derive information about the conformational properties of biochemically interesting molecules. Both techniques are used to yield information about a range of properties including:

(a) the structure of macromolecules, their secondary and tertiary structural content, and the degree of mobility of certain components;
(b) structural transitions, e.g. helix–coil transitions and the processes of unfolding and refolding of native and denatured structures;
(c) conformational changes induced by binding small molecules, e.g. substrates, coenzymes, effectors;
(d) symmetry properties of small chromophoric molecules in which optical activity is induced upon binding to macromolecules.

In considering the origins of the phenomena which give rise to the experimentally observable properties of CD and ORD it is necessary to consider some aspects of the nature of light and its interaction with matter, as well as some of the different features of asymmetry which are encountered in macromolecular systems. The concepts involved may be represented in a fully quantitative mathematical and quantum mechanical treatment, but this level of complexity is not essential for interpreting many of the experimental results. The qualitative treatment given here will illustrate that the fundamental principles are essentially straightforward and that general mechanisms conform to a few simple conceptual models.

First some considerations of macromolecular structure, illustrated with reference to polypeptide structure (Fig. 5.1).

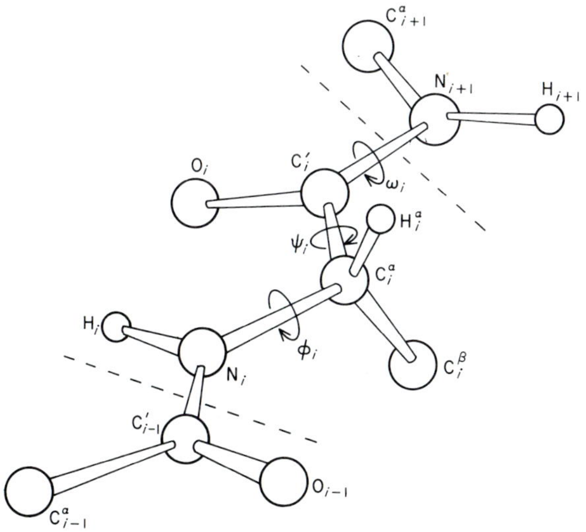

FIG. 5.1. Perspective drawing of a section of polypeptide chain. The dihedral angles (ϕ_i, ψ_i and ω_i) define the conformation. The chain is shown fully extended (ϕ_i, ψ_i, ω_i=180°), and the residue is of L configuration.

1. If asymmetric units are linked together in a stereochemically regular fashion, a *periodic* structure, namely a *helix*, results. The relative orientation of neighbouring units is identical. The conformation is given by $\{(\phi_H, \psi_H, \omega_H)n\}$ where ϕ_H, ψ_H, and ω_H are dihedral angles for the helix (see Fig. 1.1).

2. If asymmetric units are linked in a stereochemically irregular (non-identical) fashion, an *aperiodic* structure results. The relative orientation of neighbouring units is no longer identical. The conformation is given by the values $\{(\phi_1, \psi_1, \omega_1), (\phi_2, \psi_2, \omega_2), \ldots, (\phi_i, \psi_i, \omega_i), \ldots, (\phi_n, \psi_n, \omega_n)\}$.

3. If asymmetric units are linked randomly so that there is a random distribution of relative orientations between the ith and $(i+1)$th unit, a set of random structures results. Each could be written as $\{(\phi_i, \psi_i, \omega_i)\ i=1,n\}$. Any single residue will have a range of possible conformations in the population of different structures.

The three types of structure: periodic, aperiodic and the random set are all asymmetric, but they have different conformations and will therefore have distinct CD and ORD properties.

The concept of *helicity* is extremely useful in understanding the electronic origins of CD and ORD. The helical structure implies a sense of handedness (i.e. left-hand or right-hand helix) which is referred to in stereochemical terms as *chirality*. Because of this, CD and ORD are often referred to collectively as *chiroptical techniques*.

The nature of light

Light is an electromagnetic radiation and is characterized by an oscillating electric field and a perpendicular (i.e. orthogonal) magnetic field (see p. 4). The interaction of light with matter is via the effect of these fields on the distribution of electrons in the molecular framework. (Also see Chapter 2.)

In plane polarized light, the electric field vector is confined to a single defined plane. For an orientated sample, maximum absorbance occurs when the electric vector $\boldsymbol{E}$ is parallel to the direction in which there is maximum electronic displacement within the absorbing molecules (i.e. the direction of the electric transition dipole) giving the phenomenon of *linear dichroism*. The ability to absorb light is dependent upon a directional property within the molecular framework, and its relationship to the directional component (i.e. polarization) of the incident light (Fig. 5.2).

Circularly polarized light (CPL) is a less familiar concept. If two plane polarized waves are polarized in mutually perpendicular directions e.g. vertical and horizontal, such as the ordinary and extraordinary rays passing through a biaxial crystal (e.g. calcite) and the crystal dimensions are such as to give one quarter wavelength phase difference between the two (called the retardation), the resultant radiation will be circularly polarized (Fig. 5.3 a,b).

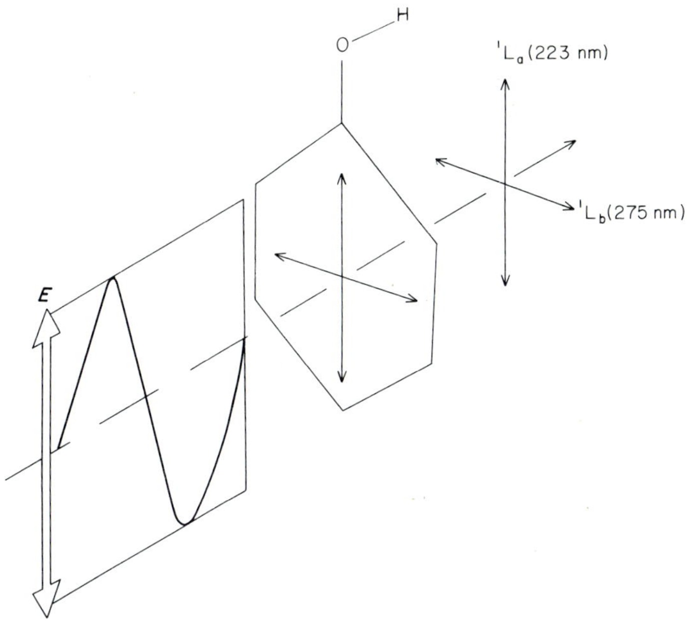

FIG. 5.2 Linear Dichroism: plane polarized light with vertical polarization, incident upon an *orientated* molecule is absorbed only by the transition whose electric moment is in the direction of the polarization. This chromophore (tyrosine) has transitions at 275 nm and 223 nm with directional properties as shown.

If the sign of the retardation is reversed, the sense of the emergent circularly polarized radiation will be reversed also (Fig. 5.3(c)). It is clear that circularly polarized radiation has helical symmetry of a defined handedness or chirality.

A helical molecular structure possesses an electronic distribution with a specific sense of hand and thus presents a different medium for interaction with either right-handed or left-handed circularly polarized light (RHCPL or LHCPL). Since the helical sense of the molecule is an intrinsic property independent of molecular orientation, the chiral property is intrinsic to the structure and is preserved even when the molecules are tumbling randomly and re-orientating in free solution. (In fixed-orientated systems additional effects will occur.) Thus a solution of asymmetric molecules will *absorb* left- and right-handed CPL to different extents.

If ε_L and ε_R are molar absorption coefficients (see p. 16) for LHCPL and RHCPL we define the *molar circular dichroism* as $\Delta\varepsilon = (\varepsilon_L - \varepsilon_R)$. Similarly, even in regions of the spectrum where there is no absorption of radiation, the solution will refract LHCPL and RHCPL to different extents. If n_L and n_R are refractive indices for LHCPL and RHCPL the circular birefringence or

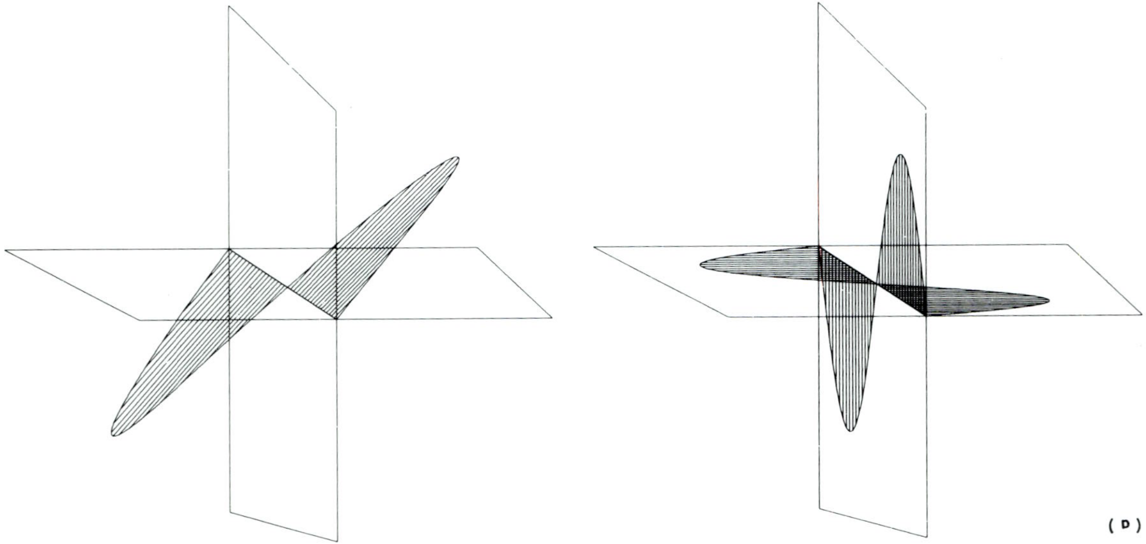

(D)

(b)

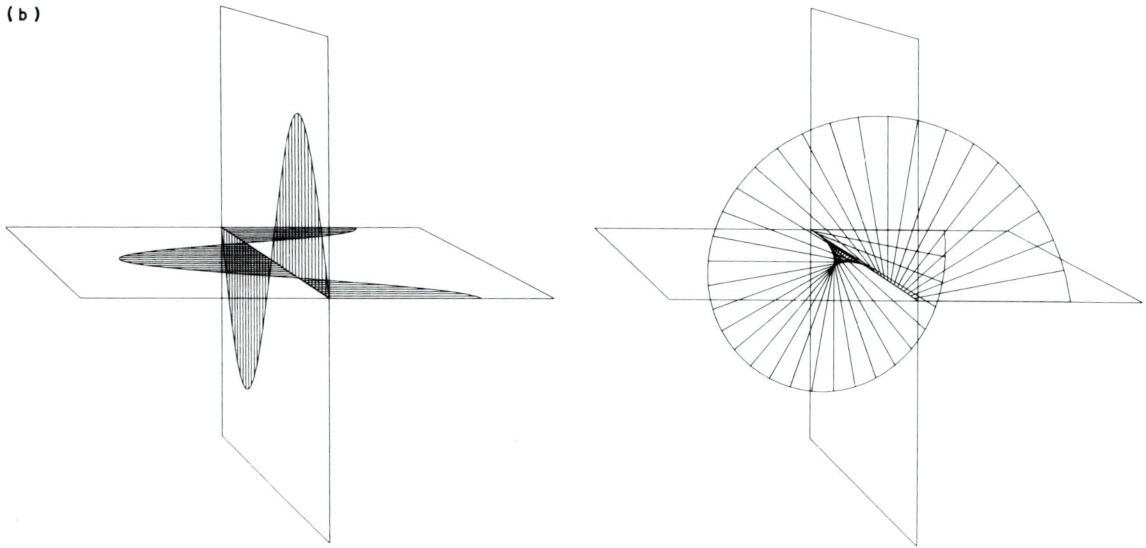

FIG. 5.3 (a) Plane and (b) right-handed circular polarization. For details see under Fig. 5.3(d).

(c)

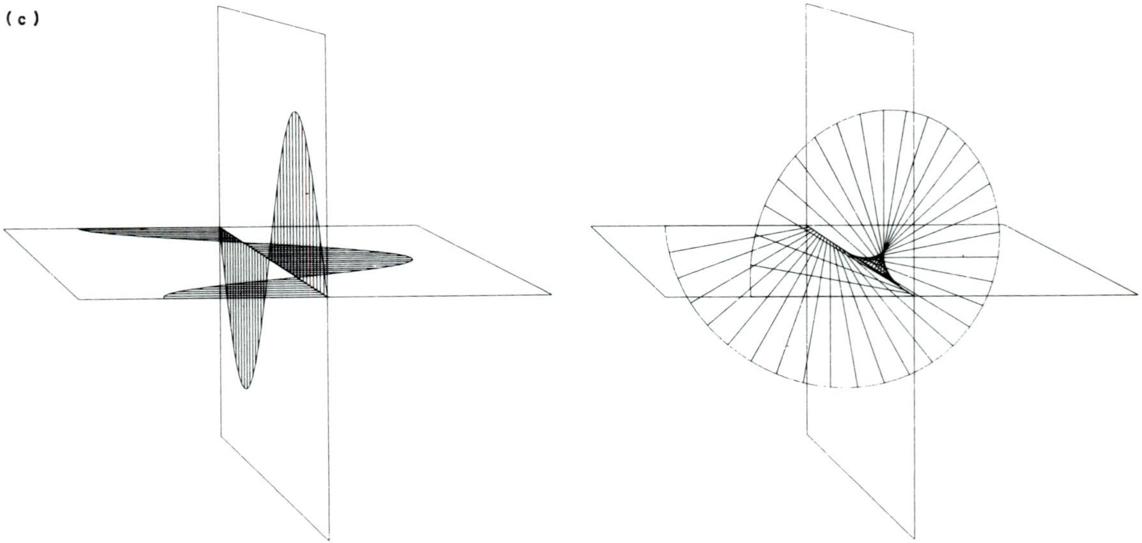

(d)

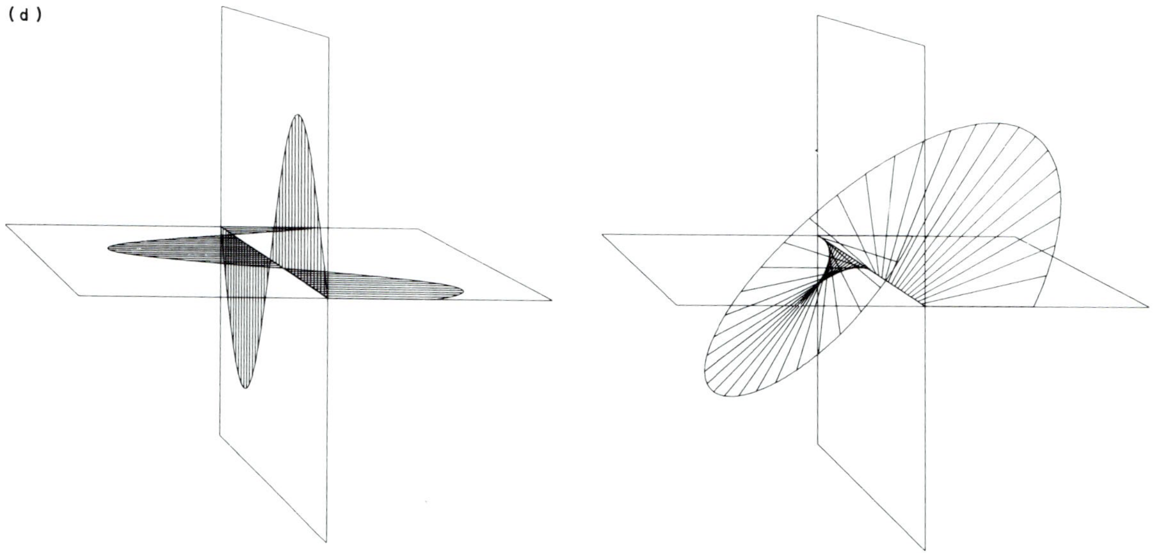

FIG. 5.3 (c) Left-handed circular and (d) elliptical polarization. The combination of two perpendicularly orientated plane-polarized curves with phase differences δ of (a) zero, (b) $+\lambda/4$, (c) $-\lambda/4$, (d) $0<\delta<\lambda/4$ generates (a) plane, (b) and (c) circular, and (d) elliptical polarization. Note also that (b)+(c)=RHCPL+LHCPL, giving plane polarized light.

optical rotation is a measure of $(n_L - n_R)$. In practice, differences of as little as one part in 10^6 are sufficient to generate readily observable effects in CD and ORD.

One further concept of polarization is important in considering CD and ORD, namely that of *elliptically* polarized light. Given the above discussion it can be described as the radiation for which the retardation has a value other than $\pm n\lambda/4$ i.e. all intermediate cases (Fig. 5.3(d)). Circular polarization (n odd) and plane polarization (n zero or even) are special cases of elliptical polarization just as a circle is a special case of an ellipse. Elliptical polarization is in fact the sum of plane and circular polarizations.

Further it follows that combining equal amounts of LHCPL and RHCPL generates plane-polarized light and this gives a satisfying phenomenological explanation of the properties of CD and ORD (Fig. 5.4). For an optically active compound, $n_L \neq n_R$ and the LH and RH circularly polarized components of an incident plane wave are transmitted with different velocities giving, on emergence, a plane-polarized wave with the plane of polarization rotated relative to that of the incident beam. i.e. optical rotation is observed (Fig. 5.4a).

Also if the incident plane polarized beam is absorbed by the optically active compound, $\varepsilon_L \neq \varepsilon_R$ and the emergent beam, composed of LHCPL and RHCPL in unequal amounts, is also elliptically polarized (Fig. 5.4(b)). Thus the molecular property of *circular dichroism*, $\varepsilon_L \neq \varepsilon_R$, naturally gives rise to the observable property of *ellipticity* and the two are directly proportional.

A historical note is relevant here. The technique of *polarimetry*, the measurement of optical rotation, gave rise in the 1960s to the ascendency of the technique of ORD, i.e. the measurement of the wavelength dependence of "dispersion" of optical rotation. A later parallel development was the establishment of measurements of ellipticity which could be combined in the same ORD apparatus. Because of the close relationship between absorbance and refraction, CD and ORD properties can be calculated each from the other using a method called the Kronig–Kramers transform and this has been an important area until the more widespread use of direct CD techniques. Present developments indicate that the greater sensitivity is available in measurements of the circular dichroic absorbance rather than ellipticity. The current literature therefore shows predominantly CD measurements in preference to the other techniques and this emphasis will be followed in this chapter.

It may be noted, however, that the phenomenon of CD occurs only within the region of absorption bands. The measurement of ORD remains a useful parameter for work in non-absorbing (i.e. transparent) regions, e.g. at high concentrations, where high sample absorbance would prevent direct CD measurements, and for monitoring effects associated with transitions in inaccessible regions, particularly in the far UV at $\lambda < 185$ nm.

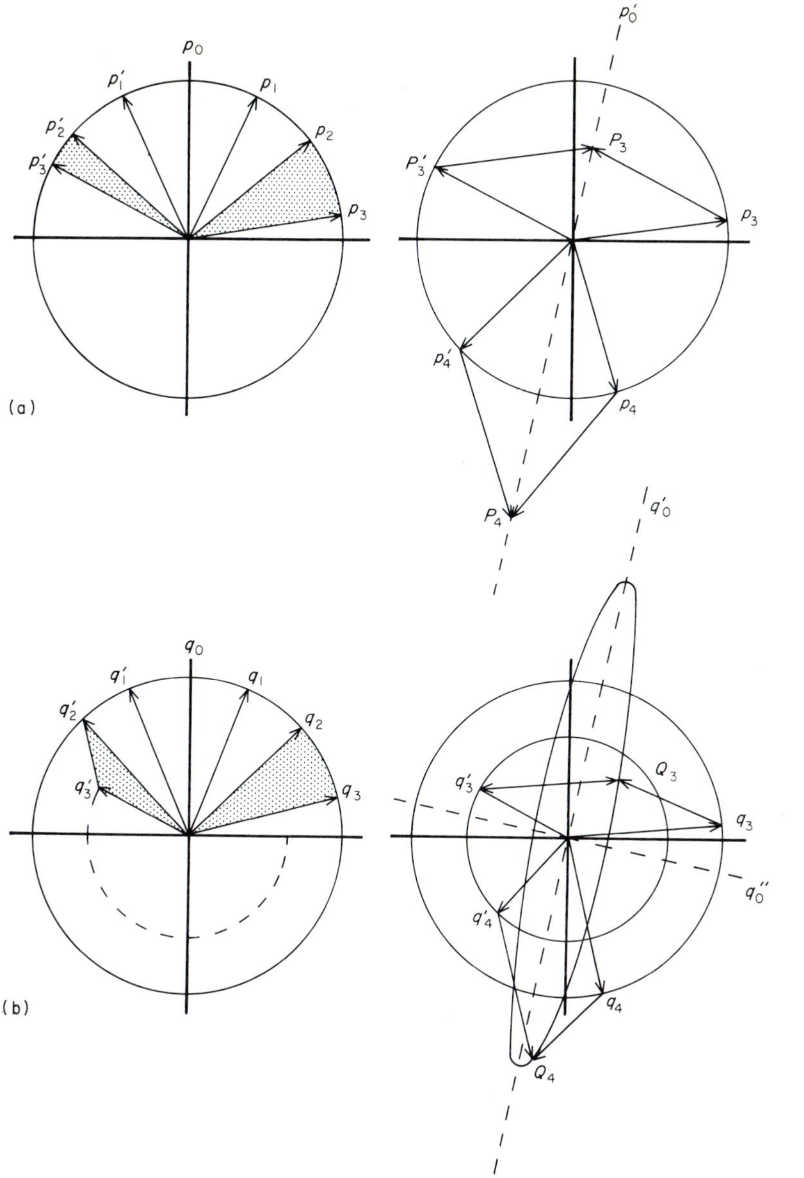

FIG. 5.4 Phenomenological explanation of optical rotation and circular dichroism. (a) Optical rotation: plane-polarized light (p_0) is composed of circular components (p_1 and p_1') which, between positions p_2 and p_3, are transmitted with different velocities, since $n_L \neq n_R$; recombination of p_3 and p_3' and subsequently p_4 and p_4' gives P_3 and P_4 orientated in the direction p_0' i.e. the plane of polarization has been rotated. (b) Circular dichroism and ellipticity: the components q_1 and q_1' between positions q_2 and q_3 are also absorbed to different extents since $\varepsilon_L \neq \varepsilon_R$; recombination of q_3 and q_3' and subsequently q_4 and q_4' gives Q_3 and Q_4 which trace out the ellipse with major and minor axes in the directions q_0' and q_0'', i.e. the emergent light is elliptically polarized.

Instrumental Considerations

Commercial instrumentation is available for the measurement of circular dichroism over the range of the near IR, visible and UV spectrum from 1000 nm to 185 nm. The technique employed in the measurement of the small differential absorption is illustrated schematically in Fig. 5.5. Light from the high-intensity source S passes through the monochromator MC (high aperture, high efficiency and good stray light rejection) and is rendered plane polarized by prism P. The polarization is at 45° to the axis of the modulator M. This may be an electro-optic crystal (Pockels cells) e.g. ammonium dihydrogen phosphate or a piezo-elastic element (e.g. silica). The optical element of the modulator acts as a variable phase retarder depending upon either the voltage across the crystal or the stress induced in the piezo-elastic element respectively. The application to the element of a high frequency signal (generally a few kHz) causes a periodic modulation of the incident beam with the components of LHCPL and RHCPL alternately at the modulation frequency ω. The magnitude of the retardation amplitude is chosen to optimize the degree of modulation at a given wavelength. In the presence of a circularly dichroic sample, the transmitted light intensity and hence the photocurrent of the photodetector PD will vary at the modulation frequency by an amount proportional to the difference in transmission for LHCPL and RHCPL. This difference will be a small AC signal superimposed on a DC component, the mean of the two transmissions. The photocurrent is resolved by a preamplifier into the DC component and an AC component and the amplitude of the latter is determined by the technique of phase-sensitive detection (or lock-in amplification) at the modulation frequency. The AC and the DC level may then be ratioed to give the circular dichroic absorbance, ΔA.

Certain features of this system resemble a simple single-beam spectrophotometer (i.e. source, monochromator, sample, photodetector, see Fig. 2.3. p. 23). However, there are important characteristics of the CD measurement system which affect experimenal performance and hence the assessment of experimental results. These all relate to the smallness of the CD signals which are to be measured. For a sample with (unpolarized) absorbance $A = 1{\cdot}0$, the circular dichroic absorbance $A_L - A_R$ will often be as small as 10^{-5} to 10^{-6}. Thus it would be impossible to do one absorbance measurement with LHCPL and one with RHCPL and then subtract the two! The modulation technique effectively allows the summation of the differential absorbance of LHCPL and RHCPL over thousands of modulation periods (i.e. time constants of several seconds). The critical factor is the signal/noise ratio which can be further enhanced by increased source brightness (not merely light power but power

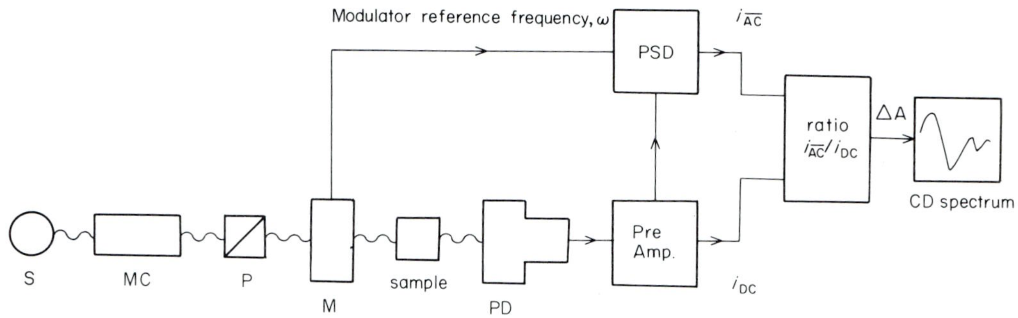

FIG. 5.5 Schematic diagram of CD Measurement System. For details see the text.

per unit area of the arc), hence the specifications of the source–monochromator combination. It may be noted that the output of conventional arc sources falls dramatically at each end of the visible and UV spectrum and the photodetector response has a similar wavelength profile. Experimentally the choice of slit widths (or an automatically servo-driven slit program) determines the light power passing through the sample, but this can be increased only by increasing the optical bandpass i.e. decreasing the spectroscopic resolution of the measurement. Thus to achieve the maximum sensitivity in CD measurements, as is frequently required in biochemical work, a compromise has generally to be struck among the following factors:

(i) length of scan and time constant;
(ii) spectroscopic resolution (bandpass);
(iii) stability to the incident radiation;
(iv) stability of the instrumental baseline.

Finally, a most important set of considerations surrounds the choice of sample concentration, buffers, solvents and pathlengths. Since the measurement depends upon transmitted light intensity, there is a strict upper limit to the absorbance of the sample (plus solvent and buffer) which can be tolerated. Optimal signal/noise is achieved at $A = 0{\cdot}8686$ and the upper limit practically is of the order $A = 1\text{–}1{\cdot}5$ (i.e. 10 to 3% transmission) depending upon wavelength. For the far UV region, short pathlengths (down to 0·1 mm) are generally required and solvents with high absorbance such as carboxylates, amides, chloride, bromide and iodide have to be avoided or used at minimal pathlengths and concentrations. Many non-aqueous solvents (dioxan, chloroform, dimethylformamide) present similar problems in the far UV.

With the ready availability of digital techniques for data processing, the possibility now exists for performing multiple scans of a sample with summation and averaging to reduce the effect of random error. The improvement in signal/noise increases as $\sqrt{N}$ where N = number of scans (see also p. 124). Such a process inevitably extends the time of an already rather slow measurement. Under these circumstances the instrumental stability may become the limiting factor and this is best monitored by careful observation of the blank (cell + buffer) over an appropriate period of time. These considerations show that the optimization of the various factors discussed is an important prerequisite to obtaining maximum sensitivity.

Digital techniques also offer considerable advantages for differential CD and for numerical analysis such as rank analysis, multiple regression and least squares techniques where the data has to be available in numerical form.

Practical Definitions and Units

As we have seen (p. 151) the molar circular dichroism is defined as $\Delta\varepsilon = \varepsilon_L - \varepsilon_R$, the difference in molar absorption coefficient for RHCPL and LHCPL. If the circular dichroic absorbance is $\Delta A (= A_L - A_R)$ for a solution of molar concentration c and pathlength l cm,

$$\Delta A = \Delta\varepsilon \, c \, l \tag{5.1}$$

from which $\Delta\varepsilon$ may be calculated (cf eqn 2.1). By comparison, ε, the molar absorption coefficient (for unpolarized light), is given by $\varepsilon = \frac{1}{2}(\varepsilon_L + \varepsilon_R)$. Units of $\Delta\varepsilon$ are the same as ε, namely $\text{M}^{-1}\,\text{cm}^{-1}$. The ratio $g = \Delta\varepsilon/\varepsilon\,(=\Delta A/A)$ is called the dissymmetry ratio and is dimensionless. Typical values are $g = 10^{-3}$ to 10^{-6}

For optical rotation, the specific rotation, $[\alpha]$, of a solution of concentration c_w g ml^{-1} and pathlength L decimeters (note 1 cm = 0·1 decimeter) with observed rotation α_{obs} (degrees) is given by

$$[\alpha] = \frac{\alpha_{obs}}{c_w \, L} \tag{5.2}$$

and the molar rotation, $[\phi]$, is given by

$$[\phi] = [\alpha] \frac{\text{MW}}{100} \tag{5.3}$$

where MW is the molecular weight and ϕ is expressed in deg cm^2 dmol^{-1}. For biopolymers the value of MW taken is the mean residue molecular weight (MRW) and is equal to the polymer molecular weight divided by the number of residues. The mean residue rotation $[m]$ is given by

$$[m] = [\alpha] \frac{\text{MRW}}{100} \tag{5·4}$$

where $[m]$ is expressed in deg cm^2 dmol^{-1}.

The ellipticity is defined as that angle whose tangent is the ratio of the minor and major axes of the ellipse (see Fig. 5.4), The units of ellipticity are the same as those of optical rotation. Thus

specific ellipticity $$[\psi] = \frac{\psi_{obs}}{c_w L} \tag{5.5}$$

molar ellipticity, $$[\theta] = [\psi] \frac{\text{MW}}{100} \tag{5·6}$$

mean residue ellipticity, $$[\theta]_{\text{MRW}} = [\psi] \frac{\text{MRW}}{100} \tag{5.7}$$

The relationship between molar ellipticity and circular dichroism is

$$[\theta] = 3300\ \Delta\varepsilon \tag{5.8}$$

Circular dichroism spectra in the literature are reported in terms of $\Delta\varepsilon$ and $[\theta]$ (and occasionally as ΔA or as "millidegrees ellipticity"). Familiarity with the above relationships is therefore essential. Note: a solution with observed ellipticity of y millidegrees would have $\Delta A = y \times 3 \times 10^{-5}$.

The units of CD conform to the SI system.

Comparison between CD and absorption spectra

Interpretation of CD spectra depends fundamentally on the knowledge and clear interpretation of the associated absorption spectra. In addition to helping establish suitable experimental conditions (see above) the absorption spectra allow identification of a particular chromophoric group (e.g. peptide, carbonyl, haem, purine, pyrimidine) providing information about the nature of the electronic transitions involved (forbidden or allowed) the electronic characteristics (π–π^* and n–π^* being most common) and the existence of vibrational fine structural details. (See Chapter 2 for a discussion of these aspects of absorption spectra.) However, CD spectra are often more complicated. Thus, a simple CD band does not necessarily correspond exactly in position and band shape with the absorption band, possibly due to different relative strengths of vibrational subcomponents in each case.

In the case of an isolated absorption band of an asymmetric chromophore (e.g. the n$\rightarrow\pi^*$ transition at 290 nm of a steroidal ketone) a correspondingly simple single CD band is generally found. Likewise a simple single ORD band is observed. The term "Cotton effect" is used for either. The characteristics of a CD spectrum are defined in terms of the wavelength of maximum amplitude, λ, the bandwidth, Δ, and the rotational strength, R, which is a measure of the area under the curve (see p. 163). The characteristic appearance of this effect is shown, idealized, in Fig. 5.6(a). A single CD band of a simple chromophore may be positive or negative, depending on the chirality. Frequently, multiple Cotton effects are found due to the close proximity within a single chromophore of a number of electronic transitions. When more than one chromophore is present the spectra become even more complicated. A characteristic effect (termed a "couplet", Fig. 5.6(b)) is often seen in the CD of a system of two interacting chromophores and this has particular value in assigning the molecular symmetry and even the stacking angle of the transition dipoles.

The characterization and interpretation of more complex spectra is most readily performed by CD. Even so, analysis of individual bands is generally not possible. The spectrum can be used as a quantitative measure of a specific conformation (e.g. the native structure of a protein or the concentration of a

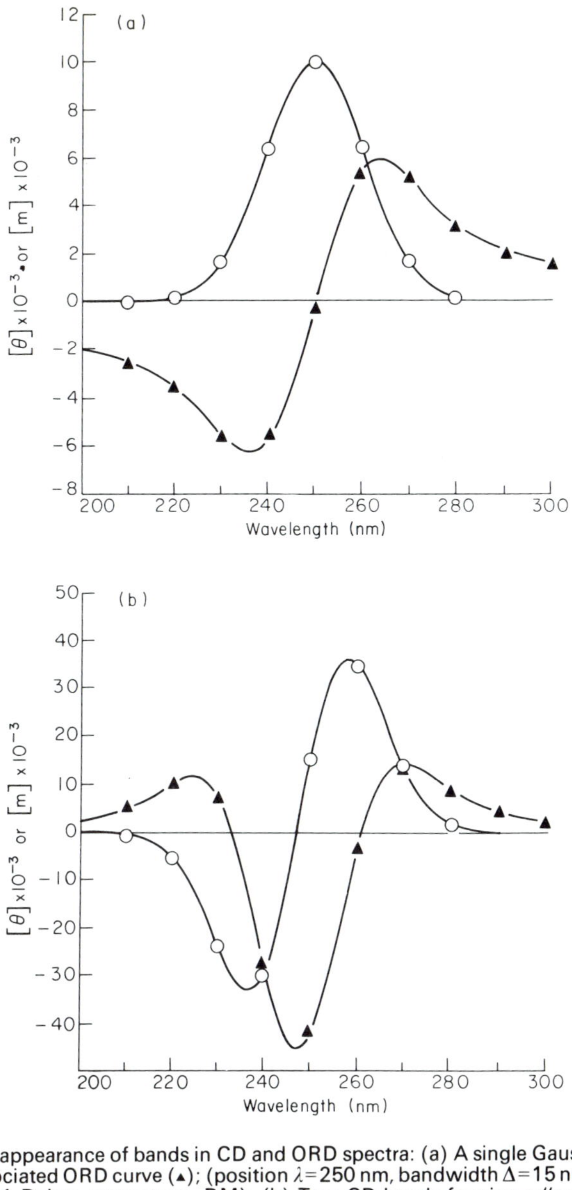

FIG. 5.6 The appearance of bands in CD and ORD spectra: (a) A single Gaussian CD band (○) and its associated ORD curve (▲); (position λ=250 nm, bandwidth Δ=15 nm and rotatory strength, R=0·1 Debye-magnetons, DM); (b) Two CD bands forming a "couplet" and its associated ORD, from equal and opposite bands split by 5 nm. (λ_1=250 nm, R_1=1·0 DM; λ_2=245 nm, R_2=−1·0 DM; Δ_1=Δ_2= 15 nm.) The characteristics of a single CD band are defined in terms of the wavelength of maximum amplitude (λ), the bandwidth (Δ), and the rotational strength (R) which is a measure of the area under the curve.

species). Knowledge of the absorption spectrum still provides essential information in characterizing properties of a non-conformational nature (such as protonic equilibria) which may also affect the appearance of the CD spectrum. The latter part of this chapter illustrates the diversity of appearance of the CD spectra of different classes of molecules found in biochemical systems.

The Electronic Origin and Conformational Dependence of CD

The electronic basis of the relationships between absorption and circular dichroism is seen in the fact that the integrated intensity under an absorption band, the *dipole strength, D*, is related to the product $\boldsymbol{\mu}_e \times \boldsymbol{\mu}_e$ whereas that under a single CD band, the *rotational strength*, is related to the product $\boldsymbol{\mu}_e \times \boldsymbol{\mu}_m$ where the quantities $\boldsymbol{\mu}_e$ and $\boldsymbol{\mu}_m$ are the electric and magnetic transition dipoles. Circular dichroism derives from the presence of electronic transitions which possess electric and magnetic transition dipoles with a common direction. This condition is satisfied in situations of molecular asymmetry, but there is no single, simple theory relating this condition to molecular structure. Rather there are several models, representing electronic mechanisms, which have been treated theoretically by the methods of quantum mechanics. These treatments cover all levels from *ab initio* calculations to highly empirical symmetry rules. However, considerable insight can be gained from consideration of the models on a qualitative basis, bearing in mind the biochemical structures for which these models serve as analogues.

Excitons

When a number of identical chromophores are organized into an array, the excitation properties of the array are no longer those of the single chromophore. The possible excited states of the array [A A A A] derive from a linear combination of states which can be symbolized as:

$$c_1\,[\mathrm{A^*\ A\ A\ A}] + c_2\,[\mathrm{A\ A^*\ A\ A}] + c_3\,[\mathrm{A\ A\ A^*\ A}] + c_4\,[\mathrm{A\ A\ A\ A^*}]$$

where A* indicates an excited group, and c_1 to c_4 are coefficients. Thus the excitation is distributed over the whole array. The chiroptical properties are then determined by the symmetry of the array, and where this is an asymmetric biopolymer, pronounced and characteristic CD and ORD properties are observed.

Helical structures

Interaction between the excited states of component chromophores in a helical array give exciton transitions with a helical and chiral character. When electronic excitation occurs over a helical path, the displacement of electron density occurs both along the helical axis (giving $\boldsymbol{\mu}_e$), and around the axis (giving $\boldsymbol{\mu}_m$, also along the axis, as shown in Fig. 5.7(a). Thus an electronic transition generates simultaneously an electric and magnetic transition dipole with a common direction, which is the condition for optical activity.

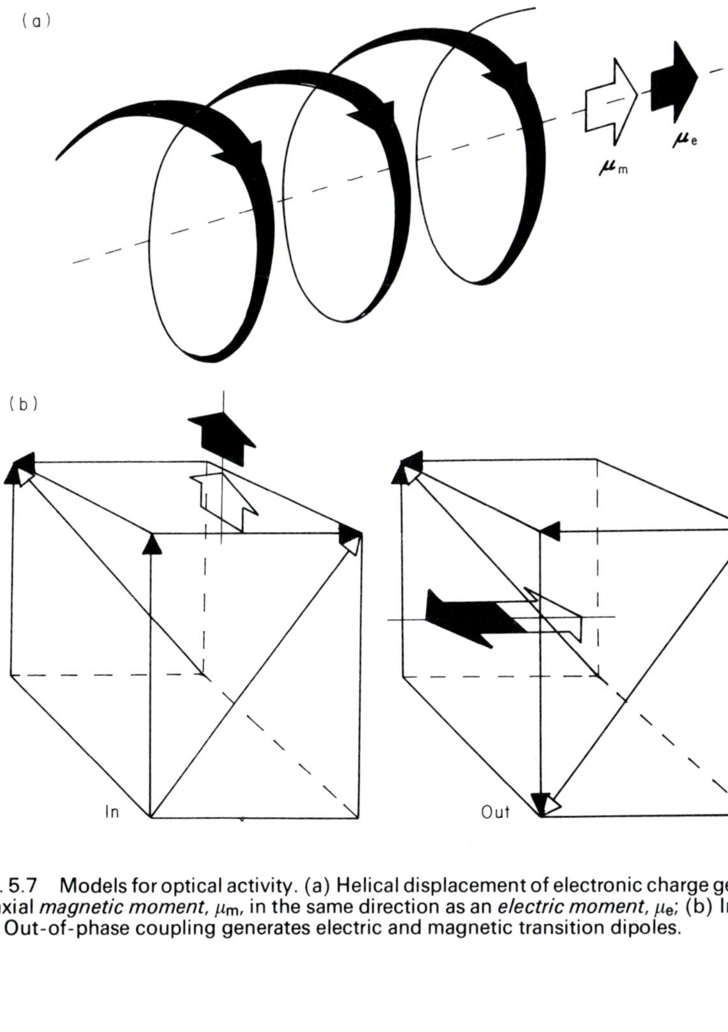

FIG. 5.7 Models for optical activity. (a) Helical displacement of electronic charge generates an axial *magnetic moment*, μ_m, in the same direction as an *electric moment*, μ_e; (b) In-phase and Out-of-phase coupling generates electric and magnetic transition dipoles.

Coupling between two chromophores

In principle this follows the same argument as the helical exciton mechanism. The asymmetric interaction of two monomers gives two coupled transitions as with a coupled oscillator. If the ground state is $A_0=[A\ A]$, excited states are $A_1=[A^*\ A+A\ A^*]/\sqrt{2}$ for the in-phase state and $A_2=[A^*\ A-A\ A^*]/\sqrt{2}$ for the out-of-phase state. The new transitions A_0 to A_1 and A_0 to A_2 have both $\boldsymbol{\mu}_e$ and $\boldsymbol{\mu}_m$ (Fig. 5.7b). This model is important because the CD appears as a "couplet" i.e. a pair of equal and oppositely signed bands, split approximately equally on each side of the monomer absorption wavelength (see Fig. 5.6(b)). Since the rotational strength of the two is equal and opposite i.e. $R_1+R_2=0$, rotational strength is conserved, and this system is called "conservative". The clearest examples are with stacked dinucleotides such as adenylyl 3′,5′-adenosine (ApA).

Intrinsically dissymmetric structures

These are structures in which the chromophore itself is twisted into a helical or chiral form. Examples are shown in Fig. 5.8, e.g. hexahelicene, twisted butadiene, polyenes contained within a larger rigid structure and the disulphide chromophore. Conformational isomers of sterically crowded molecules such as biphenyls, and di- and triphenyl methane dyes are in this category, although there will generally be rapid rotational interchange between molecules of opposite chirality in free solution. In these cases, the electronic distribution of the ground state also is spread over the chiral molecular framework, and very strong CD effects may result for chiral conformers.

Individual chromophores

Apart from the above coupling interactions, an individual chromophore may *acquire* CD by certain mechanisms.

Where the chromophore contains one electronic transition which is electronically allowed ($\boldsymbol{\mu}_e \neq 0$) and another, magnetically allowed ($\boldsymbol{\mu}_m \neq 0$), a mechanism exists for new composite mixed transitions to be produced when the chromophore is perturbed asymmetrically (the "one-electron mechanism"). The asymmetric chemical bonds constitute an asymmetric electric field and a transition with both $\boldsymbol{\mu}_e \neq 0$ and $\boldsymbol{\mu}_m \neq 0$ results. Examples are a simple peptide or carbonyl, where CD associated with the $n \rightarrow \pi^*$ transition derives from mixing with a higher energy $\pi \rightarrow \pi^*$ transition, (Fig. 5.9(a)). The relationship between the symmetry of the perturbation field and the resultant CD can often be stated as a *Sector Rule*, of which the Quadrant and Octant Rules are the simplest. Where perturbation derives predominantly from a

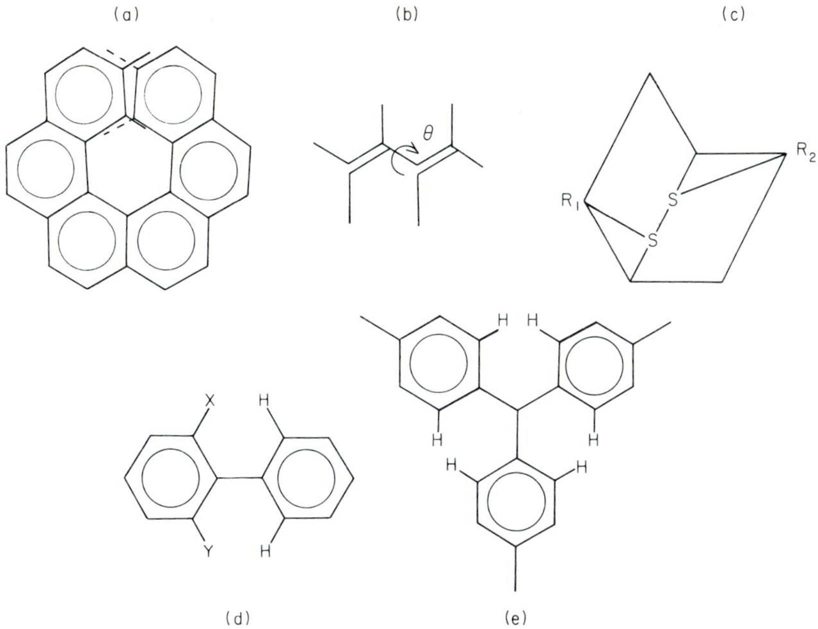

FIG. 5.8 Models for optical activity. (Top) dissymmetric chromophores: (a) hexahelicene, (b) the twisted diene, (c) disulphide; bottom: chromophores which can adopt dissymmetric conformations: (d) sterically crowded biphenyl, (e) triphenyl methane type.

single atom or group, the rules find important use on an empirical basis for closely related structures such as steroidal ketones, peptides, etc.

A single chromophore with an electrically allowed transition can acquire CD by interaction between $\boldsymbol{\mu}_e$ and the asymmetric electron distribution of the molecule. This $\boldsymbol{\mu}_e$ interacts with the polarizability of the rest of the asymmetric molecule. (Fig. 5.9(b)). Polarizability is the extent to which the electron distribution can be perturbed by an electrostatic field, and has directional properties related to the chemical structure (see also Chapter 4). In fact it is also related to the possible higher energy transitions, although these may be in the very far UV. Formally the mechanism approximates to interaction between $\boldsymbol{\mu}_e$ and all the other high-energy $\boldsymbol{\mu}_e$'s giving a sign and magnitude to the CD determined by summation over all possible interactions. Thus an aromatic residue such as tyrosine or tryptophan, chromophoric substrates or coenzymes may acquire CD by interaction with the "bulk asymmetry" of the macromolecule. Coupling mechanisms may operate simultaneously.

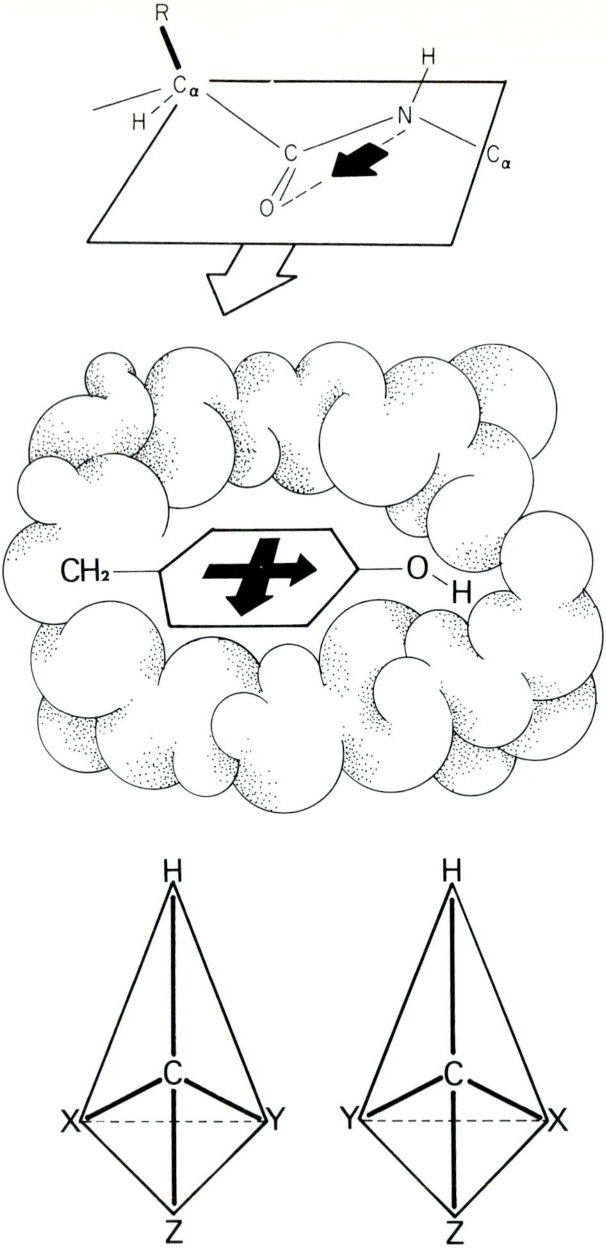

(a) (b) (c)

FIG. 5.9 Models for optical activity. (a) *Peptide residue*: the $\boldsymbol{\mu}_e$ of the $\pi \rightarrow \pi^*$ transition and $\boldsymbol{\mu}_m$ of the $n \rightarrow \pi^*$ transition mix under the influence of the asymmetric centre at C_α. (b) The *polarizability* of the groups surrounding the chromophore provides an asymmetric environment with which the $\pi \rightarrow \pi^*$ transitions of the tyrosine side chain interact. (c) *The asymmetric carbon atom*: the two stereoisomers have the lower three substituents ordered in different handedness, XYZ clockwise or anticlockwise respectively when viewed from above.

The asymmetric carbon atom

This represents one of the simplest models for optical activity. It may be noted that the two isomers of the model H—C(X, Y, Z) do indeed have a sense of chirality (Fig. 5.9(c)) and hence some helical character. When none of the groups is obviously chromophoric, polarizability interactions would account for the existence of optical activity. This would be readily observed by ORD; CD would only be observed at much shorter wavelength (see p. 156). Where one group is obviously chromophoric, we have the possibilities discussed above for interpretation in terms of the mixing of polarizability mechanisms.

Problems in assigning mechanisms

The principal problem in any molecule containing more than one chromophore is that several mechanisms are operating simultaneously. Summation of the different mechanisms is only rarely valid. Thus a $\pi \rightarrow \pi^*$ transition might be involved in exciton interactions with equal energy (degenerate) transitions, coupling interactions with higher energy (non-degenerate) transitions, interactions with molecular polarizability, and mixing effects with local $n \rightarrow \pi^*$ transitions. The integration of the mechanisms into a single theoretical model is a complex and demanding problem. Theory is most successful when one or two mechanisms predominate. Fortunately this situation is found sufficiently often for CD to be usefully interpreted in terms of molecular conformation. General reviews treat the cases of helical polypeptides and polynucleotides in some detail. The predominant interactions for polypeptides are coupling and mixing mechanisms and for polynucleotides are coupling and polarizability interactions. The resultant CD properties have considerable diagnostic value and this is discussed in the subsequent sections.

Peptides and Proteins: Determination of Secondary and Tertiary Structure by CD

Multicomponent analysis

Early work on the ORD and CD of the synthetic homopolypeptides, particularly poly-L-glutamic acid and poly-L-lysine established the distinctive spectroscopic properties of the α-helical and β-sheet conformations. In addition, a characteristic CD spectrum was found for either polymer in the fully charged state (Fig. 5.10(a)). Since this state has the hydrodynamic properties of an expanded and random coil, it was taken to represent a random arrangement of the chromophores also. More recent work with several polymers in different denaturants (neutral salts, guanidine, urea, etc.) shows a different typical spectrum which is thought to derive from a more nearly random structure. The charged polymers probably adopt a specific "extended helix" conformation with the optical properties principally determined by relatively local interactions within the helix segments and the more macroscopic properties determined by an open random coiling of these segments.

These observations in both their original and modified forms have acted as a great stimulus to the analysis of the conformation of globular proteins in solution from the CD (or ORD) spectra. If the approximation is made that the properties of the component structures are additive, an individual spectrum can be represented as follows:

$$S_\lambda = x_1 [\alpha]_\lambda + x_2 [\beta]_\lambda + x_3 [R]_\lambda \tag{5.9}$$

In this equation, $[\alpha]_\lambda$, $[\beta]_\lambda$, and $[R]_\lambda$ represent the spectra of α-helical, β-sheet and random or aperiodic structure (Fig. 5.10(a)). Using a linear least-squares procedure, which performs the multiple regression of the spectrum S_λ with the three components $[\alpha]_\lambda$, $[\beta]_\lambda$ and $[R]_\lambda$, the values of the coefficients x_1, x_2 and x_3 are found which give the best approximation to the observed spectrum, S_λ.

The simplicity of this operation is rather deceptive. In reality, establishing the validity of the result requires considering the assumptions inherent in writing eqn (5.9) and of the limitations of the mathematical procedure itself. Equation (5.9) contains the assumptions that the secondary and tertiary structures can be fully represented by the three types of conformation and that the reference spectra are truly appropriate. Both theory and experimental results with oligopeptides show that the property per residue within an α-helix increases with increasing helix length. Thus to represent all helical residues by $[\alpha]_\lambda$ is probably to overestimate their contributions, because of the occurrence of short helical lengths in many proteins. Similarly, CD spectra of model polypeptides in β-conformation are rather sensitive to environment and theory confirms that the properties are rather critically dependent on the exact relationship of hydrogen-bonded chains. Thus $[\beta]_\lambda$ is not well defined. The recent recognition of the widespread occurrence of the β-turn has raised the question of the contribution to be assigned to this feature. Finally, the assignment of the spectrum of the denatured conformation $[R]_\lambda$ to the specific, but aperiodic, tertiary structure is an approximation which would hold exactly only if the aperiodic component contained all of the possible nearest neighbour interactions found in the denatured structure. The occurrence of *cis* as well as *trans* peptides in the denatured state is a further complication.

A rather different empirical approach has therefore evolved to obtain a different group of reference components based on the CD of a set of globular proteins whose secondary–tertiary structure had been determined by X-ray crystallography. Assigning the structure of each protein as containing known

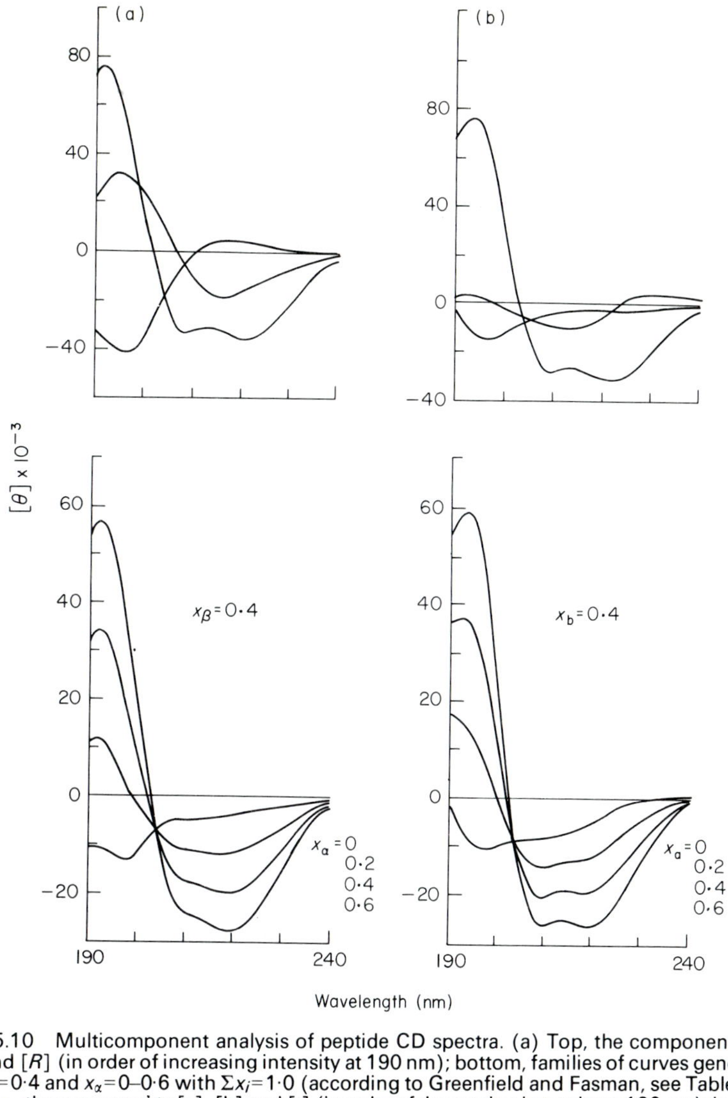

FIG. 5.10 Multicomponent analysis of peptide CD spectra. (a) Top, the components [α], [β] and [R] (in order of increasing intensity at 190 nm); bottom, families of curves generated for x_β=0·4 and x_α=0–0·6 with Σx_i=1·0 (according to Greenfield and Fasman, see Table 5·1). (b) Top, the components [a], [b] and [r] (in order of decreasing intensity at 190 nm); bottom, families of curves generated for x_b=0·4 and x_a=0–0·6 with Σx_i=1·0 (according to Chen, Yang and Martinez, see Table 5).

fractions x_a, x_b and x_r (of α-helix, β-sheet and aperiodic structure respectively), an equation similar to eqn. (5.9) can be written for each protein, spectrum S_λ in terms of the unknown spectral contributions $[a]_\lambda$, $[b]_\lambda$, and $[r]_\lambda$.

$$S_\lambda = x_a\,[a]_\lambda + x_b\,[b]_\lambda + x_r\,[r]_\lambda \qquad (5.10)$$

From the set of spectra, the best values of the spectral components $[a]_\lambda$, $[b]_\lambda$ and $[r]_\lambda$ can be determined by least-squares procedure. These new reference spectra represent the average properties of the α, β and aperiodic regions of the proteins of known conformation (Fig. 5.10(b)). In fact, component $[a]_\lambda$ is quite similar to $[\alpha]_\lambda$, as might be expected from the distinctive and intense property of the α-helix, while $[b]_\lambda$ and $[r]_\lambda$ are substantially different from $[\beta]_\lambda$ and $[R]_\lambda$. The form of $[a]_\lambda$, $[b]_\lambda$ and $[r]_\lambda$ might be expected to depend to some extent on the standard proteins which are used. The value of the method lies in applying it to proteins of relatively similar overall conformation.

Critique of the method

Irrespective of whether the analysis is carried out for x_1, x_2 and x_3 using eqn (5.9) in terms of $[\alpha]_\lambda$, $[\beta]_\lambda$ and $[R]_\lambda$ or eqn (5.10) in terms of $[a]_\lambda$, $[b]_\lambda$, or $[r]_\lambda$, certain decisions are necessary in the multiple regression analysis.

1. In the experimental spectrum to be analysed, instrumental noise causes increased uncertainty in values in the important region below 220 nm. This may be reduced by careful averaging but may also require a weighted least-squares procedure, with each point weighted by the inverse of its standard deviation, determined in a number of scans.

2. In multiple regression, the component spectra should be dissimilar. Ideally they will be uncorrelated and preferably orthogonal. There is little choice in this because of the desire to associate them with the physically meaningful components of α, β and aperiodic structure. If two components are correlated, their coefficients will be correlated also.

3. Physical reality requires all of x_1, x_2 and x_3 to be in the range $0 < x_i < 1{\cdot}0$, and ideally that $x_1 + x_2 + x_3 = 1{\cdot}0$.

Only in the case that $\sum x_i > 1{\cdot}0$ can remedial action be taken simply; a correction of $x_i' = x_i / \sum x_i$ can be made if, for example, there is some doubt about the absolute value of the experimental spectrum. Otherwise, applying the constraints that $0 < x_i < 1{\cdot}0$ and $\sum x_i = 1{\cdot}0$ means that alternative numerical solutions are arbitrarily neglected. Such constraints require that non-linear least squares techniques must be used.

Given the number of choices to be made, it is important that the result of the analysis should not be strongly dependent on the approach used. The variation found by different choices of wavelength range of the data, weighted/unweighted, constrained/unconstrained fits, gives some idea of the true resolution of the method. The choice of basic components should reflect the experimental data, e.g. a synthetic polypeptide and a globular protein might well be treated differently.

Results of multicomponent analysis

Given these limitations, and the implicit warning against the over-interpretation of the results, it is still true to say that multicomponent analysis is a most useful, if approximate, measure of protein conformation. The CD spectrum remains the most practical experimental measure of the general secondary–tertiary characteristics which can be performed rapidly on submilligram quantities of material. In cases where the experimental spectrum resembles one of the component spectra, an unambiguous assignment can be made. In this way, the virtually α-helical structure of paramyosin was recognized (from ORD) as early as 1961. The relatively high α-helical content of haemoglobins and myoglobins was consistent with X-ray crystallographic results and the high β-content of the immunoglobulins was recognized first from the typical CD.

For globular proteins, the method based on [*a*], [*b*] and [*r*] from reference proteins is generally preferred. Results showing intermediate α-helical content in the range 20–50% are often ambiguous, although the likelihood of substantial amounts of β-structure implied by these results appears to be reasonably consistent with the more recent X-ray results. Several globular proteins have been successfully analysed and Fig. 5.11 shows the reference spectra and the computed fits for a number of examples.

A further development of the method allows inclusion of an additional parameter characterizing the mean α-helical length which gives a better fit of the CD spectra particularly in the far UV.

In general, results with the multicomponent analysis as described are not expected to be better than ±5%. The fact that the best fits still deviate significantly from the observed spectra suggests the possibility of more than just three components and an obvious additional source of intensity resides with the aromatic residues (see p. 185) which are neglected in the above treatment.

An example of the application of multicomponent analysis is shown in Table 5.1. The far UV CD of bovine glutamate dehydrogenase was measured from 250 nm and the mean of four scans was taken. This was analysed as above using three sets of reference data. The results ($100 \times x_i$) are shown for the constrained analysis with $\sum x_i = 1{\cdot}0$, and the fits are illustrated in Fig. 5.12. It is clear that the range of the data is very important: 240–210 nm can be fitted satisfactorily but extending the lower limit to 200 nm decreases the fit very seriously as judged visually (Fig. 5.12) and numerically (see χ^2, Table 5.1). For 240–210 nm the range of the parameters is no better than ±5% between the three sets of reference data and a reasonable mean result of 38% α, 12% β, 50% *R* is obtained. It should be noted that the standard deviation of any x_i refers to the internal consistency of that particular calculation and should not be taken

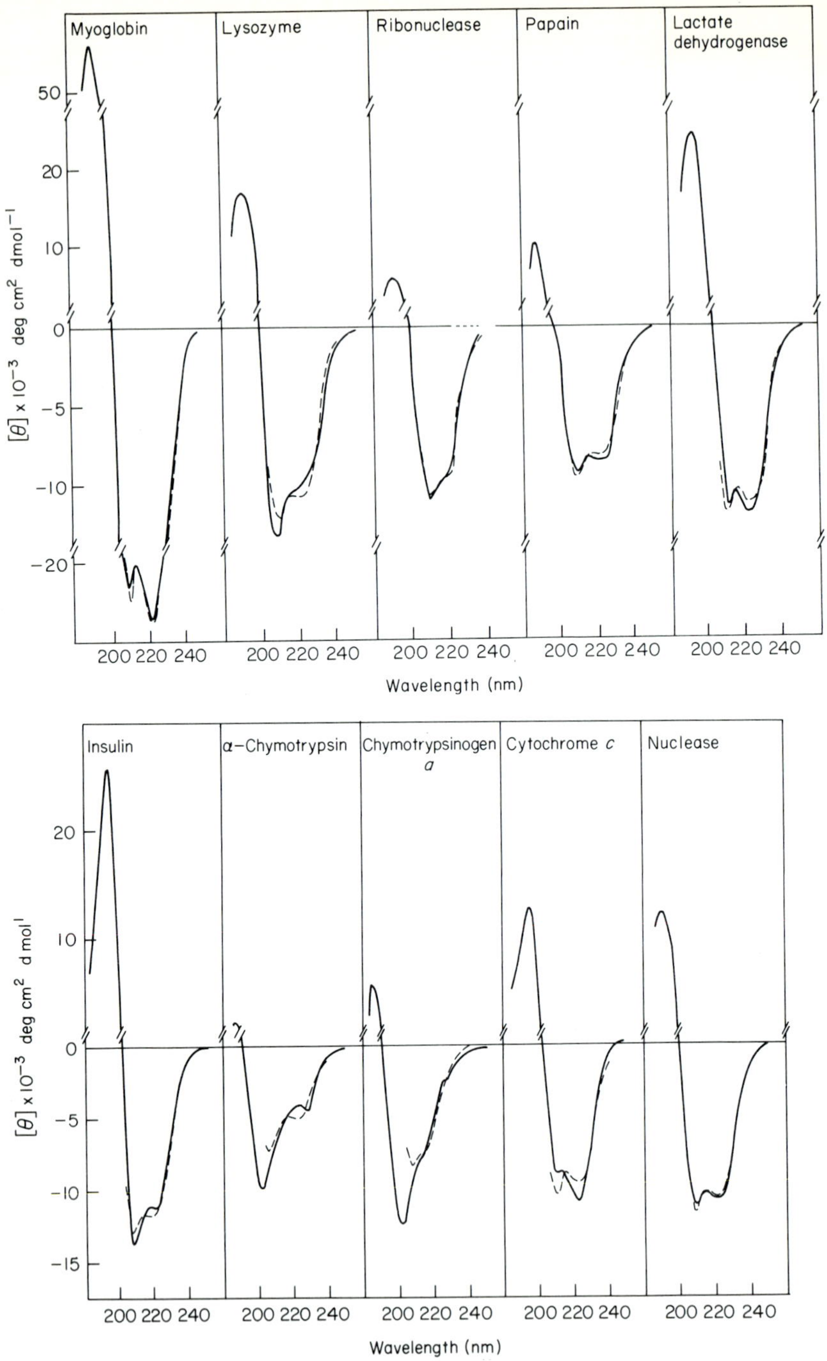

FIG. 5.11 Applications of multicomponent analysis: Comparisons of experimental CD (solid line) with calculated CD (broken line) for (a) five reference proteins from which [*a*], [*b*] and [*r*] were derived and (b) five additional proteins, using [*a*], [*b*], [*r*] (Data of Chen, Yang and Martinez, see Table 5.1).

TABLE 5.1 Multicomponent analysis of GDH peptide CD

Data 240–210 nm	α	β	ρ	χ^2
GF	34 (1·3)[b]	19 (2·7)	47 (3·0)	0·40
CYM	35 (0·9)	7 (3·3)	58 (3·4)	0·28
CYC	44 (0·9)	11 (3·4)	45 (3·5)	0·32

Data 240–200 nm	α	β	ρ	χ^2
GF	16 (0·9)	58 (1·3)	26 (1·5)	1.65
CYM	35 (0·8)	16 (3·1)	49 (3·2)	9·25
CYC	39 (0·8)	42 (2·0)	19 (2·2)	5·15

$$\chi^2 = \frac{10^{-6}}{N_f} \sum_{i=1, N_{pts}} ([\theta_i^{obs}] - [\theta_i^{calc}])^2 \text{ where}$$

N_f = no. of degrees of freedom = $(N_{pts} - N_{parameters} - 1)$

Figures in parentheses are 100 $[x_i\ (\pm\sigma_i)]$ where σ_i is the standard deviation of x_i in the fitting procedure.
GF: Greenfield, N. J. and Fasman, G. D. (1969). *Biochemistry* **8**, 4108.
CYM: Chen, Y. H. Yang, J. T. and Martinez, M. (1972) *Biochemistry* **11**, 4120.
CYC: Chen, Y. H., Yang, J. T. and Chau, K. H. (1974). *Biochemistry* **13**, 3350.

as an assessment of the accuracy of the result without consideration of the χ^2 to which it applies. Better fits than these can be obtained with unconstrained fitting, i.e. giving lower χ^2, but since these involve some negative values of x_i or $\sum x_i$ very far from 1·0, little trust can be placed in them.

The CD spectra of *S*-carboxymethyl insulin A chain in various solvent mixtures was also used to test various multicomponent analysis procedures. For data taken from 240 nm to 210 nm there is a pronounced monotonic increase in intensity with increasing proportion of organic solvent (trifluoroethanol) consistent with the adoption of up to 50% α-helical structure. Below 210 nm the data were not fitted satisfactorily, possibly due to contributions at low wavelengths from aromatic transitions (see Fig. 5.12 caption). Such effects, particularly in the near UV, have their own diagnostic uses (see p. 185) but detract from the use of the analysis with small peptide systems. Also the possible occurrence of β-turns in small peptides as well as proteins represents another component which may contribute to the CD in this region. There is clearly insufficient resolution in any CD spectrum to allow the expansion of the basic set of components indefinitely: inclusion of more parameters will always improve the fit but will not necessarily add to the physical interpretation.

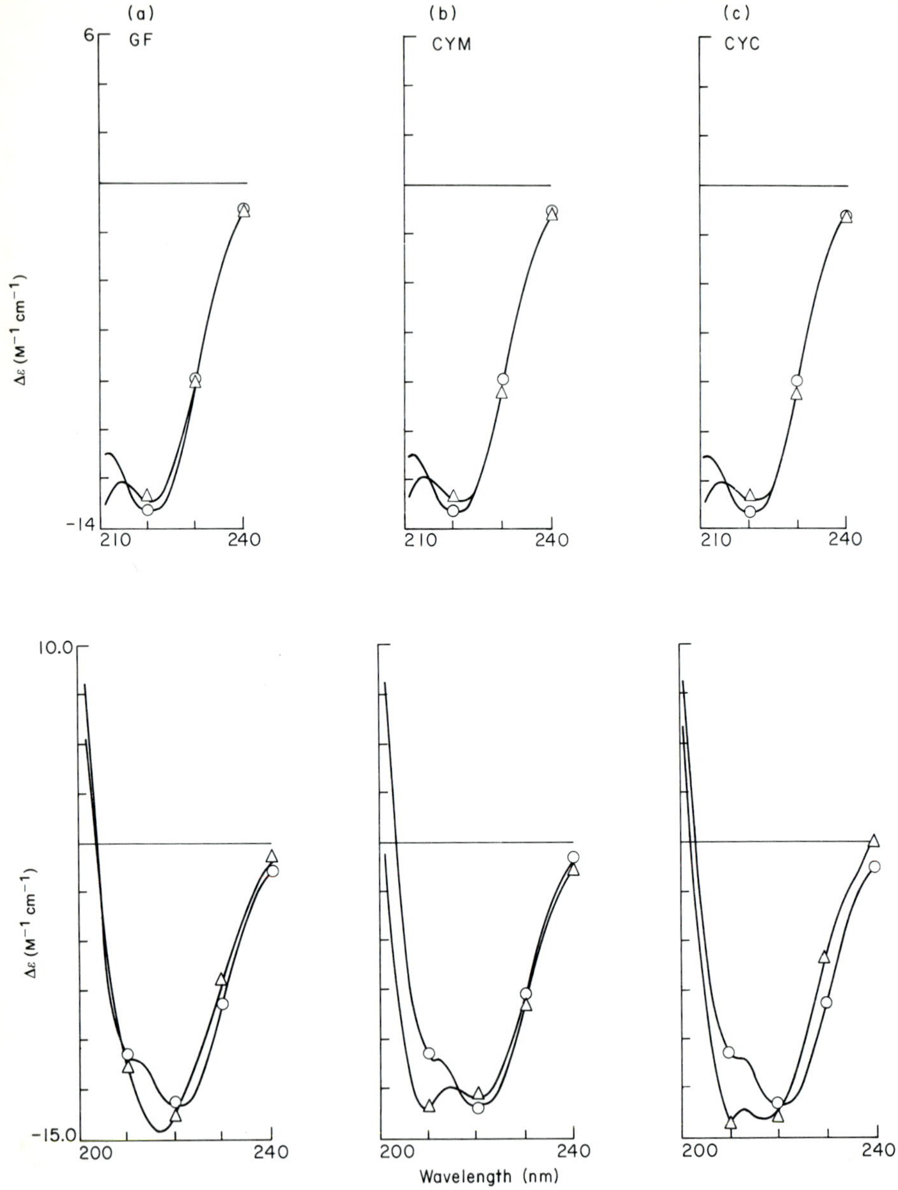

FIG. 5.12 Protein CD spectra, peptide region. Multicomponent analysis of the far-UV peptide CD for bovine glutamate dehydrogenase using three sets of reference spectra according to (a) Greenfield and Fasman (GF), (b) Chen, Yang and Martinez (CYM) and (c) Chen, Yang and Chan (CYC). For references, see Table 5.1. Top, data from 240 to 210 nm. Bottom, data from 240 to 200 nm. (○), Experimental; (△), Calculated. A similar problem is discussed for insulin A by L. A. Holladay (1977). *Biophys. Chem.* **7**, 41–49.

Peptides and Proteins: Structural Transitions of Environmentally Sensitive Molecules

Some examples will be presented from protein work of rather large conformational changes which can readily be observed using measurements of the CD in the far UV (peptide) region as a function of gross changes in the environment. Thus, pH, temperature, organic solvents and neutral salts all affect macromolecular conformation and such studies provide valuable information about the range of stability of many peptide and protein systems.

Small peptides

Peptide hormones are often short oligopeptides whose conformational properties in aqueous solution are ill defined. Since their potent biological activity is associated with the specific stereochemical interaction at a receptor site, it is important to investigate the conformational possibilities of such molecules. Using non-aqueous solvents which enhance intramolecular interactions such as hydrogen bonding and electrostatic salt bridges, it is often possible to convert an apparently structureless peptide into a specific structure. A solvent frequently used is trifluorethanol (TFE) which is less polar that water and might, to some degree, mimic the environment at a membrane interface.

An interesting example is the potent pressor hormone angiotensin II(AII), an octapeptide with sequence Asp—Arg—Val—Tyr—Ile(or Val)—His—Pro—Phe. Comparative studies have been performed with the hormone and derivatives in which each position of the peptide was substituted with an aliphatic residue.

Circular dichroism spectra were evaluated in trifluorethanol and in water as a function of pH (Fig. 5.13). In aqueous solution the octapeptide appears highly flexible and no specific conformation is found. In TFE the spectrum is clearly of a β-type conformation, possible due to a folded conformation with intramolecular hydrogen bonding. In the derivatives, the substitutions in positions 1–5 have relatively little effect (Fig. 5.13(a)), but positions 6,7, and 8 are much more critical (Fig. 5.13 (b)). Thus the presence of D-His^6 (or Gly^7) allow an α-helical conformation to develop. This suggests that Pro^7 of the native AII has a definitive role in maintaining the potential for β-structure. the *C*-terminal Phe^8 is important for conformation (as well as for biological function). The β-structure develops only partially in the Ile^8 substitution suggesting that the *N*-terminal end has folded normally. The orientation of Tyr^4 is markedly changed in the Ala^6 derivative as judged by the strong aromatic effect at 230 nm, although in Ile^8-AII it shows no abnormal effect

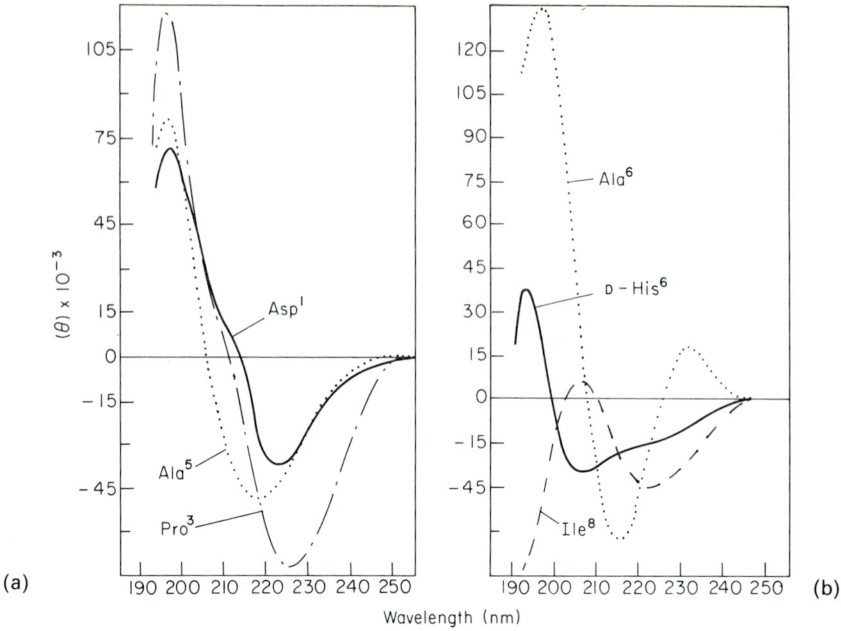

FIG. 5.13 Peptide CD of structural analogues of angiotensin II in trifluoroethanol. (a) Analogues adopting a β-type conformation; (b) analogues in which substituents are in positions 6 to 8 affect this conformation. (Data of Lintner, K., Fermandjian, S., Fromagsot, P., Khosla, M. C., Smeby, R. R. and Bumpus, F. M. (1977). *Biochemistry* **16**, 806–812.)

(Fig. 5.13(b)). The picture is that of a molecule in which β-structure is relatively stable in the *N*-terminus and up to the His[6] residue. It may be noted that the β-structures show intense CD; intermolecular association in pure TFE could possibly contribute.

A rather different behaviour is shown by lipotropin (β-LPH), the 91-residue peptide which contains within it the sequences of β-melanocyte-stimulating hormone (β-MSH) (residues 41–58) and β-endorphin (C fragment, residues 61–91) of which residues 61–65 constitute the pentapeptide Tyr—Gly—Gly—Phe—Met ([methionine]-enkephalin). Circular dichroism measurements show that neither β-LPH, nor its major fragments, have appreciable secondary structure in water (Fig. 5.14). For the 91-residue peptide, this lack of structure is unusual. As the proportion of TFE in the medium increases, progressive changes occur in structure, which are largely complete by 25% TFE. The maximum effects inducible by TFE are comparable to the ordered structures predicted by the sequence rules of Chou and Fasman. The conformation of β-LPH and derivatives may be important in directing the specificity of endopeptidase cleavage of the prohormone. This in turn would require the

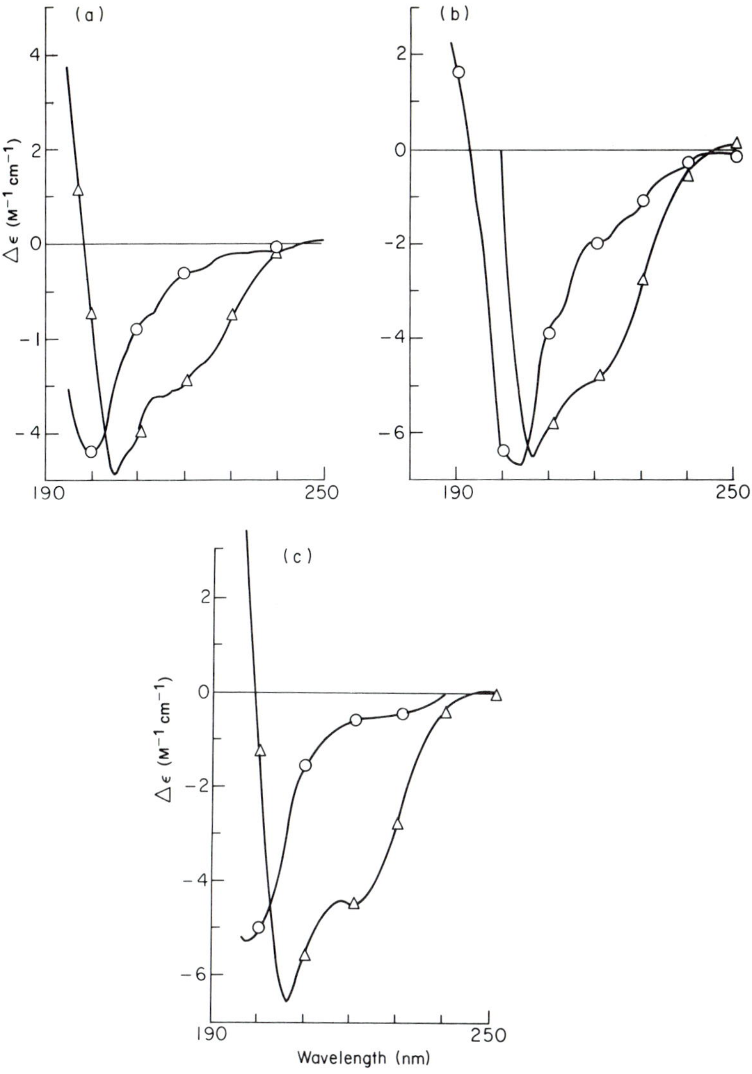

FIG. 5.14 Peptide CD of β-lipotropin and fragments. (a) β-LPH; (b) γ-lipotropin; and (c) β-endorphin; (○) H_2O and (△) 50% trifluoroethanol. Increased helical content is induced by low concentrations of TFE.

cleavage to occur while the prohormone was in some relatively apolar environment and hence in a folded conformation. The fact that β-LPH and adrenocorticotrophic hormone (ACTH) both derive from a common precursor (MW=31 000), supports the idea of independent domains for both peptides. The solvent sensitivity of the small biologically active opiate peptide

β-endorphin suggests that it, too, requires an apolar environment for its activity, and its intermolecular interactions could be important in destabilizing the conformation complementary to its receptor. The effect on transfer to aqueous solution observed here is common to many short peptide systems and could be a general principle allowing termination of the biological action by proteolysis.

A further example of interesting effects of non-aqueous solvents is the case of the ionophore gramicidin A, a linear peptide with the sequence: formyl—(L)-Val—Gly—(L)-Ala—(D)-Leu—(L)-Ala—(D)-Val—(L)-Val—(D)-Val—[(L)-Trp—(D)-leu]$_3$—(L)-Trp—ethanolamine. This effectively has an alternating L-D-L-D chirality, and is hydrophobic. It is thought to form an ionophoric channel, spanning the apolar region of the membrane. Circular dichroism spectra of three distinct conformations have been characterized, of which two are effectively mirror images as judged by the inversion of the CD (Fig. 5.15(a)). Three species are separable chromatographically, in dioxan–water (110:1) and may be interconverted by treatment with ethyl acetate. The mirror image spectra are interpreted as deriving from helices of opposite handedness with predominantly parallel β hydrogen bonding between two molecules in a dimeric cylindrical structure, approximately 3 nm long with a pore of 0·3 nm diameter. (The third conformation is predominantly antiparallel β as judged by a characteristic 1680 cm^{-1} infrared amide I band—see p. 134.) A monomeric species is also found in more polar solvents such as ethanol or dimethyl sulphoxide. The models proposed for these structures differ from the model head-to-head left-handed π_{LD} helices, also proposed as pore-structures for this ionophore.

The CD spectra alone do not allow an unambiguous assignment of structure. Interpretation of the CD is complicated by contributions from higher energy transitions of tryptophan at 230 nm; the different environments of the tryptophan residues may be seen in the CD at 280 nm (Fig. 5.15(b)). The CD evidence does show the capacity of the molecule to form conformations of different chirality, the multiplicity of potential conformational states, and the slow rates of interconversion in non-polar solvents. One notable difference of the models is that the peptide carbonyls of the π_{LD} were proposed to interact with the ion while in the double helical structures they are involved in inter-chain hydrogen bonding. Direct evidence on conformation in the transmembrane channel is still required.

Conformational changes in histones

Following the isolation of individual histones in pure form, conformational properties of these proteins and interactions between them have been studied extensively by CD. Given the highly polycationic nature of their amino acid

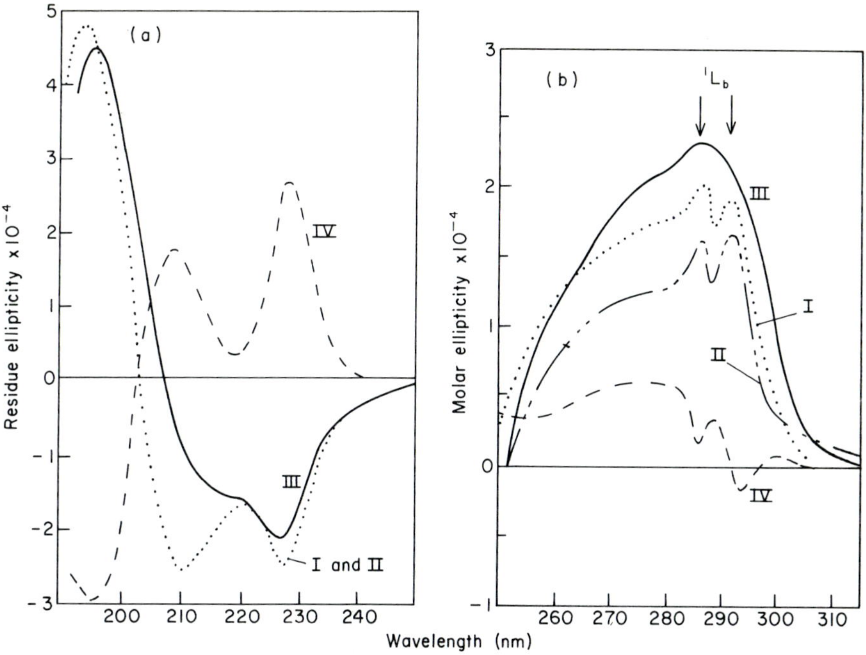

FIG. 5.15 Circular dichroism of four different conformational species of Gramicidin A in organic solvents. (a) peptide region (propan-2-ol); (b) aromatic region (dioxan). (Data of Veatch, W. R., Fossel, E. T. and Blout, E. R., *Biochemistry* (1974). **13**, 5249–5256.)

composition, it is not surprising to find that in aqueous solution at neutrality and low ionic strength they show little evidence of ordered structure. However, transitions from such unfolded conformations to folded ones, presumed to be analogous to, or even identical with, the native conformations, have been demonstrated by CD for several histones. These effects have contributed significantly to current ideas of nucleohistone structure and to the functional role of histones in chromatin structure. Thus histone H1 undergoes a salt-induced conformational change which is apparently located in the central region of the molecule. This effect is also shown by isolated subfractions of H1 and comprises an increase of approximately 13–16% of residues in α- and β-structure, a folding phenomenon which can also be induced by increasing the pH to near 9. The subfractions of H1 show different affinities for complexing with the non-histone chromosomal proteins HMG1 and HMG2 and there is no appreciable change in CD on complex formation.

Related, but characteristically different behaviour, is shown by the inner histones H2a H2b H3 and H4. All of these show a definite folded conformation induced by conditions within the physiological range (Fig. 5.16). Analysis by multi-component methods (see pp. 169–172) shows that H3 and H4 undergo significant conformational changes. Both fast and slow changes are observed: In H3, in the faster phase, approximately 14% α-helix forms; after the slow phase, lasting a few hours, 8% helix and 25% β-structure is found. The slow phase is both salt- and temperature-sensitive and involves aggregation. All the histones show a similar level of ordered structure in their folded forms. Perhaps of greater importance is that complex formation between pairs of histones to form the strong complexes H3·H4, H2a·H2b, H2b·H4 is accompanied by enhanced ordered structure as indicated by more intense CD at 220 nm. Analysis shows that the enhancements are only small in structural terms, e.g. approximately nine residues (per tetramer) of $(H3 \cdot H4)_2$ (Fig. 5.16a); 15 residues (per dimer) of (H2a·H2b) and eight residues (per dimer) of (H2b·H4) adopting α-helical conformation in these strong complexes, and aggregation and β-sheet formation of individual histones is inhibited. Neither H2a nor H2b complexes sufficiently strongly with H3 to prevent its self aggregation and increase in β structure. Although the CD changes involved are quite small their careful characterization provides strong evidence for the complex involving the interactions H2a–H2b–H4–H3 and requiring specific conformational changes of the component histones in its formation.

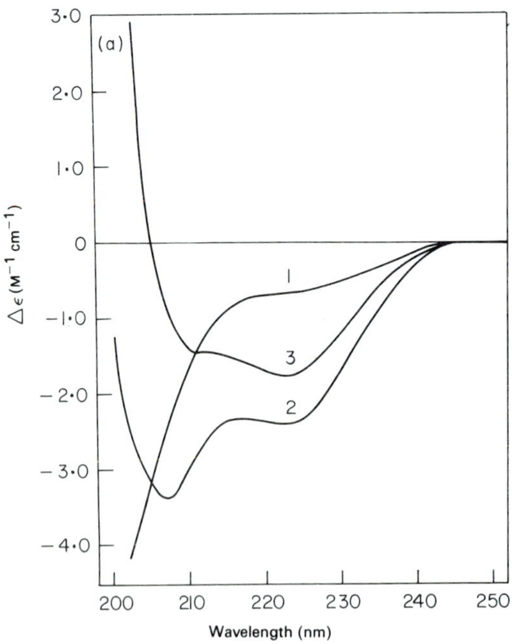

FIG. 5.16 (a)

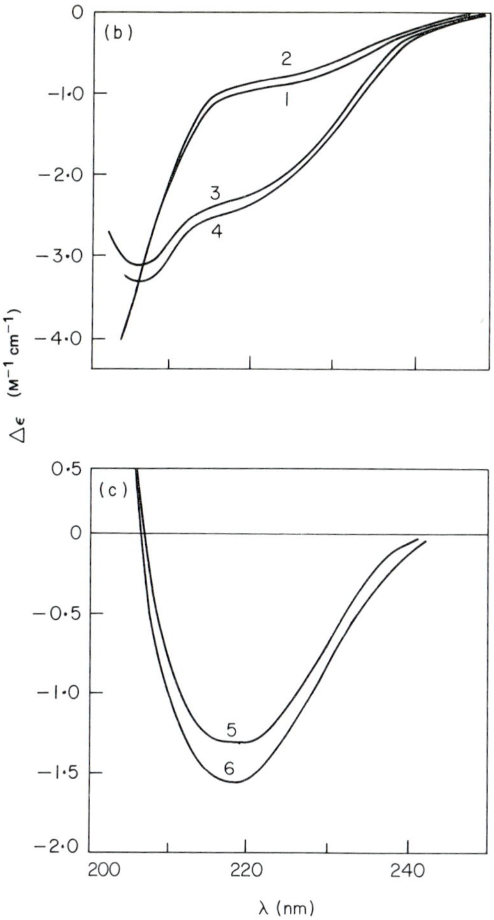

FIG. 5.16 Conformational changes in histones. (a) H3+H4 (5 μM each) in (1) 1mM HCl; (2) 20 mM phosphate pH 7; the difference spectrum (3) is greater than for H3 or H4 alone; (b) H3H2a (10 μM in each) in (1) 1mM HCl; (3) 15 mM phosphate pH 7; H3+H2b (10 μM each), in (2), 1 mM HCl; (4), 15 mM phosphate, pH 7. (2) and (4); (c) the differences (5) = (3) − (1); (6) = (4) − (2) are the same as for H3 alone. (Data of D'Anna, J. A. Jr and Isenberg, I. (1974). *Biochemistry* **13**, 4992–4997.)

Conformational changes in synthetic polypeptides

Circular dichroism is the most direct parameter of the large-scale conformational changes which characterize the transitions of synthetic polypeptides as a function of temperature, solvent and salt concentration. Some of the complex behaviour of these polymers is summarized in Table 5.2 and shows the marked effect of n, the number of methylene groups in the side chain R. Thus

Table 5.2(a) shows the strong tendency for polymers based on L-Glu to form RH α-helices; the analogous polymers based on L-Asp behave very differently, although the structures differ only in one CH_2 group.

The variation in properties of polymers related to poly-L-lysine is also markedly dependent on the length of side chain (Table 5.2(b)). Thus the midpoint of the helix–coil transition on protonation decreases smoothly with n, as does the proportion of methanol stabilizing the helical form. However, anomalous behaviour of poly-L-lysine is shown in that addition of SDS to the protonated (coil) form generates β-conformation for this polymer alone. This

TABLE 5.2 (a) $R=(CH_2)_nCOOR'$

Polymer	n	R′	Solvent	Conformation
Poly (benzyl L-glutamate)	2	$—CH_2C_6H_5$	—	RH α-helix
Poly (benzyl L-aspartate)	1	$—CH_2C_6H_5$	—	LH α-helix
Poly (L-glutamic acid)	2	—H	pH 4·6	100% RH helix
Poly (L-aspartic acid)	1	—H	pH 4·6	20% RH helix

(b) $R=(CH_2)_nNH_2$

Conditions	$n=3$	4	5	6	7
pH (Helix coil)	—	10·1	9·6	9·0	8·7
%MeOH (Mid-point)	—	88	72	52	27
Effect of SDS on coil form	α	β	α	α	α

conformation can also be generated from the α-form by heating. Different behaviour again is shown by the protected polymers with $R=(CH_2)NHCBz$ (CBz=carbobenzoxy group). Heating in chloroform/dichloracetic acid causes a transition from coil to helix for $n=4$ (lysine) and $n=3$ (ornithine) but the inverse (helix to coil) for $n=2$.

It is emphasized that these different conformations are adopted by polymers of the same optical configuration which is L. The function of the side chain in determining conformation is thus a major factor in these regular homopolymers and indicates the complexity of the problem in the folding of globular proteins into their specific secondary and tertiary structures.

The ease of determining the thermodynamic parameters of the helix–coil transition as a function of some variable, say temperature, has led to an interesting method for trying to quantitate the contributions of individual residues to the stability of helical structures. Thus the effect of introducing the residue in question into the parent polymer poly (L-alanine) can be used to assess the helix-forming potential of this residue from changes in position and slope of the midpoint of the thermal transition as monitored by CD.

Sequence copolymers have been synthesized to provide model compounds for a number of natural systems. Peptides of alternating D and L residues are models for gramicidin A; poly-(tri) peptides have been studied for their immunogenic activities; proline-containing poly-(tri)peptides of the sequence (Pro·HOPro·X) are models of collagen and poly-(penta)peptides have been synthesized as analogues of elastin. It is an essential part of such studies to use CD criteria to see how closely the model compares with the natural system in adopting a "native" conformation and in undergoing the appropriate environmentally induced transitions. In general terms the presence of a repeating sequence suggests the possibility of a regular periodic structure. If the sequence itself contains a specific structure, the linking together of this in a regular fashion will produce new helical forms with distinctive optical properties. Again CD is the most readily applied experimental tool.

Peptides and Proteins: Conformational Information from CD of Aromatic and Disulphide Side Chains

The side chains of phenylalanine, tyrosine and tryptophan absorb in the near UV (see p. 29) and, in simple chiral amino acid derivatives, exhibit rather weak CD in the characteristic wavelength range around 280 nm (Fig. 5.17). However, the residues of Trp and Tyr contribute more strongly to the near UV CD of native globular proteins (cf. Figs. 5.12 and 5.15). This is most clearly seen in the case of proteins containing only tyrosine e.g. insulin and pancreatic ribonuclease (RNAse) (Fig. 5.18) and bovine pancreatic trypsin inhibitor (Fig. 5.21). Since, in model compounds, immobilization of the side chain in the conformation of a given rotameric isomer causes 10–100 times intensification of the CD, it is reasonable to argue that "buried" tyrosine side chains are responsible for the high CD in proteins. Surface tyrosines restricted by specific interactions will also be intensified. The electronic interactions of coupling and polarizability mechanisms (see p. 166) will also operate. It may be noted that "frozen" rotameric conformations with oppositely signed CD are possible (for the same L stereoisomeric configuration) and this may give cancellation. Also an immobilized residue with zero CD is entirely possible. The important point is that strong effects *may* be produced and are frequently encountered.

When tryptophan is also present a more complex situation arises. Model compound studies show that, in immobilized Trp derivatives, a variety of types of Trp spectra are possible (Figs. 17(c),(d)) in which the two electronic transitions "$'L_a$" and "$'L_b$" have different relative strengths, and possibly different signs. The superposition of a number of oppositely signed effects gives overall a rather weak CD spectrum but the information content is in principle relatively high, due to the characteristic fine structure of the individual

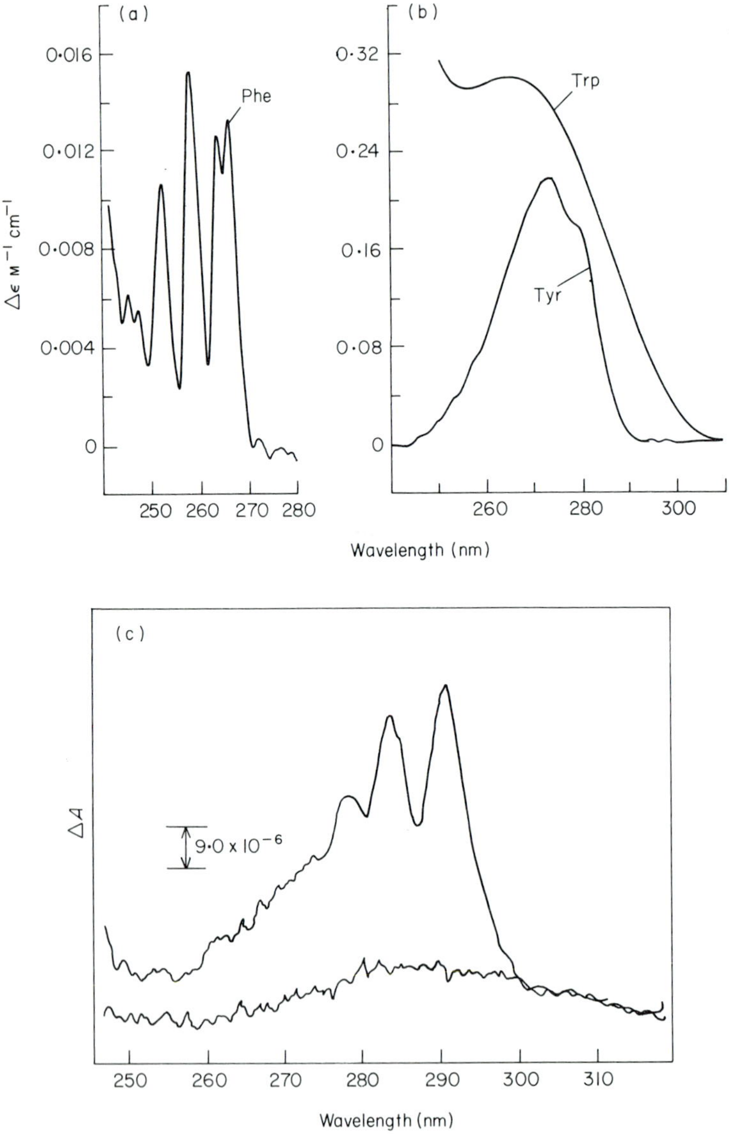

FIG. 5.17 (a), (b) and (c).

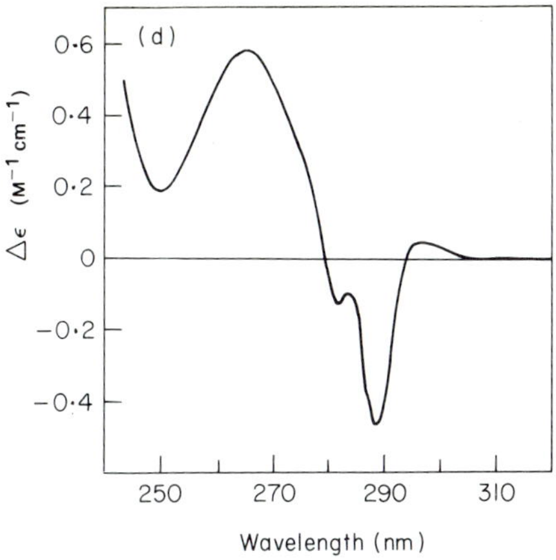

FIG. 5.17 Circular dichroism of Aromatic Chromophores in the near-UV. (a) Phenylalanine (b) tyrosine and tryptophan (c) positive ^1L_b type CD of the L-tryptophan chromophore of *N*-acetyl-L-Trp-ethyl ester at 77 K (d) negative ^1L_b type CD of the L-tryptophan chromophore of cyclo (L-His-L-Trp) at 297 K in water pH 10.5. (c) and (d): data of Strickland and co-workers—see Bibliography).

components and the rather high environmental sensitivity of the Trp chromophore which causes spectral shifts of individual residues.

Thus the near UV CD spectrum of a protein is often too complex to resolve into a large number of different overlapping and cancelling effects. It is, however, a very fine criterion of the native structure of the protein. With normal CD techniques, much fine structure can be detected and recent increases in instrumental sensitivity and data processing capabilities reveal finer details which can be used as criteria for the structural state of a given system. This is illustrated in Fig. 5.19(a) showing the degree of resolution available in the near UV CD spectrum of bovine glutamate dehydrogenase. This spectrum may be used quantitatively for diagnostic purposes in ligand binding, folding, unfolding, etc. Figure 5.19(b) shows the corresponding region for liver alcohol dehydrogenase, which shows again a characteristic aromatic spectrum.

Detection of conformation changes in proteins

A large number of controlling processes are postulated to operate by conformational changes and CD has been widely used for monitoring changes in protein conformation on binding a non-absorbing ligand. The interpretation of changes in CD following such binding parallels the corresponding technique of ultraviolet absorption difference spectroscopy, (see p. 57–61),

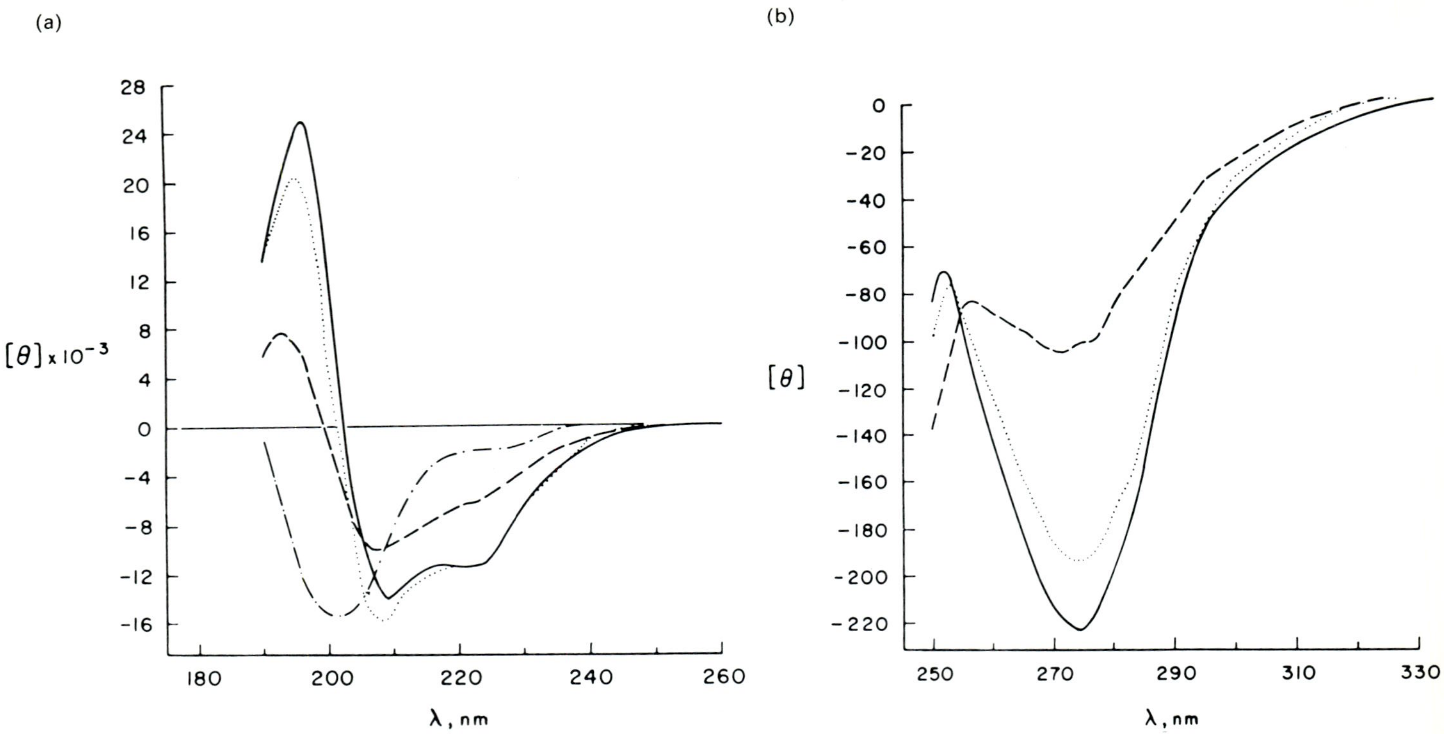

FIG. 5.18 Aromatic CD in near-UV of tyrosine-containing proteins. (a) Insulin: pH 3·0 at various temperatures (right) near-UV CD, (—) 27°C, (. . .) 45°C, (– –) 70°C (left) the corresponding far-UV with the random poly-L-Lysine spectrum (–·–·–) for comparison. Even at 70 °C, a degree of ordered structure is retained. (Data of Ettinger, M. J. and Timasheff, S. N. (1971). *Biochemistry* **10,** 824–830.)

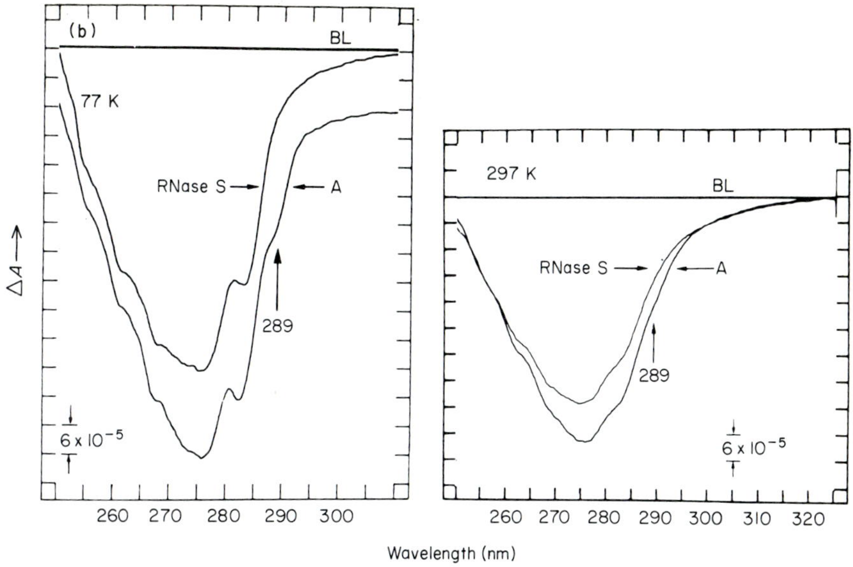

FIG. 5.18 Aromatic CD in near-UV of tyrosine-containing proteins. (b) Pancreatic RNAse A and S at 77 K and 297 K. The already intense CD of tyrosine residues is not greatly enhanced on cooling, i.e. they are already highly immobilized; in RNAse S the environment of a single Tyr has become more exposed to solvent, shown by the blue shift from 289 nm of its vibrational fine structure at 77 K. (Data of Horowitz, J. and Strickland, E. H. (1971). *J. Biol. Chem.* **246**, 3749–3752.)

although in CD the preferred experimental approach involves the recording of two spectra separately and taking their difference numerically rather than using double-beam techniques. Thus if a CD difference is seen to correspond to just a single residue, conformational change is not necessarily involved since the direct effect of the ligand (bound asymmetrically) could be reponsible. If a CD difference is seen to involve several spectroscopic types or residues then the simplest explanation is that the environment and relative orientation of a number of chromophores and other structural components has changed. Where the aromatic chromophores themselves are interacting there can still be ambiguity. Thus the CD of an interacting Trp and Tyr pair will not necessarily be the simple sum of the two and if, say, the Tyr property was changed due to interaction with a ligand (i.e. Trp+Tyr+ligand) the difference could involve the Trp transitions also. Thus the interpretation of CD difference systems in the aromatic region requires careful consideration of a number of possible effects often acting simultaneously.

The disulphide chromophore

Studies with rigid model compounds show strong CD in the region 260–300 nm associated with the —S—S— chromophore and the absorption maxima depend strongly on the dihedral angle. Much ingenious work has attempted to correlate the chirality of the —S—S— bond with observed signs of the CD. However the spectral properties are the sum of a number of effects; in addition to the intrinsic dissymmetry, the group as a chromophore can participate in coupling and polarizability mechanisms, as well as being perturbed asymmetrically by the groups R_1 and R_2 (Fig. 5.8(c)). Thus in cystine, where both

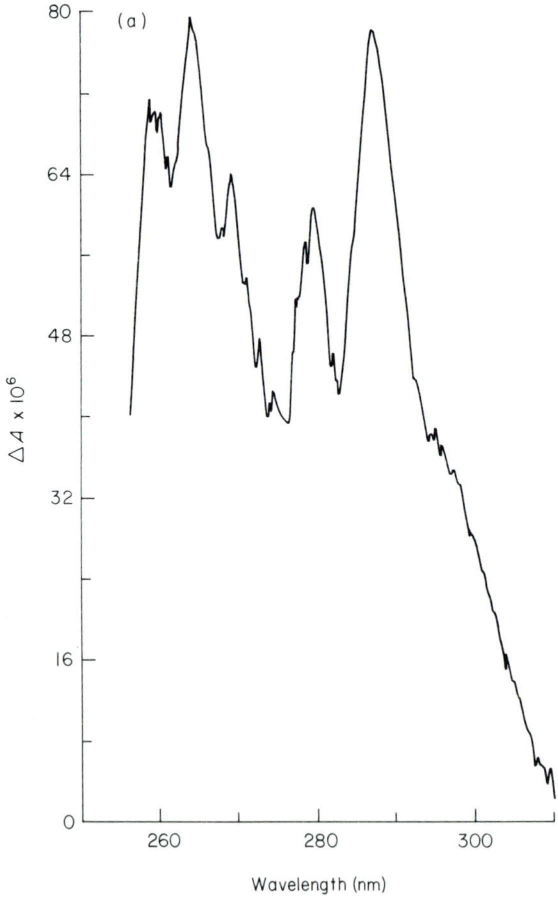

FIG. 5.19 (a)

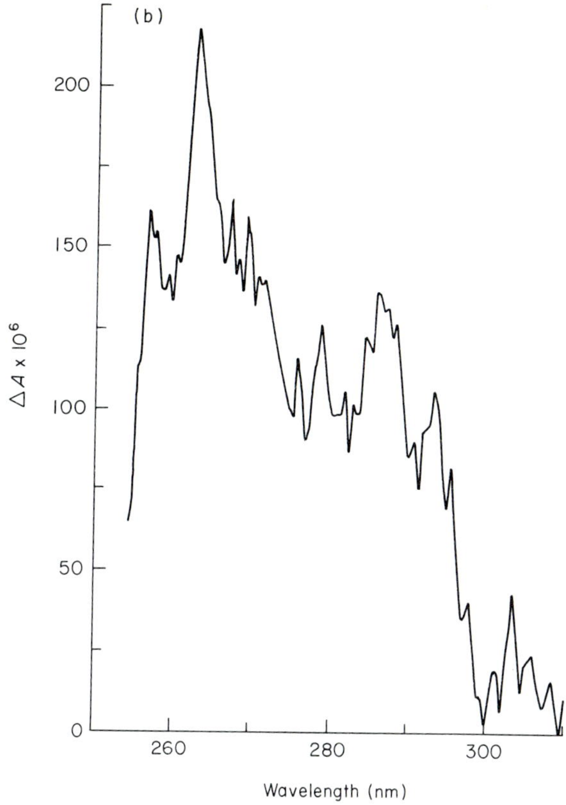

FIG. 5.19 Aromatic CD in near-UV of two dehydrogenases. (a) bovine glutamate dehydrogenase 10 mg ml^{-1}; 1mm path (average of four scans); measured at 1 nm bandwidth, time constant=16 s (Jasco J41C); (b) horse liver alcohol dehydrogenase 4.5 mg ml^{-1}; 5 mm path (average of five scans); at variable bandwidth, time constant=10 s (Jouan Mk II). Instrumental baselines are subtracted and summations performed numerically by computer. Noise levels: (a) $\delta\Delta A \sim \pm 1 \times 10^{-6}$; (b) $\delta\Delta A \sim \pm 8 \times 10^{-6}$.

chiral forms are present equally in aqueous solution, simple asymmetric substitution accounts for $\Delta\varepsilon_{257} = -0{\cdot}7 \text{M}^{-1}\text{cm}^{-1}$ (Fig. 20 (a)(b)). Only when $\Delta\varepsilon$ exceeds $\pm 2{\cdot}0\ \text{M}^{-1}\ \text{cm}^{-1}$ is a preferred chirality reasonably certain. In proteins a broad and featureless CD is sometimes found and the —S—S— group is often the only candidate for broad CD effects observed in very weakly absorbing regions at $\lambda > 300$ nm (Fig. 5.20). Higher energy effects at 210–220 nm may contribute a characteristic positive effect in proteins, which has an apparent maximum at 240–250 nm. However, this may not always be resolvable from high-energy aromatic effects.

Folding processes in globular proteins

Circular dichroism has been widely used to monitor properties of both aromatic and peptide chromophores during protein folding/unfolding reactions in solutions of denaturants such as urea or guanidine hydrochloride. The enzyme penicillinase loses its aromatic CD at a low denaturant concentration, where peptide conformation is virtually unchanged (Fig. 5.21). This is evidence in favour of an intermediate (hydrodynamically expanded) structure in which the asymmetric environment of these residues has been destroyed, presumably by increased side chain mobility. Aromatic CD is also absent after removal of denaturant from CNBr-cleaved penicillinase, which allows reformation of a non-covalent complex from three major peptide fragments. Although the backbone conformation is effectively that of the unmodified enzyme, the aromatic residues have greater conformational freedom than in the native enzyme.

A different type of information has been sought in refolding studies of bovine pancreatic trypsin inhibitor BPTI (Fig. 5.22) using the intense aromatic CD associated with the tyrosine residues. Bovine pancreatic trypsin inhibitor can be unfolded and refolded from denaturants reversibly and undergoes a reversible thermal transition on heating; in both cases the native CD is effectively fully restored. The time course of re-establishment of the native structure by reoxidation of the fully reduced denatured protein has been followed in great detail by quenching the reoxidation reaction at different times and identifying chemically the intermediate reaction products in

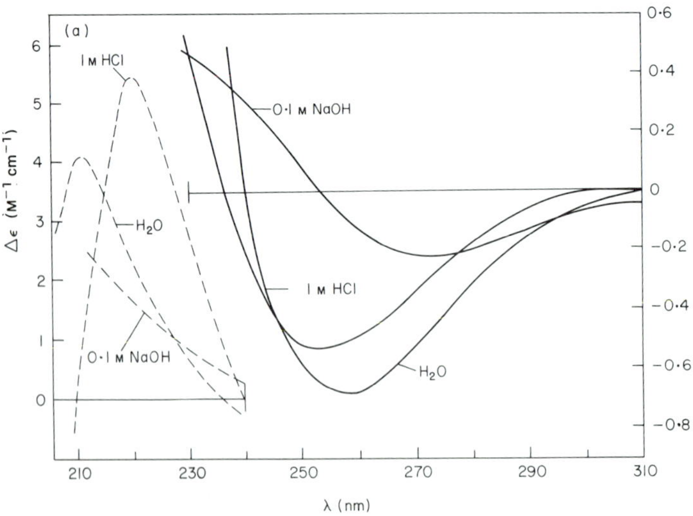

FIG. 5.20 (a)

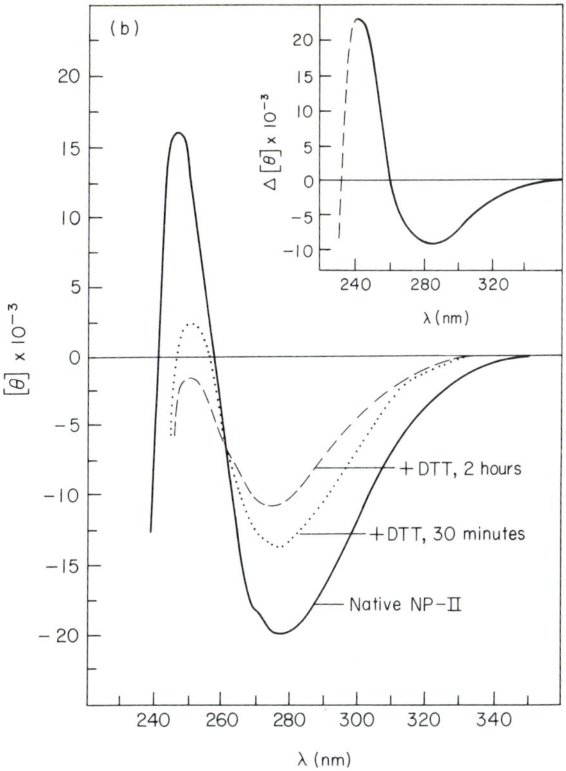

FIG. 5.20 Circular dichroism of cystine residues. (a) L-Cystine in acid, neutral and alkaline solution; alcoholic solvents such as *n*-butanol or isopropanol (with L-cystine diethylester hydrochloride) cause an inversion at $\lambda > 250$ nm (data of Casey, J. P. and Martin, R. B. (1972) *J. Am. Chem. Soc.* **94**, 6141–6151); (b) The effect of dithiothreitol (DTT) (1:1) on the CD spectrum of neurophysin II (NP-II) as a function of time at pH 8. The inset shows the 2 h difference spectrum due to loss of the disulphide in NP-II. (Data of C. J. Menendez-Botet and E. Breslow (1975). *Biochemistry* **14**, 3825–3835.)

terms of the pattern of their disulphide bonding. Some of these intermediates have been examined by CD (Fig. 5.22(b)). It is seen that the fully reduced carboxymethylated form has effectively zero tyrosine CD indicating a complete randomization of Tyr side chain conformations. The intermediate disulphide (Cys-30–51) already shows a pronounced negative effect (possibly Tyr-35) which may well be in its native environment, while the incorrect disulphide (Cys-5–30) gives less intense CD. The fully reoxidized form has CD comparable with that of native BPTI.

Measurements of the aromatic CD have been particularly useful in the study of immunoglobulin (Ig) conformation. This system poses many questions about the interaction of domains within a single heavy or light chain as well as the conformational parameters involved in reforming active Ig structure from the isolated chains. An example is shown from studies of IgD, a human myeloma immunoglobulin, in Fig. 5.23. In one IgD, the summation of conformations of Fab and Fc to give a CD indistinguishable from the parent IgD argued against striking conformational effects between Fab and Fc following the proteolytic cleavage. By contrast the Fab of another patient showed an intense low wavelength CD band which was not affected by the presence of the Fc fragment but which rapidly disappeared on further tryptic cleavage of Fab. This suggests that it is a characteristic marker for interactions between other domains with Fab and studies with the smaller fragments of Fab suggest that it resides in the light chain and is produced by interaction with a heavy chain domain. The wavelength suggests Tyr or disulphide. Interestingly the effect has similar intensity to that of Tyr in Ala^6–Angiotensin II which is adjacent to β-structure. Since β-structure predominates in immunoglobins (hence the characteristic $\Delta\varepsilon_{217}$ of about -1 to $-1{\cdot}3\ \text{M}^{-1}\,\text{cm}^{-1}$) the

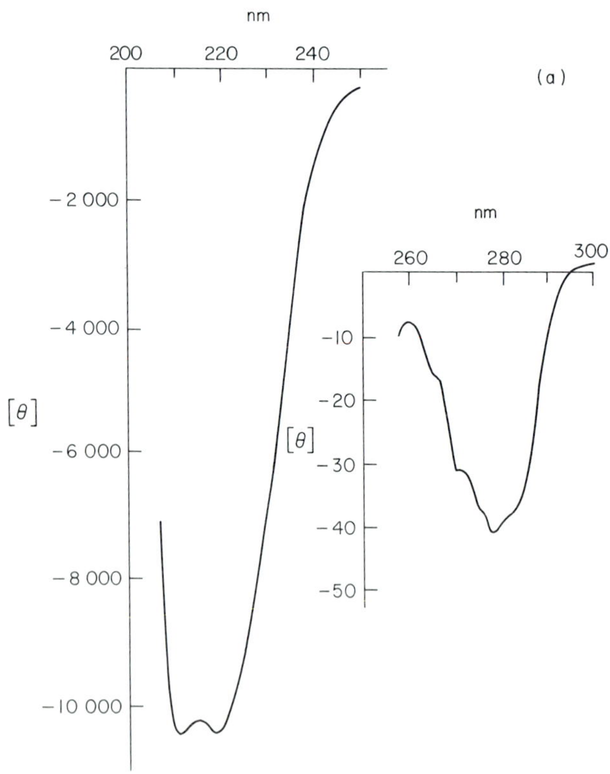

FIG. 5.21 (a)

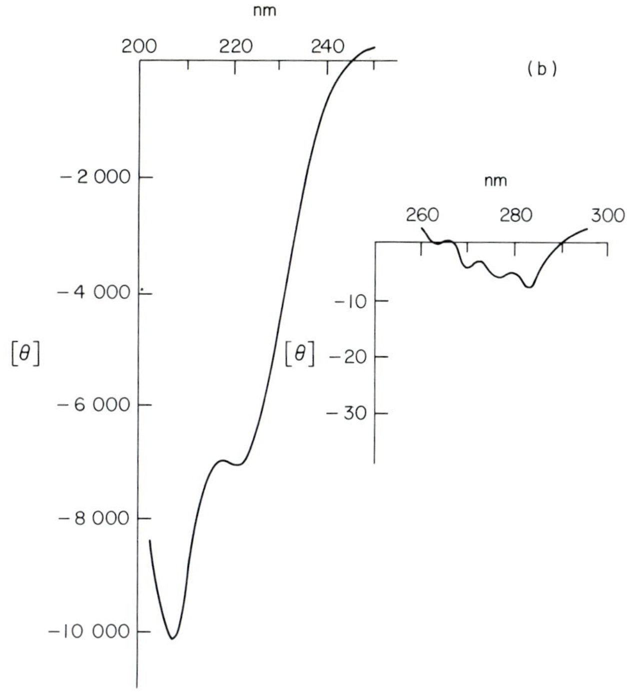

FIG. 5.21 Unfolding of Penicillinase (*Staphylococcus aureus*). near and far UV–CD: (a) at pH 7·0; (b) in 0·84 M guanidine hydrochloride which reduces the aromatic CD while having little effect on peptide conformation. (Data of Carrey, E. and Pain, R. (1978). *Biochim. Biophys. Acta* **533**, 12–22.)

effect could be due to a similar structure. Its absence from other IgD's suggests that a primary sequence difference in the light chains could be responsible.

These examples illustrate the wide range of conformational questions which can be approached through CD studies. Invariably, supporting evidence is sought by other techniques, NMR of smaller peptides, fluorescence and absorption spectroscopy, plus studies of the state of macromolecular association by gel chromatography and ultracentrifugation.

Nucleic Acids and Their Components

Mononucleotides

The strong absorption bands of the common nucleosides and mononucleotides (see p. 32) show rather weak CD around 260 nm which is generally negative for purine nucleotides, and positive for pyrimidine nucleotides in

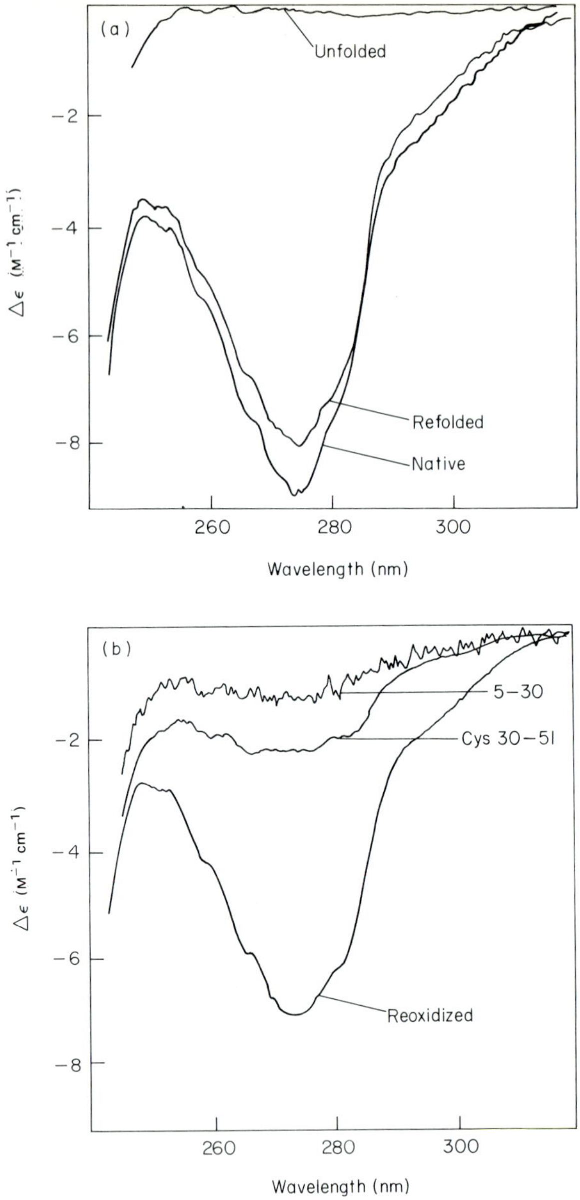

FIG. 5.22 Refolding studies of pancreatic bovine trypsin inhibitor (BPTI) using near UV–CD. (a) native BPTI, BPTI refolded from denaturants, and fully reduced carboxymethylated BPTI; (b) BPTI reoxidized from the fully reduced form; BPTI intermediate with CYS 30–51 reformed; "incorrect" intermediate with CYS 5-30 reformed. (Data of Creighton, T. E. and Bayley, P., unpublished.)

aqueous solution (e.g. see adenosine Fig. 5.24(a)). These molecules have β-glycosidic conformation; the α-anomers show CD of reversed sign. However, model compound studies show that this classification is deceptively simple. Derivatives of adenine, for example, in which the glycosidic angle is held rigidly by covalent ring formation from N6 to C5′, can show greatly enhanced CD. Experiments and theory show that the rotational strength is highly dependent on the glycosidic angle (χ) and may even invert in certain cases. Electronically this is because the perturbation of the transitions of the base show different symmetries in the rotamers defined by different values of χ. Thus the weak CD in the normal mononucleotides reflects the rotational freedom in χ with averaging out of both positive and negative contributions. It also follows that it is possible to have a fixed conformation with giving effectively zero CD (i.e. a nodal position). The effect of different sugar conformations (3′*endo*, 2′*endo*, etc.) is likely to be to restrict the available range of χ.

The clearest interpretations of mononucleotide properties can be expected only when rather intense CD is found. The assignment of χ (even in the rather loose sense of *syn* and *anti* conformers) may not be unambiguous. Treatments such as protonation and deprotonation, which can change the sign of the CD, alter the basic properties of the chromophore, and hence the orientation within the base of the $\pi \rightarrow \pi^*$ transition dipoles. Such changes are specific to the individual bases; signs, intensities and dependence on χ can all vary in an individual way. Nonetheless, a change from *anti* to *syn* has been established for guanosine on protonation (Fig. 5.24(b)) as seen by the similarity in its CD at pH 1 compared with formycin and other guanosine derivatives of known *syn* conformation.

Circular dichroism evidence on conformation in these rather flexible systems generally requires strong corroborative evidence from other sources; NMR has been most successful, providing detailed equilibrium and kinetic information (see p. 263, Example 2).

Changes in state of association of mononucleotides are readily observed by CD. The couplet observed with thioguanosine derivatives in isopropanol is direct evidence for intermolecular interaction, (Fig. 5.25(a)). This spectrum resembles that of the di- and oligonucleotides discussed below.

Di- and oligonucleotides

In contrast with the rather confused situation with mononucleotides, CD has been a useful parameter in studying the dimeric (and oligomeric) assemblies of nucleotides. Base-stacking in a simple dinucleotide such as D-adenylyl(3′5′)D-adenosine(D-ApA) is necessarily asymmetric. The optical properties are those of the coupled oscillator or dimeric "exciton" model (see p. 164); intense CD is observed in the form of a couplet of overlapping equal positive

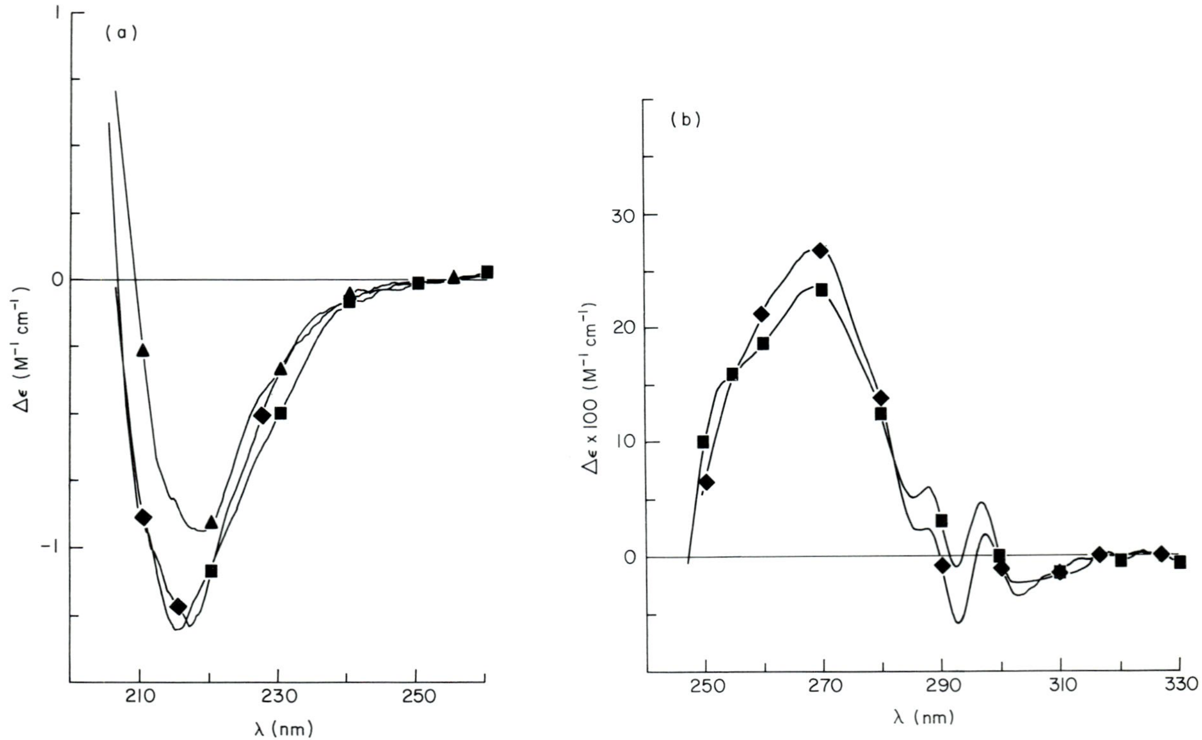

FIG. 5.23 Circular dichroism studies of immunoglobulin-D. (a) Far UV and (b) near UV CD spectra of t-Fc (◆), t-Fab (▲) and p-Fc (■) fragments of IgD-Ha. (t=trypsin- and p=papain-derived fragments.) (Data as for Fig. 5.23 (e)).

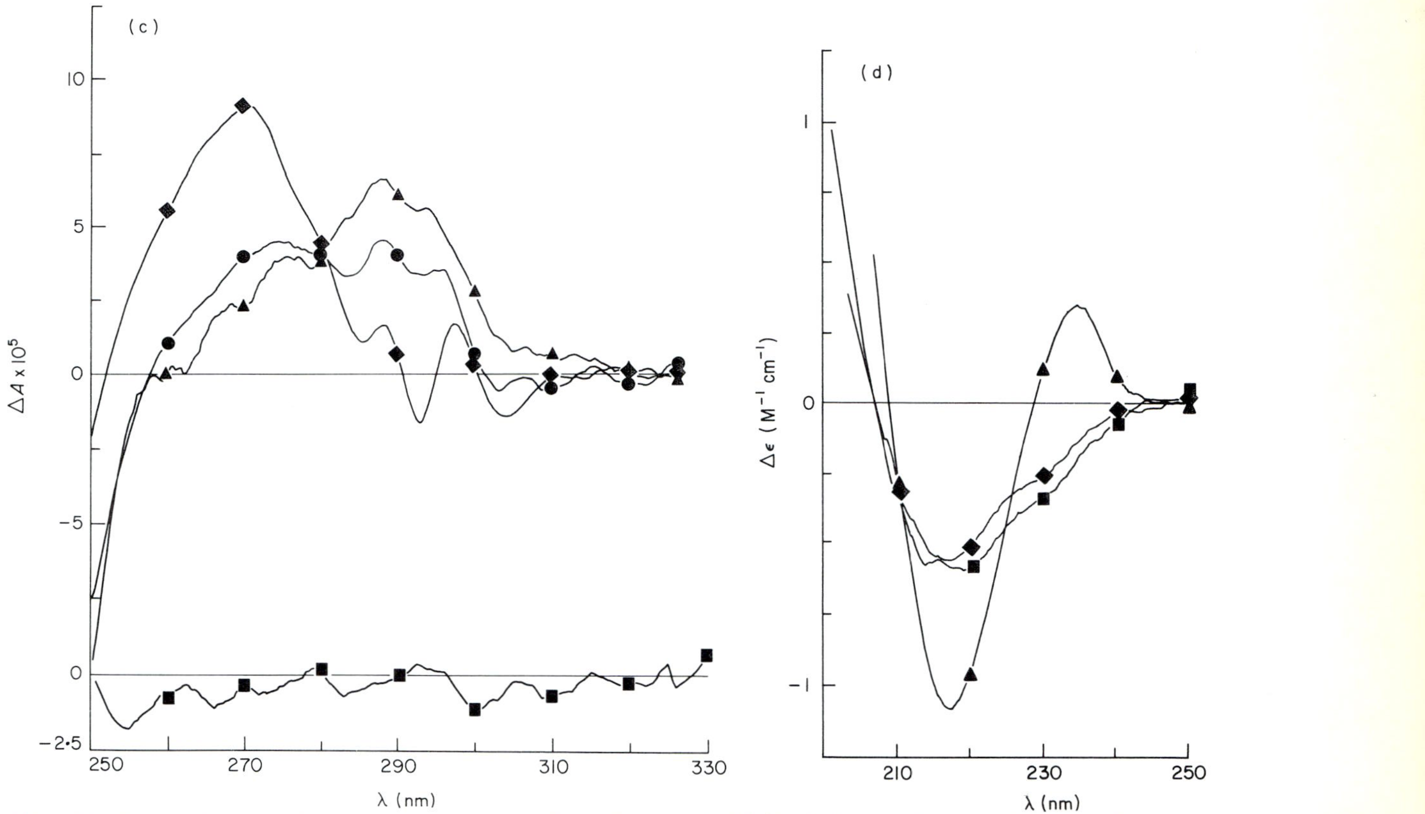

FIG. 5.23 Circular dichroism studies of immunoglobulin-D. (c) The near UV CD of IgD-Ha (●) is the sum of t-Fc and t-Fab shown by the weighted difference spectrum (■). (d) Far UV CD spectra of t-Fc (◆), t-Fab (▲) and p-Fc (■) of IgD-Ai showing the distinctive band at n=235 nm. (t=trypsin- and p=papain-derived fragments.) (Data as for Fig. 5.23 (e)).

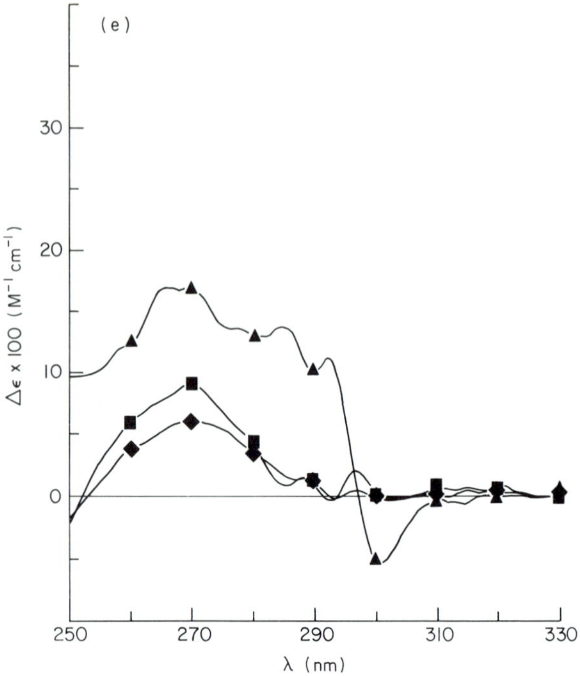

FIG. 5.23 Circular dichroism studies of immunoglobulin-D. (e) Near UV CD spectra of t-Fc (◆), t-Fab (▲) and p-Fc (■) of IgD-Ai (t=trypsin- and p=papain-derived fragments). (Data of Jefferis, R., Matthews, J. B. and Bayley, P. M. (1977). *Immunochemistry* **14**, 393–396.)

and negative bands displaced about the absorption maximum (Fig. 5.25(b)). This is one of the most reliable CD effects diagnostic of chromophore–chromophore interaction. For D-ApA, the long wavelength positive band indicates a stacking in which the transition dipoles interact with right-handed symmetry. The L-ApA clearly has the inverse symmetry. For the D-ApA this implies a structure equivalent to the common B-type helix of DNA. The couplet effect has been widely used to study the melting profiles of simple oligonucleotides, comparisons with hypochromicity as an index of stacking, derivation of the thermodynamics of base-stacking interaction, degree of stacking, and cooperativity in oligonucleotides (see also p. 65). Extending the length of the stacked structure, (i.e. in oligo-A) intensifies the CD of the fully stacked form, but the characteristic line-shape is unchanged. This observation indicates that large changes are associated with stacking, with rather less distinction between single-strand stacked and double-strand stacked structures by CD.

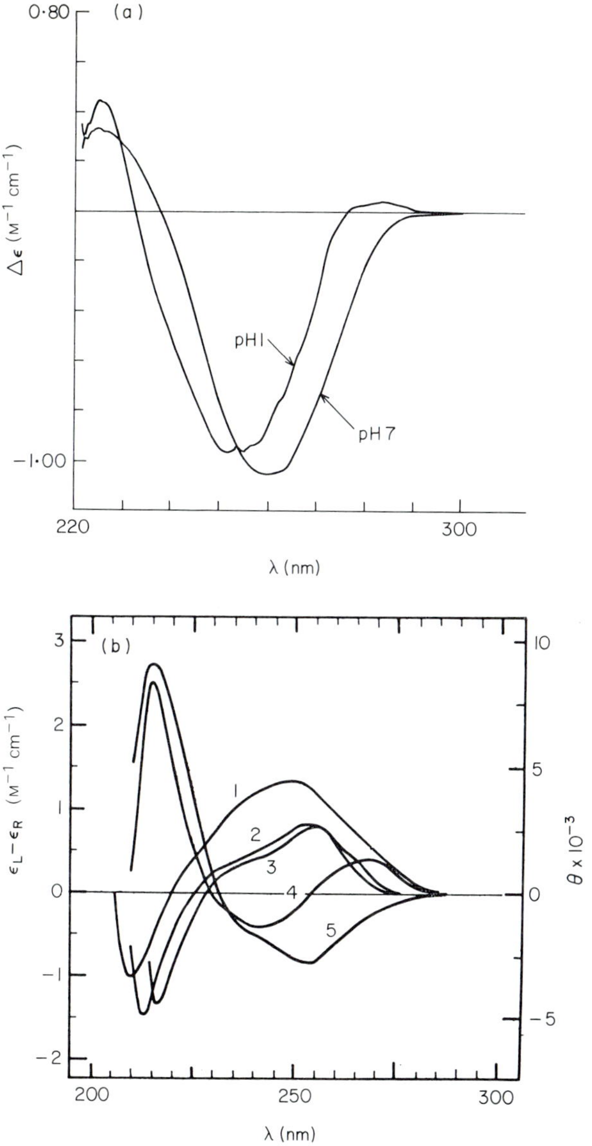

FIG. 5.24 Circular dichroism of purine nucleosides: *syn*- and *anti*-conformers. (a) Adenosine at pH 7 and pH 1 (b) Guanosine at pH 7 and pH 1, with derivatives adopting the *syn*-conformation. (1. formycin B; 2. guanosine pH 1: 3. 7-methyl guanosine, 4. 8-bromoguanosine; 5. guanosine pH 7; data of Guschelbauer, W. (1972) Jerusalem Symposium on Quantum Chemistry and Biochemistry **IV**, 297—310).

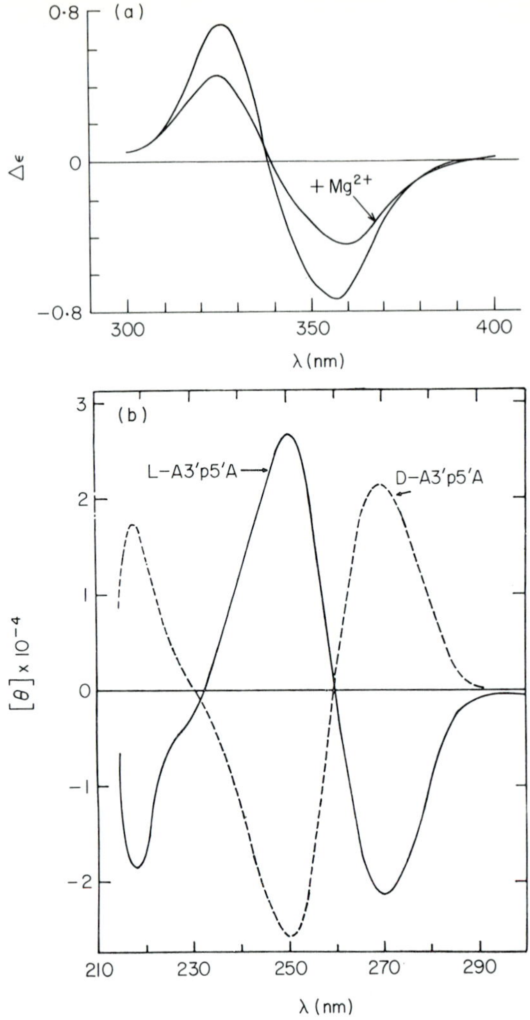

FIG. 5.25 Circular dichroism of stacked nucleotide systems. (a) thio-GTP (57·5 μM) in 95% isopropanol (1mM acetate buffer pH 4.5+0·1 mM dithiothreitol; 20°C) plus 5 mM $MgCl_2$. (Eccleston, J. F. and Bayley, P.M. unpublished.) (b) L-A3′p5′A and D-A3′p5′A; pH 7, 21·5°C (data of Tazawa, I., Tazawas, S., Stempel, L. M. and Ts'o, P. O. P. (1970). *Biochemistry* **9**, 3499–3514).

Polynucleotides

Much of the impetus of this work derives from studies on synthetic polynucleotides, initially with homopolymers and more recently with sequence-specific polymers. To some extent, the very diversity of the material precludes simple generalizations about the relationship between structure and CD. The CD is determined by the interaction of a large number of conformational parameters and provides a valuable and sensitive indicator of a conformational state (e.g. an ordered structure or a multi-strand complex). Almost by contrast, DNA from a wide variety of sources shows a typical CD profile in aqueous solutions with intense bands, positive and negative, about 260 nm, in an approximately conservative system. These CD properties undergo striking changes in solutions of high salt concentration and in non-aqueous solvents, involving loss of the conservation and shifts in the band positions. These effects are interpreted as changes in polymer structure from B-form to A-form, and B-form to C-form, due to base tilting and changes in number of residues per turn, analogous to changes which also occur with DNA films under conditions of low hydration (Fig. 5.26).

Ribonucleic acid spectra show a wide variety of forms. Double-stranded RNA (known from X-ray studies of rice dwarf virus) shows an intense CD band, mainly positive, with maximum close to 260 nm (Fig. 5.27(a)). The lack of conservatism is similar to that in A-type DNA and indicates a polynucleotide conformation in which bases are tilted relative to the helix axis (cf. above). The CD of the leading negative band at 295 nm is related to G+C content. The characteristic weak mycophage RNA is shown in Fig 5.27(b,c). Melting of the ordered structure results in a diminution of the band and a shift of up to 15 nm to the red. RNA from other sources (ribosomal and *t*RNA in particular) show similar shifted spectra (Fig. 5.27) which appear to be some combination of double strand and denatured spectra. Attempts have been made to separate the contribution of the two forms and to assess the average base composition of single- and double-stranded regions. To date this has been only partially succesful, but recent progress with the analysis of DNA CD suggests that the problem is not intractable.

The CD profiles of DNA and various RNAs have been used in many studies e.g. thermal stabilities, binding of metal ions, particularly the role of intercalation by mutagenic and therapeutic drugs and dyes (proflavin, acridines, ethidium, anthracyclines, etc.), interactions of DNA with histones and synthetic histone-like polypeptides, reconstitution experiments and many other systems in which the polynucleotide conformation is of fundamental importance. The Bibliography at the end of this chapter (p. 234) supplies more references to this work.

Sequence analysis of nucleic acids

General methods of analysis of the CD spectrum of nucleic acids depend upon the fact, observed in model components, that the CD of an oligonucleotide depends upon its nucleotide sequence. Further, the CD of small oligo-

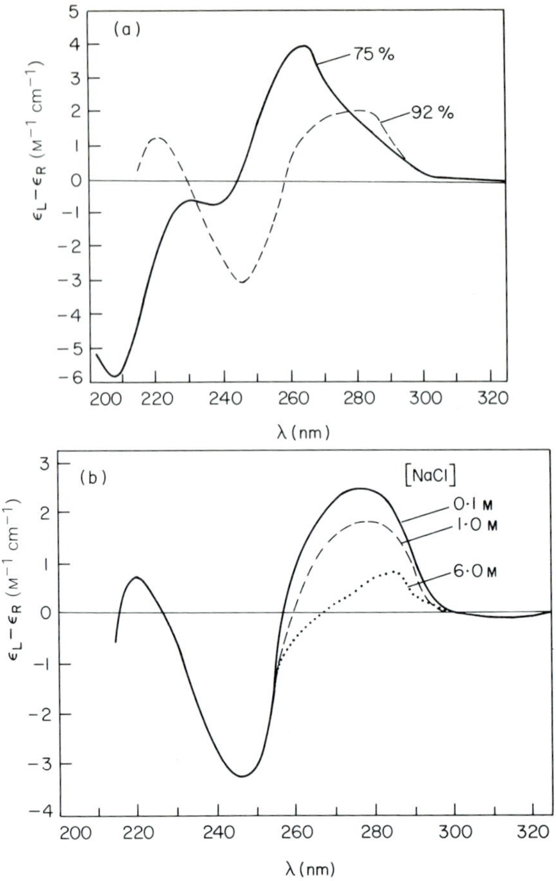

FIG. 5.26 Circular dichroism of conformational transformations of DNA. (a) CD of film of DNA (calf thymus; Na salt) at humidity of 92% (B form) and 75% (A form). (b) CD of DNA (calf thymus) in NaCl (0·1 M, 1·0 M and 6·0 M); films of Li-DNA show similar changes on reducing relative humidity (B form→C-form). (Data of Tunis-Schneider, M. J. B. and Maestre, M. F. (1970). *J. Molec. Biol.* **52**, 521–541.)

meric ribonucleotides (e.g. the trinucleotides) could be represented as the sum of the nearest-neighbour interactions. By building up a library of the CD spectra of the component dinucleotide interactions (16 for the four bases) attempts have been made to represent the CD of the larger polyribonucleotides and *t*RNA as a sum of the first-neighbour interactions. This first-neighbour hypothesis has, to date, not been fully successful with these rather complex RNAs, partly due to the structural features discussed above. However, progress with the application of the hypothesis to analysis of the CD of DNA is now yielding useful biochemical information.

Individual DNAs from different sources show a similar broad pattern of CD intensity and band shape. Intuitively this pattern looks like the superposition of a large number of (base-specific) couplets whose CD contributions differ in position, intensity and band shape according to base content. The approach has been to use, as basis spectra, the CD of a number of sequence-specific synthetic polynucleotides containing repeating di- or tri-nucleotide sequences (Fig. 5.28). Since the nearest-neighbour frequencies of these polymers are by definition, known, a set of equations can be written from which the contributions from individual dinucleotide interactions (and their Watson–Crick complementary base pairs) can be evaluated. For the four bases A, T, G, C the ten dinucleotides required to define the full sequence are A–A, A–T, A–G, A–C, T–A, T–G, T–C, G–G, G–C, C–G, (plus their complements). A least-squares procedure gives the set of CD components which best represents the chosen polymer set. Some of these are also shown in Fig. 5.28. These components are now used in fitting the CD spectra of samples of DNA, using as initial test, DNA whose nearest-neighbour frequencies have been determined enzymatically. The fit is optimized in terms of the minimum standard deviation of experimental and computed CD spectra for a number of sets of basis spectra. These conditions are used to establish the best set of basis spectra. Figure 5.29(a) shows the excellent fits obtained for DNA from calf thymus and various bacterial sources with (A + T) content ranging from 57 to 29%. A systematic decrease of fitting at higher (A + T) suggests that the basis functions have not yet been perfectly defined. The agreement between the two sources of sequence information is represented in the histograms of Fig. 5.29(b). The nearest-neighbour estimates are most accurate for the lower (A + T) content, but are in any case in good semiquantitative agreement. Problems exist with repeating sequences e.g. A–A–A where interactions between more distant bases may contribute. In fact homopolynucleotides are excluded from the best set of polymer data.

Analysis of four (A + T)-rich satellite DNAs from *Drosophila* show a relatively simple distribution of nearest neighbours, in good agreement with experimental results. Analysis of satellite DNA from the crab shows the high predominance of the poly [*d*(A–T)*d*(A–T)] sequence for three species, but the fourth (*Cancer*) is substantially different in base content and sequence. Both systems are illustrated in Fig. 5.29(c,d).

Overall, the results show that (A + T) content can be assessed to within a few percent, and semi-quantitative sequence information can be derived. The accuracy of the method should improve as minor systematic errors are eliminated from the reference spectra, which currently lead to an overestimate of *d*(G–C)·*d*(G–C) and *d*(C–G)·*d*(C–G) and an underestimate of

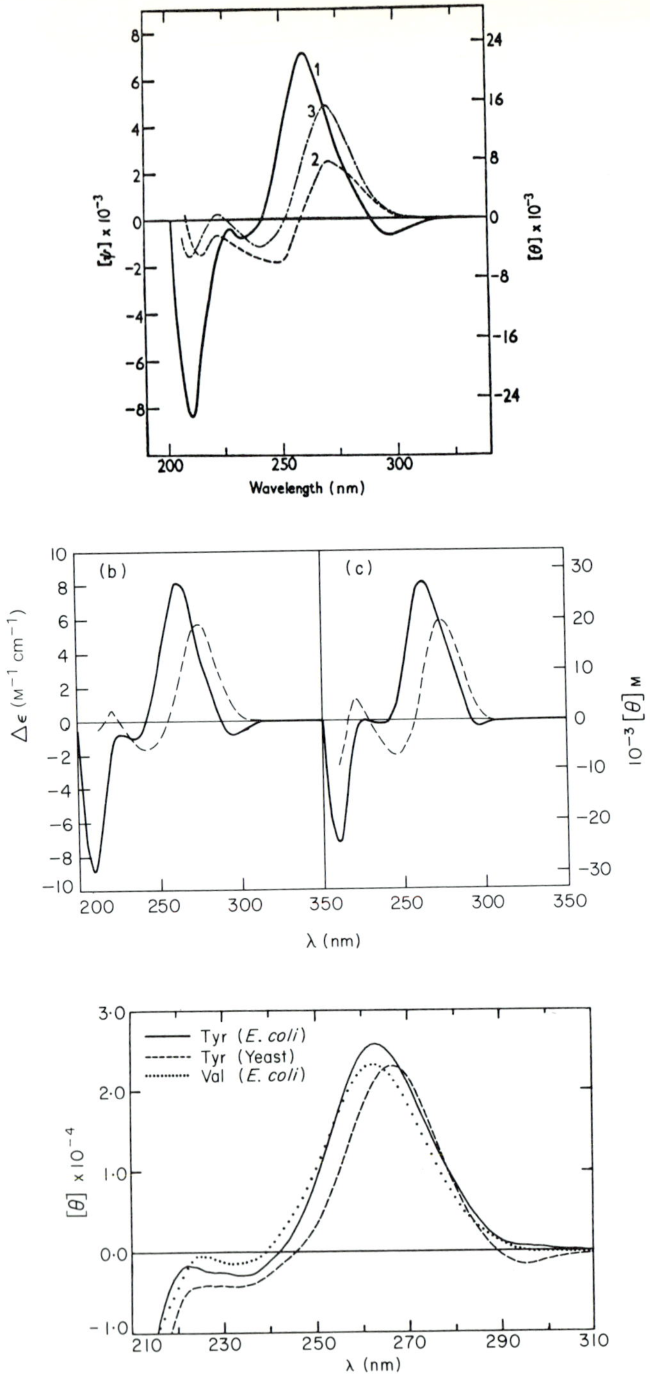

[ψ] x 10⁻³
[θ] x 10⁻³
Wavelength (nm)
1
2
3
(b)
(c)
Δε (M⁻¹ cm⁻¹)
10⁻³ [θ] M
λ (nm)
Tyr (E. coli)
Tyr (Yeast)
Val (E. coli)
[θ] x 10⁻⁴
λ (nm)

(a)

(d)

FIG. 5.27

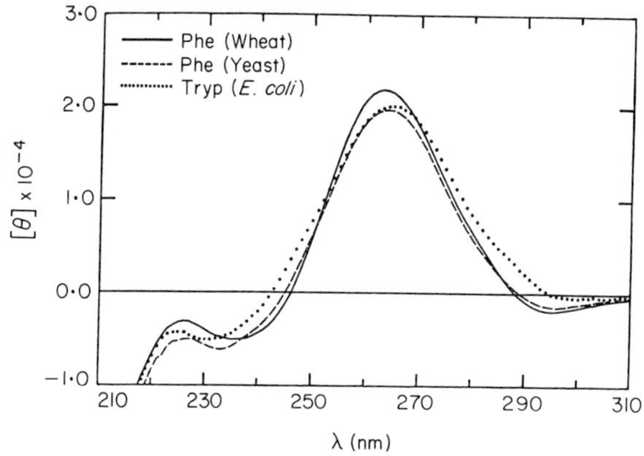

(e)

FIG. 5.27 Circular dichroism of double-stranded RNA (a) to (c) and transfer RNA (d) and (e): (a) Rice dwarf virus RNA at (1) 24 °C (2) 94 °C (3) 24 °C after slow cooling from 94 °C. (T. Samejima, H. Hashizume, K. Imahori, I. Fujii and K. Miura. (1968) *J. Molec. Biol.* **34,** 39–48.) (b) and (c) Mycophage RNA fractions at (∼) 25 °C (⁽‿⁾) 25 °C after heating to 95 °C in 1% formaldehyde. (Cox, R. A., Kanagalingam, K. and Sutherland, E. S. (1970). *Biochem. J.* **120,** 549–558.) (d) and (e) Various species of native *t*RNA at 25 °C in 10 mM tris-1 mM $MgCl_2$. (Blum, A. D., Uhlenbeck, O. C. and Tinoco, I. Jr. (1972). *Biochemistry* **11,** 3248–3256.)

d(A–A)·*d*(T–T). The use of a wider basis set of DNA composition and other polymer combinations will also assist in the refinement.

Induced Optical Activity

Binding of simple symmetric chromophores

When a chromophoric molecule is symmetrical and hence has no CD, the phenomenon of induced optical activity or induced CD may be observed following specific binding to an asymmetric environment such as is afforded by a binding site on a protein or nucleic acid. For a simple, planar molecule such as proflavin binding specifically to DNA (into which proflavin intercalates) coupling mechanisms with the purine and pyrimidine bases are possible. Also, coupling between adjacent proflavin molecules is indicated by the increasing CD with increasing *r* value (Fig. 5.30a). For a molecule like the trypanocidal drug Berenil (Fig. 5.30(b)), which does not intercalate, but presumably interacts with external groups of the DNA by virtue of the two amidine residues, the intense CD suggests that this extended molecule may bind in a twisted helical conformation, possibly in the larger DNA groove. The non-isodichroic system of curves shows that several chiral forms contribute.

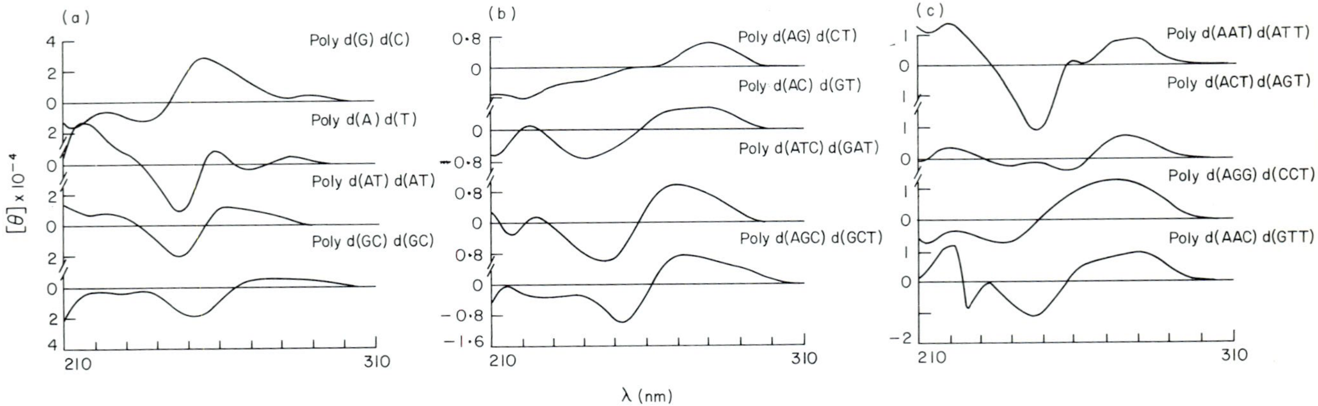

(a)
(b)
(c)
Poly d(G) d(C)
Poly d(A) d(T)
Poly d(AT) d(AT)
Poly d(GC) d(GC)
Poly d(AG) d(CT)
Poly d(AC) d(GT)
Poly d(ATC) d(GAT)
Poly d(AGC) d(GCT)
Poly d(AAT) d(ATT)
Poly d(ACT) d(AGT)
Poly d(AGG) d(CCT)
Poly d(AAC) d(GTT)
[θ] x 10⁻⁴
λ (nm)
210
310

FIG. 5.28

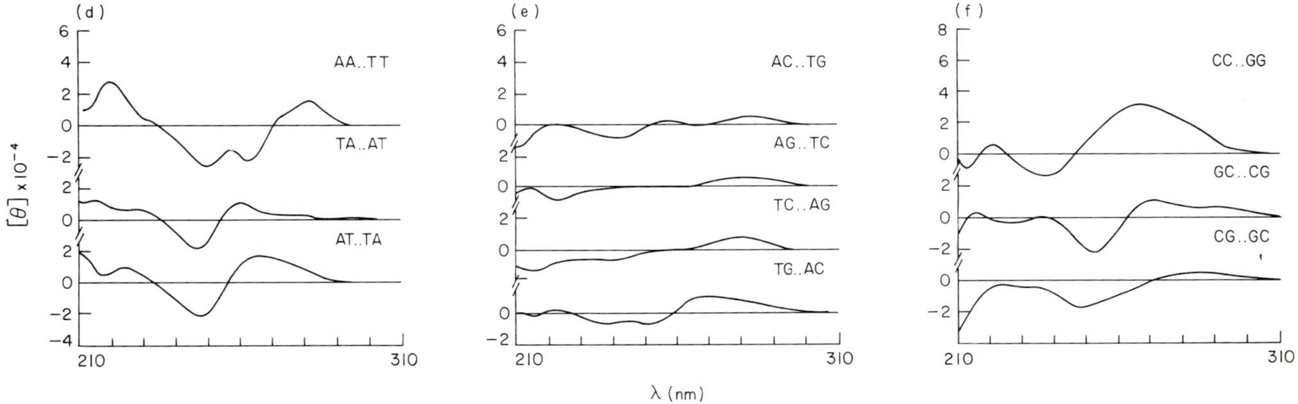

FIG. 5.28 Circular dichroism of various synthetic polydeoxyribonucleotides and component spectra derived from them for nearest-neighbour analysis: (a) synthetic polydeoxyribonucleotides used as basis spectra; (b) component spectra for specific nearest neighbours (plus their complements of opposite polarity). (Data from Gray, D. M., Hamilton, F. D. and Vaughan, M. R. (1978). *Biopolymers* **17**, 85–106.)

For the diphenyl methane dye, Auramine 0, which acts as a specific inhibitor of liver alcohol dehydrogenase, it is likely that the CD of the bound conformation (Fig. 5.31) is due to one of the inherently dissymmetric propeller forms present in rapid equilibrium in free solution. Similar considerations suggest that the (less specific) binding to DNA of the triphenyl methane bacteriocidal dye, crystal violet, involves the selection of one specific chiral form into the bound state, giving intense CD.

An interesting case of induced CD is shown by the binding of Cibacron Blue to a number of NADH-linked dehydrogenase enzymes. This dye, which is the chromophore of blue dextran, has been suggested as having specificity for the dinucleotide fold of the coenzyme binding site by adopting a conformation similar to that of the bound nucleotide. Figure 5.31(b) shows that the CD of the bound form is undoubtedly different for three enzymes suggesting specific binding in each case. Of course, interactions other than those within the chromophore are possible and indeed the CD of complexes of the three enzymes with NADH itself shows characteristic differences. Other dyes with

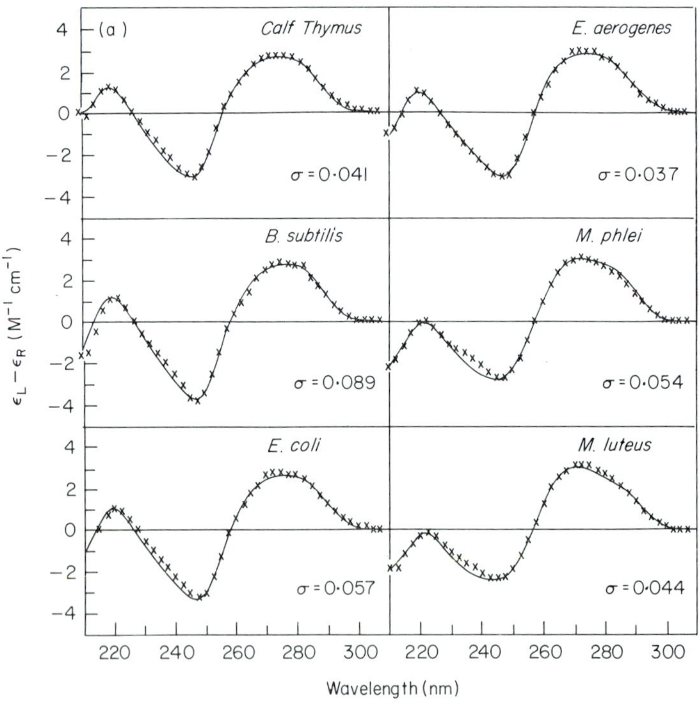

FIG. 5.29 Circular dichroism of various natural DNA samples and their nearest-neighbour analysis. (a) calf thymus, bacterial and mycobacterial DNA samples. (Data as for Fig. 5.28.)

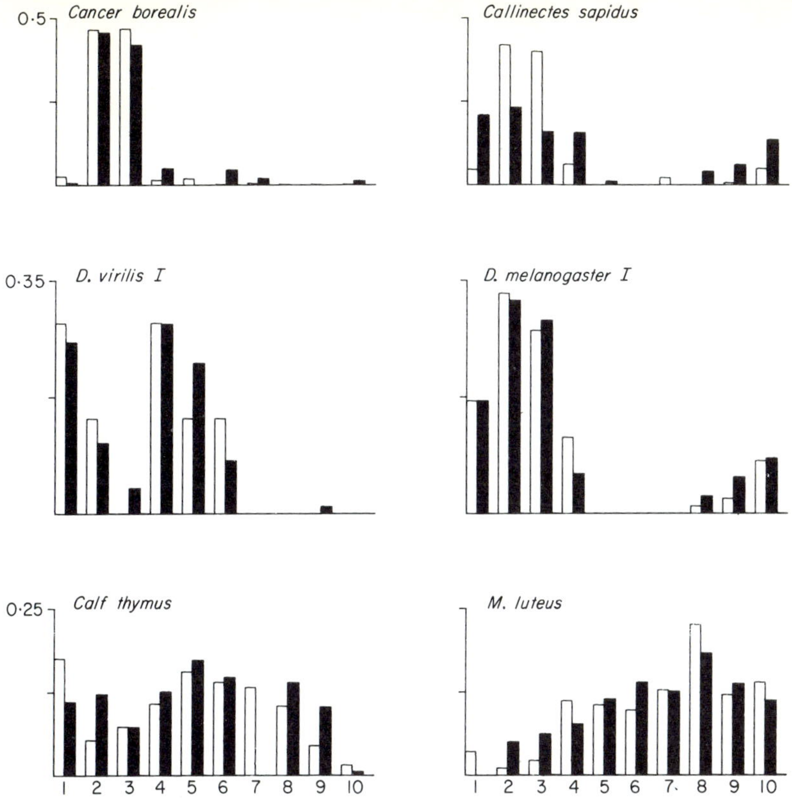

FIG. 5.29 Circular dichroism of various natural DNA samples and their nearest-neighbour analysis. (b) Histograms of nearest-neighbour frequencies: open columns determined enzymatically; filled columns, from CD spectra. The numbers correspond to the ten dinucleotides cited on p. 205. (Data as for Fig. 5.28.)

affinities for coenzyme binding sites are Congo Red and many fluorescein derivatives.

Thus, these essentially symmetrical and optically inactive dyes can afford considerable useful information about their binding sites on a variety of macromolecules from CD properties of appropriately characterized complexes.

Binding of nucleotides and coenzymes

In considering the binding of more natural chromophoric molecules such as nucleotides and coenzymes, additional considerations are introduced by the fact that such molecules are themselves optically active. One exception is pyridoxal phosphate (vitamin B_6) the cofactor of many decarboxylases and aminotransferases. Formation of the enzyme–Schiff's base with the apoenzyme of aspartate aminotransferase generates intense CD in the absorption

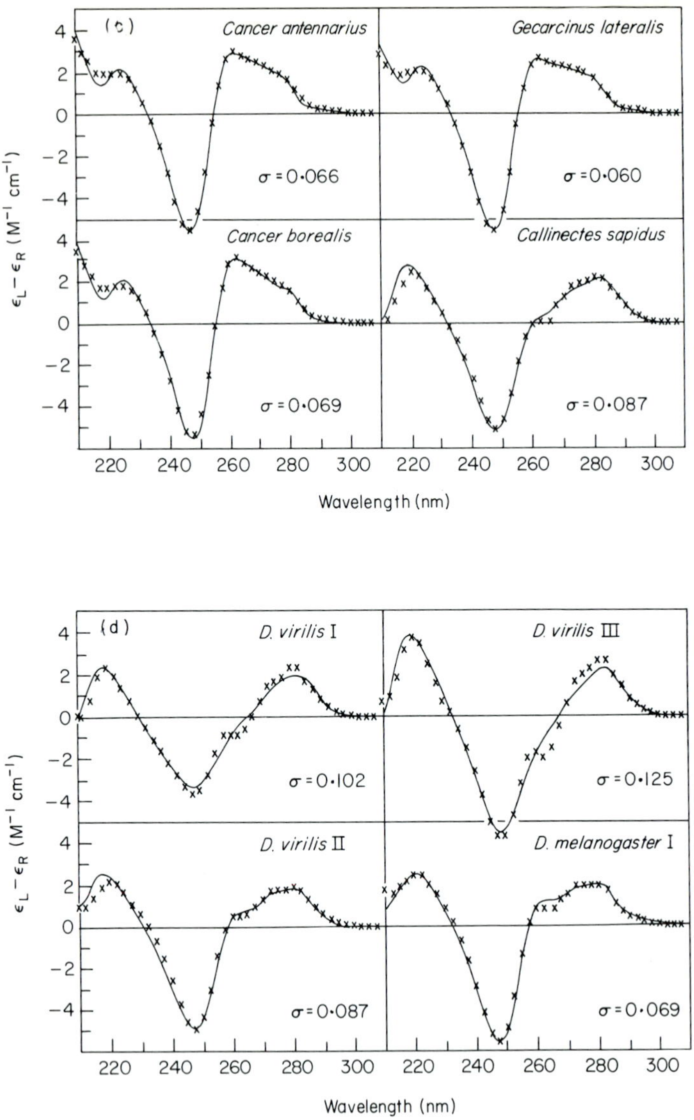

FIG. 5.29 Circular dichroism of various natural DNA samples and their nearest-neighbour analysis: (c) Four crab (A+T)-rich satellite DNA samples—measured spectra (——) and calculated spectra (X). (d) Four *Drosophila* DNA samples. (Data as for Fig. 5.28.)

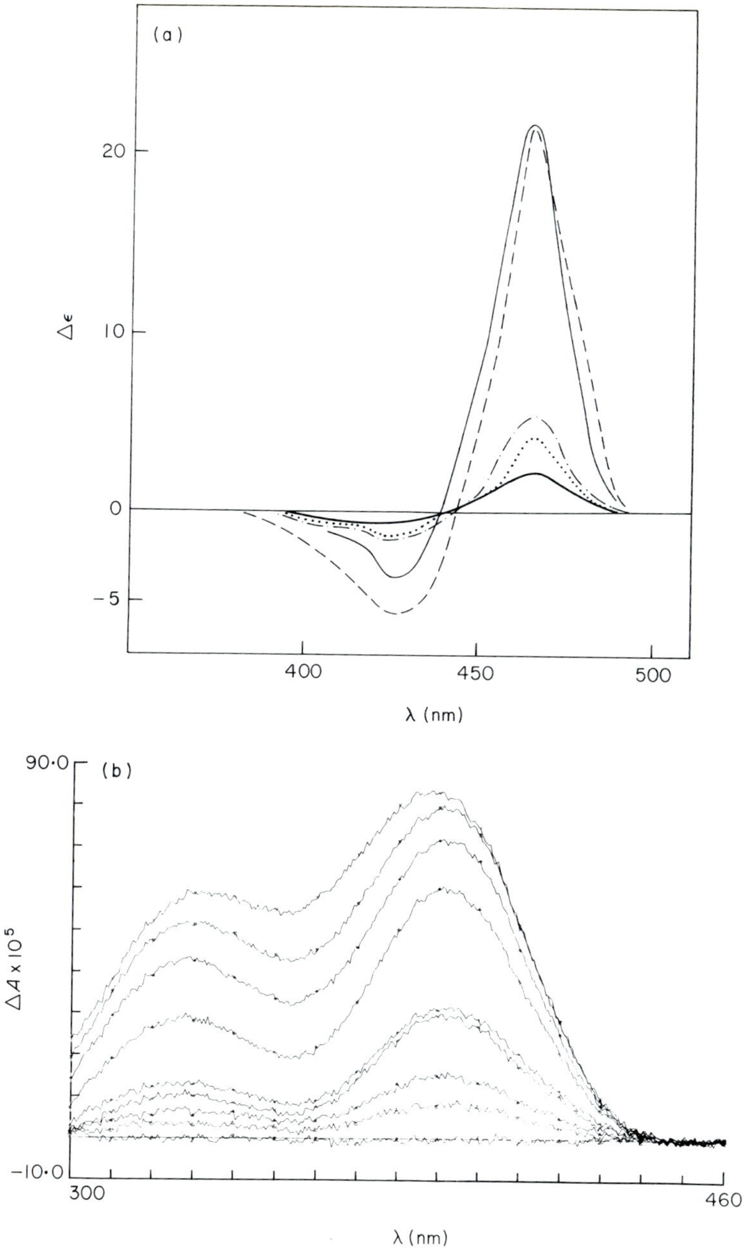

FIG. 5.30 Induced CD of chromophores, bound to DNA (calf thymus) as a function of r (moles dye bound per mole DNA phosphorus) or r′ (total dye per total DNA); (a) proflavine + DNA for r 0·20, 0·18, 0·14, 0·07 and 0·035, decreasing at 466 nm (data of Dalgleish, D. G., Fujita, H. and Peacocke, A. R. (1969). *Biopolymers* **8**, 633–645); (b) berenil (pp′ diamidino, diazoaminobenzene; 21·5 μM) + DNA in 0·1 M tris, pH 7·5 for r′ = 0·026. 0·044, 0·066, 0·133. 0·33, 0·33, 0·66, 1·33 decreasing at 390 nm, i.e. the reverse effect (data of Martin, S. R. unpublished).

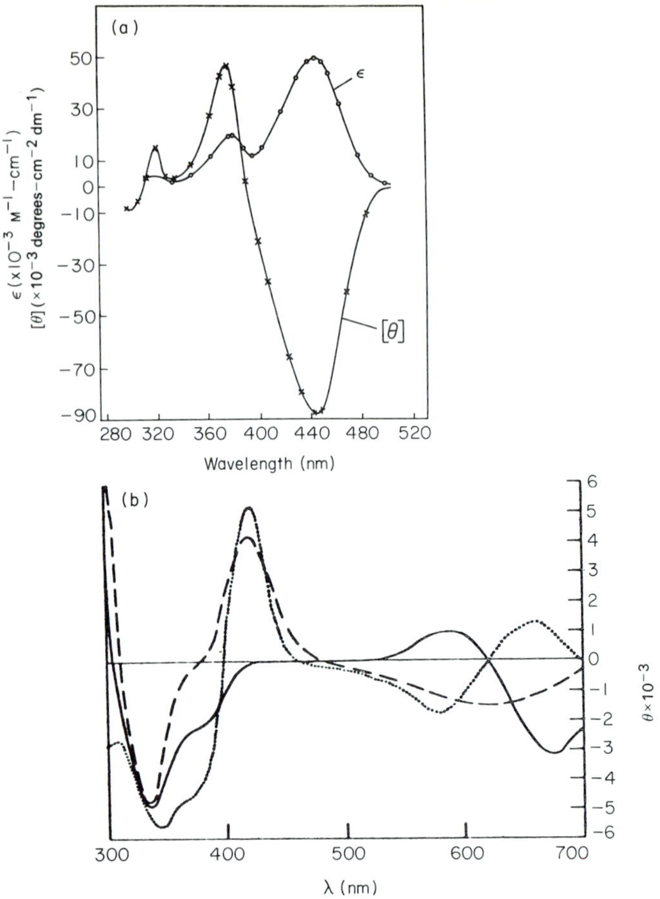

FIG. 5.31 Induced CD of chromophores bound to dehydrogenases: (a) absorption and CD spectra of Auramine 0 (47·6 μM) with liver alcohol dehydrogenase (23 μM) pH 7.4, (data of Wicken, T. S. and Woody R. W. (1973). *Biochemistry* **12**, 3459–3466); (b) CD of Cibacron Blue (CB) with three dehydrogenases—GAPDH -- LADH and .... cMDH, (data of Edwards, R.A. and Woody, R. W. (1977) *Biochim Biophys. Res. Commun.* **79**, 470–476).

bands of the protonated and unprotonated complex ($\lambda \sim 440$ nm and 365 nm respectively Fig. 5.32.) In addition to the usual coupling mechanisms, a twisted π-electron system could occur, possibly due to the restraints on both, forming the Schiff's base linkage and allowing optimal contact of all the substituent groups of the pyridine nucleus with the binding site.

For the optically active mononucleotides such as the common di- and triphosphates, the interpretation of the CD of the bound nucleotide draws on extensive information about the conformational and spectroscopic parameters of individual purine and pyrimidine derivates. Model compound studies and theoretical investigations have established that the observed CD is dependent upon the glycosidic angle χ, as discussed above (pp. 195–197). The CD at a given single value of χ can generate considerable intensity. It therefore

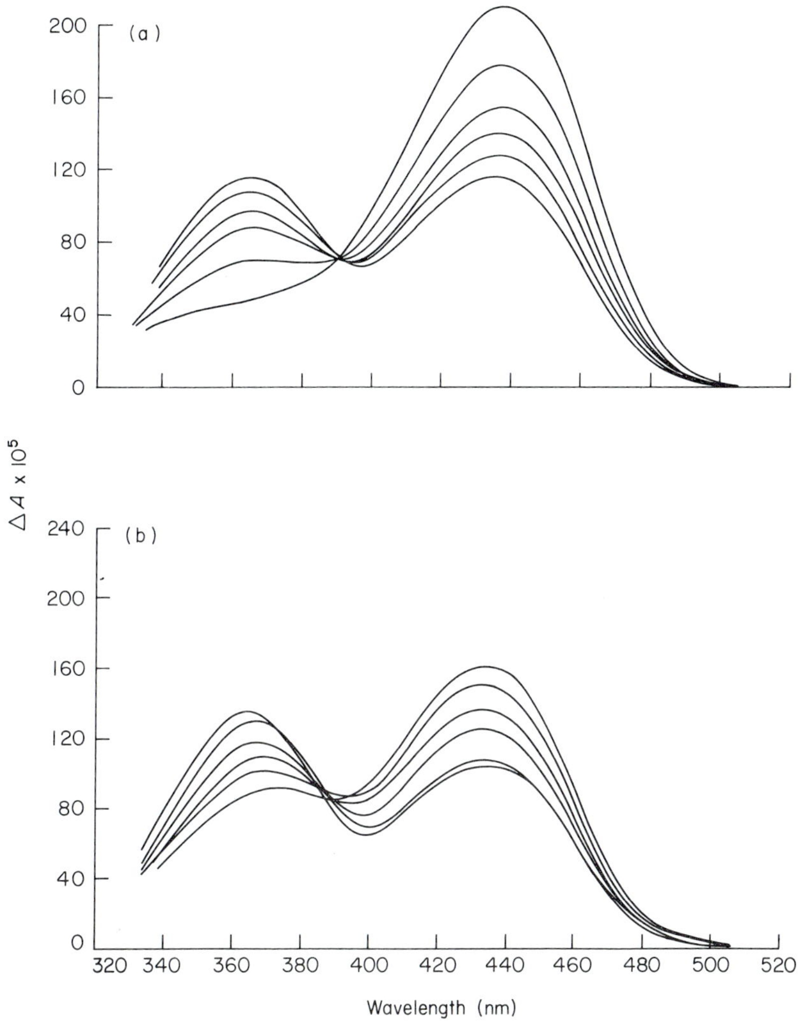

FIG. 5.32 Circular dichroism of the pyridoxal phosphate cofactor of aspartate aminotransferase (α-isomer, 0·5 mg ml^{-1}in 20 mM phosphate, pH 6.5); the effect of aliphatic and aromatic dicarboxylates. (a) α-Oxoglutarate concentrations 0, 0·02, 0·06, 0·11, 0·29, 1·7 mM. (b) *o*-Phthalate concentrations 0, 6, 20, 38, 68, 107 mM. The dicarboxylate causes protonation and increasing ΔA_{440}, in a isodichroic system (a), but not in (b). (cf. Bayley, P. M. and Harris, H. E. (1975). *Eur. J. Biochem.* **56,** 455–465.)

follows that the very fact of immobilization of the nucleotide on binding, for example to an enzyme active site, will often produce enhanced CD in the nucleotide chromophore. In addition the groups of the protein may also contribute. Such mechanisms appear to be responsible for the enhancement of the pyrimidine nucleotide CD observed when cytosine and uridine derivatives bind to RNAse (Fig. 5.33(a)).

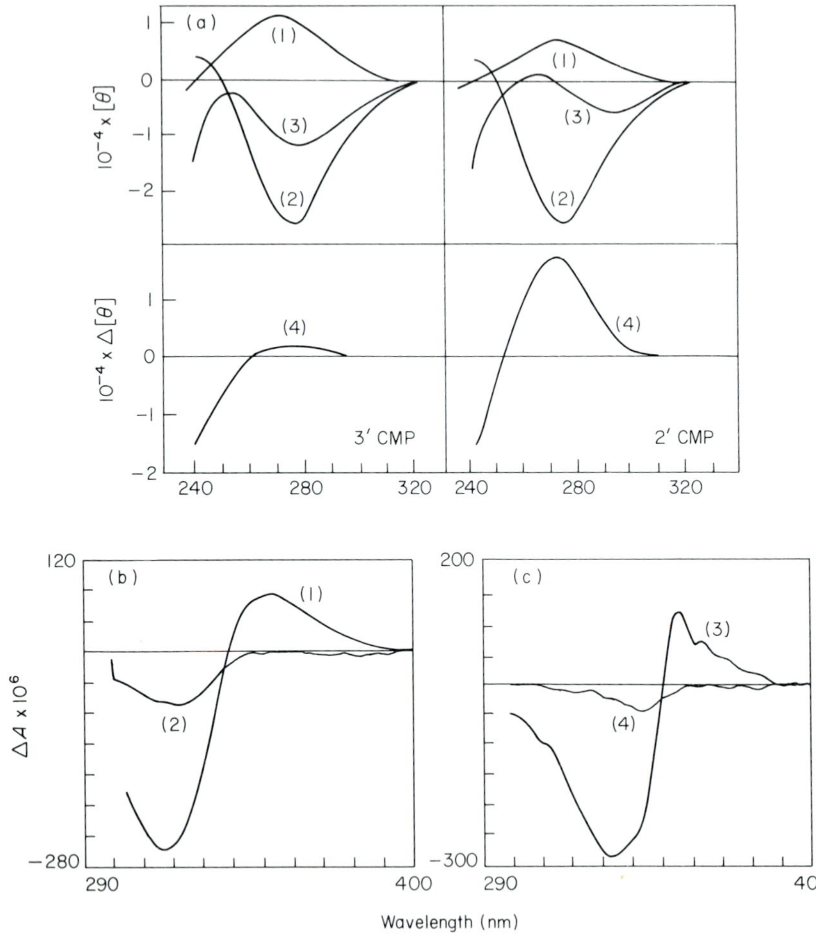

FIG. 5.33 Circular dichroism of nucleotides bound to certain enzymes: (a) 2′ and 3′ CMP bound to pancreatic ribonuclease A, (1) nucleotide, (2) RNAse A, (3) complex, (4) difference (complex–nucleotide), (data of Dudkin, S. M., Karabashyam L. V., Karpeisky, M. Y., Mikhailov, S. N., Padyukova, N. S. and Sakharovsky, V. G. (1977). *Biochem. J.* **167,** 749–757); (b) and (c) thionucleotides bound to myosin S1 fragment. (1) free thio ITP (50 μM); (2)+S1 (57 μM) pH 6·5, 5 °C; (3) free thio GTP (58 μM); (4)+S1 (83 μM) plus trinucleotide regenerating system pH 6·5; 5 °C, (data of Eccleston, J. F. and Bayley, P. M. unpublished).

An even greater enhancement is seen when the useful chromophoric thioguanosine or thioinosine nucleotides are bound to the myosin ATPase subfragment S1. In addition to intensification of the main CD band (Fig. 5.33(b)), the leading shoulder (probably an $n \rightarrow \pi^*$ transition of the thio-G chromophore) appears inverted, possibly due to a specific interaction with the sulphur atom. Similar effects are seen in S1 plus various thio-GTP analogues indicating a similar conformation for a number of complexes analogous to those in the ATPase mechanism.

Direct observation of purine CD nucleotide in the presence of protein generally requires careful difference CD techniques. Figure 5.34 shows the unusual intensification of the ADP CD on binding to the regulatory site of bovine glutamate dehydrogenase. From the CD titration, an enhancement of 25 is deduced, together with a dissociation constant of 2 μM. This effect is found with a range of adenine-containing ligands and, given the possible involvement of a nucleotide binding site involving ordered α and β regions, extensive interactions between the transitions of adenine and the exciton transitions of the ordered polypeptides appear likely as the source of the intensity. Using difference techniques, it has been possible to follow directly the binding of the coenzyme NAD^+ to GDH. The CD effect at 260 nm also derives from the adenine chromophore. Using the CD of bound "adenine" from Fig. 5.34(a), plus free enzyme and free NAD^+ (Fig. 5.34(b)) it was possible by linear regression technique to calculate the CD spectra, in excellent agreement with the experimental data, for the titration of GDH with NAD^+ (Fig 5.34(c), (d)). From the proportions of free and bound coenzyme, a dissociation constant of 0·7 mM was calculated. This and other considerations suggest that NAD^+ in the absence of substrates or inhibitors binds preferentially at the regulatory adenine nucleotide site.

Extensive studies of reduced coenzyme binding to dehydrogenases indicate a range of CD properties of the dihydronicotinamide moiety. Again for GDH, the existence of two sites capable of binding NADH was shown by oppositely signed CD effects of NADH at the active and regulatory sites. In the case of dehydrogenases known to contain the "nucleotide binding domain" there appears, so far, no obvious correlation in the dihydronicotinamide CD properties. In fact, given the more fundamental role in binding of the adenine moiety, it would be more reasonable to expect a correlation with this property but generally the data are not yet available.

An interesting case of coenzyme CD is shown by NADH binding to liver alcohol dehydrogenase Fig. 5.35(a). The most striking effect is an intense CD band in the dihydronicotinamide absorption. This is also found for the binary complexes involving NAD-X where X is a strongly nucleophilic group such as hydroxylamine, cyanide or pyrazole (cf. X = hydride, giving NADH). Similar intensities are found with complexes of a range of coenzyme analogues and these intensities in general exceed those found for binary complexes of other dehydrogenases. It is relevant to note that although the dihydronicotinamide ring can apparently adopt a planar conformation (e.g. in certain crystal structures), non-planarity could also be produced, either by rotation of the carboxamide group or twisting of the ring structure. Carboxamide rotation might be less likely since the carboxamide group is not essential for the effect, acetyl, aldehyde or thio-amide also being effective (cf. Fig. 3.5(b)). Ring twisting would generate a chiral form with intense CD which would therefore account for the observed properties and possibly be prototypic of the non-planar transition states which have been suggested for the enzyme mechanism. It may be noted that the dihydronicotinamide ring does not appear to lie close to the α-helical or β-sheet structures of the coenzyme site. Also there is a total absence of the adenine CD in this case, although the adenine moiety is

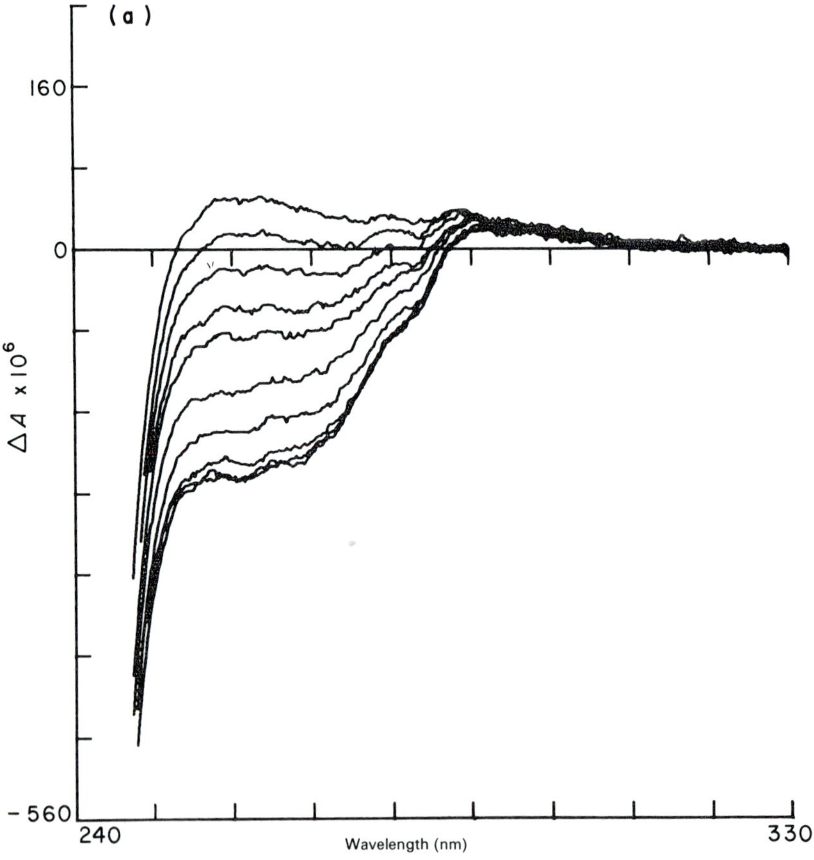

FIG. 5.34 Analysis of NAD binding to bovine glutamate dehydrogenase 0·1 M phosphate; pH 7, 20 °C: (a) the titration of GDH (100 μM) with ADP forming 1:1 complex.

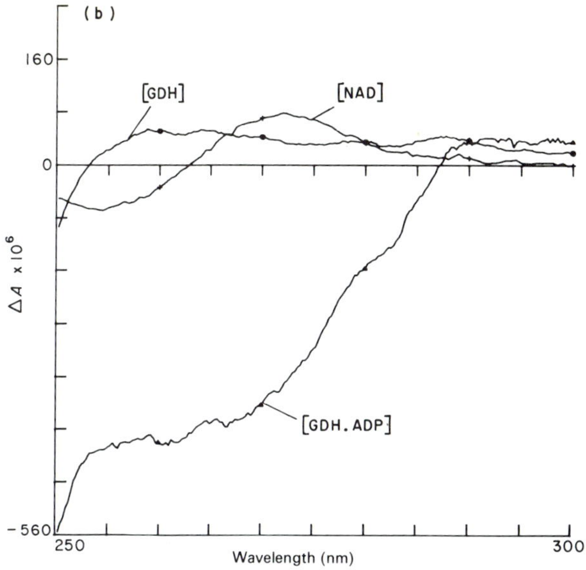

FIG. 5.34 Analysis of NAD binding to bovine glutamate dehydrogenase 0·1 M phosphate; pH 7, 20 °C: (b) the components for linear regression; NAD (400 μM free), GDH (100 μM free) and GDH+ADP (200 μM, 1 : 1 complex).

strongly bound and must be strongly immobilized. Presumably cancellation of the effects or the freezing of the adenine glycosidic angle close to an optical node is responsible.

Haem chromophores

A further class of chromophoric ligands which have been studied intensively by CD are the haem chromophores as found in haemoglobin and myoglobin (see Fig. 5.36). Useful spectroscopic probes are afforded by both the weak (electrically forbidden) α and β transitions at 500–650 nm as well as the intense (electrically allowed) Soret transition close to 400 nm (see p. 36). The positions of these bands are of course dependent upon the state of oxidation and ligation. The reformation of native myoglobin from apomyoglobin and haem has been verified by the restoration of the CD spectrum in both visible and UV wavelength regions. The various transitions have also been used to monitor association–dissociation reactions in haemoglobin. While such reactions might intuitively be thought to proceed via conformational changes in the subunits, changes in the CD spectra on dissociation do not necessarily justify such a conclusion. This is because, given the intensity of the Soret transitions ($\varepsilon \approx 10^5$ M^{-1} cm^{-1}) the $\boldsymbol{\mu}_e$ associated with these transitions interact not merely with other groups within the subunit, but also with aromatic residues and possibly haem groups in other subunits. Thus changes

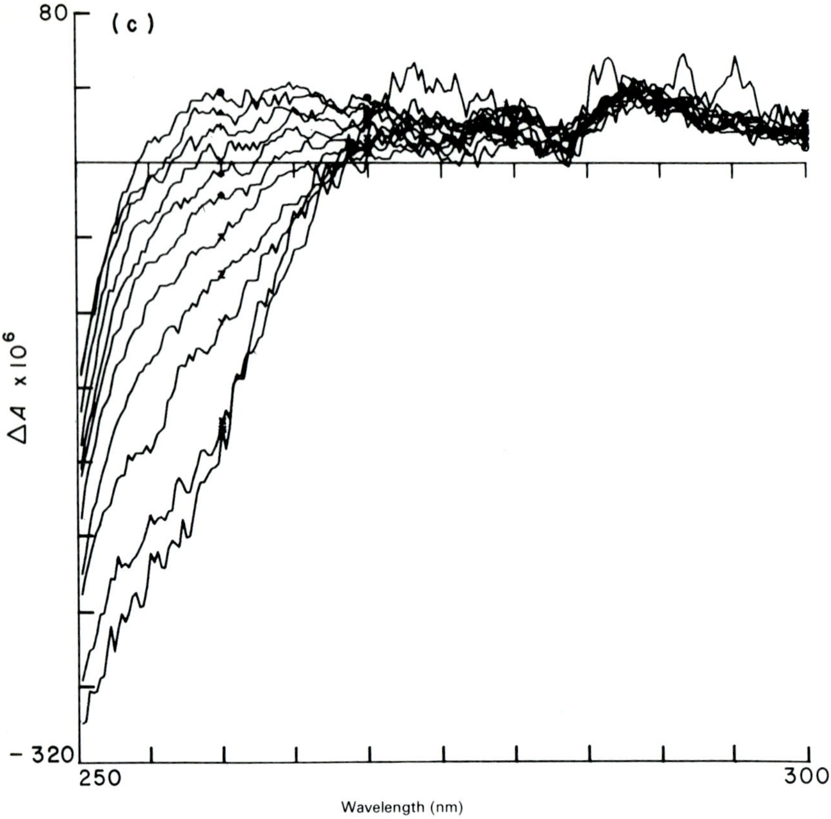

FIG. 5.34 (c)

in quaternary structure may be monitored. It seems likely that the 260 nm transitions are sensitive to the local environment of the haem and hence may be used additionally as monitors of changes in secondary–tertiary structure. Interactions between the Soret transitions and the helical exciton transitions also appear possible from theoretical calculations, presenting an additional structural feature which can contribute to the observed CD.

Studies of Supramolecular Structure

The use of CD in studying large molecular aggregates has produced interesting results in two areas of biological importance, namely membranes and ordered aggregates of chromophores.

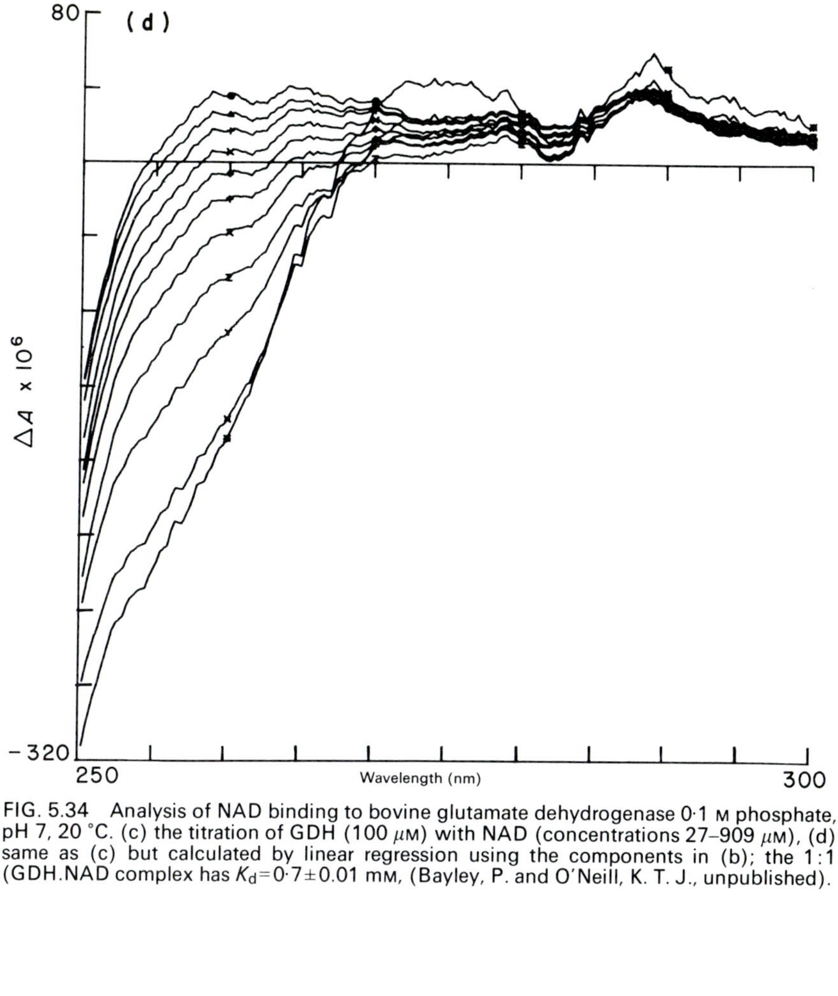

FIG. 5.34 Analysis of NAD binding to bovine glutamate dehydrogenase 0·1 M phosphate, pH 7, 20 °C. (c) the titration of GDH (100 μM) with NAD (concentrations 27–909 μM), (d) same as (c) but calculated by linear regression using the components in (b); the 1:1 (GDH.NAD complex has K_d=0·7±0.01 mM, (Bayley, P. and O'Neill, K. T. J., unpublished).

Membrane proteins

A primary question has been whether the conformational properties of different models of membrane protein structure can be distinguished by their CD spectra. Two main problems exist: (a) experimental spectra require correction for distortion effects due to differential scattering and "shadowing" at relatively high turbidities; (b) the interpretation of the corrected spectra. Considerable progress has been made in applying suitable corrections for the attenuation of the observed CD of membranes, and conditions minimizing the distortions have been identified. Enlarged solid angle for collection of scat-

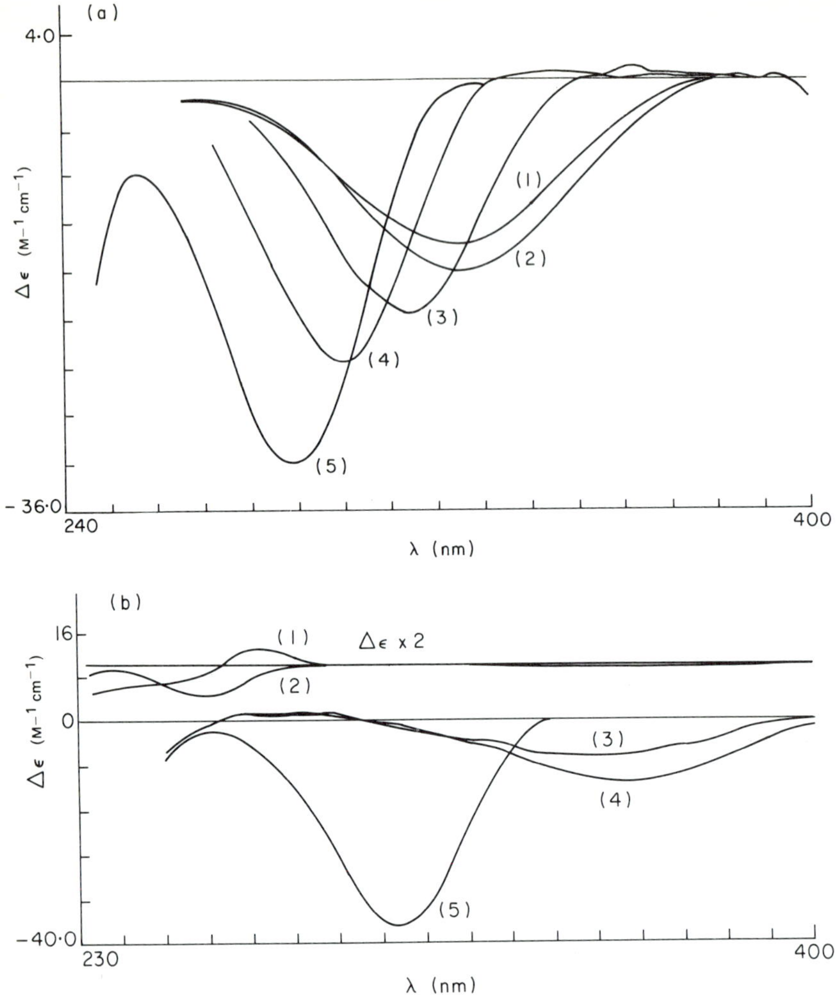

FIG. 5.35 Circular dichroism of NADH and derivatives, bound to liver alcohol dehydrogenase. (a) (1) NADH (2) NADH+excess isobutyramide (3) NAD–CN complex (4) NAD–hydroxylamine complex (5) NAD–pyrazole complex. (b) Acetylpyridine (AP) analogues of NAD (1) AP–NAD (free) (2) AP–NADH (free) (3) AP–NADH (bound) (4) AP–NADH (bound+isobutyramide) (5) AP–NAD-pyrazole (bound). (Bayley, P. and Martin, S. R. unpublished).

tered light is desirable. Sonication of the membranes reduces artefacts and the use of very short pathlength cells (e.g. 0·1 mm) reduces scattering. Measurements on membrane proteins in detergents, phospholipid dispersions, chlorophylls, rhodopsin, bacteriorhodopsin and other large assemblies have been made, virtually free from artefacts. A spectrum of intact erythrocyte ghosts is

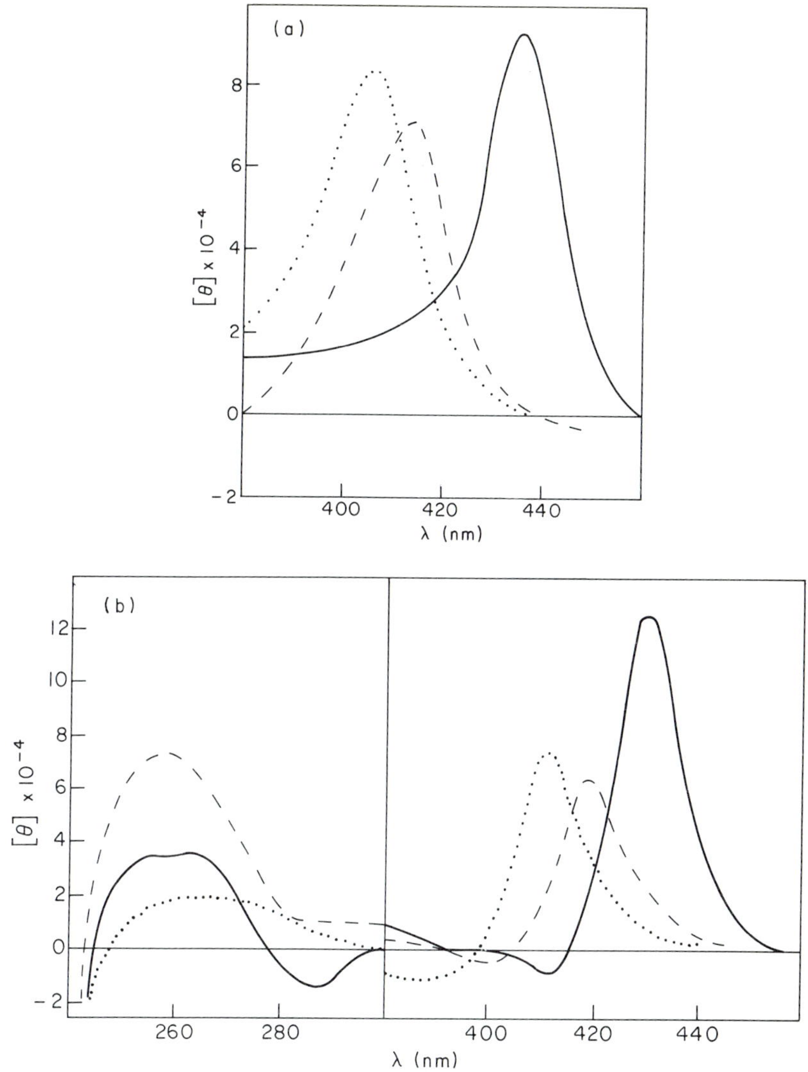

FIG. 5.36 Circular dichroism of haem proteins in the Soret and near-UV regions: (a) myoglobin; (b) haemoglobin, (– – – –) oxy (· · · · · ·) met (——) deoxy, (data of Nagai, M., Sugita, Y. and Yoneyama, Y. (1969). from Hsu M.-C. and Woody, R. W. (1971). *J. Am. Chem. Soc.* **93**, 3515.)

shown in Fig. 5.37(a) and the protein spectrin which forms 30% of the ghost protein is shown in Fig. 5.37(b).

Interpretation of these results has also progressed significantly, together with the understanding of the diversity of membrane proteins and organization of membrane structure. The question of the conformation of individual proteins requires the protein to be isolated in pure form, and the CD measurements to be made on the sample, solubilized in a suitable mild phospholipid or detergent medium. Interpretation of the corrected spectra in terms of multicomponent analysis allows a reasonably confident assignment of overall conformation. Thus the spectrin of human red blood cells (fortunately soluble in aqueous solution) shows a large helical content. Similarly the apoproteins of high density lipoprotein complexes are highly helical, as are the proteins of bacteriorhodopsin and mammalian rhodopsin. Questions relating to changes in membrane protein conformation appear to be more difficult to answer than those relating to soluble proteins, but it may be hoped that, with the enhanced sensitivity of CD measurements now available, e.g. in the aromatic region, these problems may also be treated.

Ordered aggregates of chromophores

These present rather different problems from the above. The assembly of chromophoric groups in an array which is regularly and asymmetrically ordered and which extends over distances which are large compared to the wavelength of light, produces unusually intense effects, due to differential scattering of RH and LH–CPL. This appears as a difference in transmitted

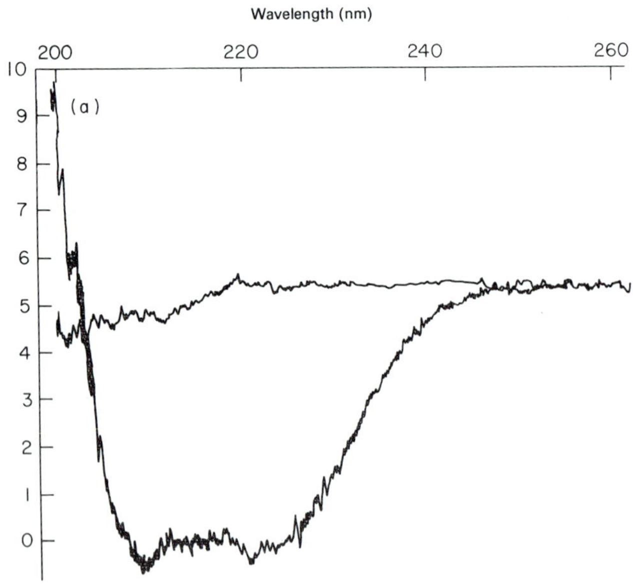

FIG. 5.37 (a)

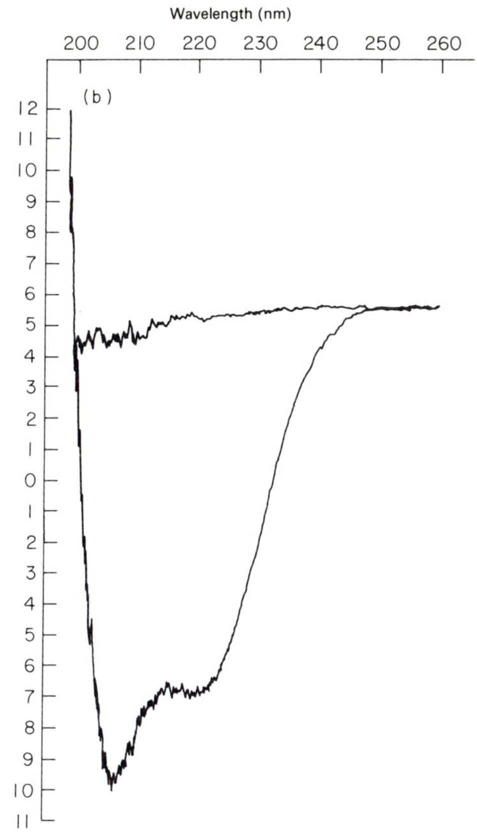

FIG.5.37 Far-UV CD of membrane proteins: (Jasco J41C Spectrometer). (a) Experimental CD of erythrocyte ghosts 0·65 mg ml^{-1}, 5mM tris buffer, pH 8. 0·1 mm path; (Data of Beavens; unpublished). (b) Experimental CD spectrum of erythrocyte spectrin 0·2 mg ml^{-1} 1 mm path; in 5 mM tris buffer, pH 8. The horizontal traces represent the baseline in each case. Vertical scale: one unit is $\Delta A = 6 \times 10^{-5}$.

RH and LH–CPL and is not easily distinguished from true circular dichroic absorbance. The effects are associated with the absorption bands and are potentially very intense. Thus the formation of micellar arrays of amphotericin-B, an (asymmetric) polyene antibiotic, appears to itensify the CD by more than 100-fold, presumably in the micellar state (Fig. 5.38). Such effects are unusually sensitive to changes in physical conditions (e.g. temperature, apolar solvents) which can affect the integrity of the micelles. Parallel monitoring of the state of aggregation, e.g. by light scattering, is a valuable check on such systems.

A related but distinctive case is that of the extended supra-molecular arrays formed between polynucleotides and positively charged synthetic polypeptides. Such polyelectrolyte interactions may result in extremely large assemblies which are readily observed in the electron microscope. The CD properties allow simple monitoring of the formation of these structures using the CD in the region of the nucleic acid absorption. Scattering effects are probably responsible and the extraction of useful structural information is not easy. However, the close parallel with natural organized nucleoprotein assemblies is notable and CD work on chromatin itself has been reported.

Thus there appears to be no essential restriction on the size of assemblies which can be studied by CD, although the problems and additional considerations of such systems do present some limitations to the amount of structural information which can be derived.

Recently Developed Techniques

Notwithstanding the already vast literature of CD studies, the field continues to expand actively. Recent extensions and developments of CD techniques are now discussed briefly.

High-sensitivity measurements

Increases in CD resolution have become available with improvements in electronics and modulation techniques. Machine resolution of better than $\delta\Delta A = 1 \times 10^{-6}$ is available routinely; such sensitivity requires time constants of the order of 16 s or more. The limiting factor then becomes machine stability over the period of scanning. Computer systems with facilities for prolonged multiple scanning are available and (though subject to the same ultimate limitation) show the expected improvements in effective sensitivity (see Fig. 5.39). The data-handling of *difference CD* by computation is becoming a standard procedure, and has been referred to on p. 189.

Low-temperature studies

Low-temperature CD of model compounds in transparent rigid media at liquid nitrogen temperatures has provided important information about aromatic acid residues in proteins and appears a very useful adjunct in studies of molecular mobility. Experiments in low-temperature fluid solution have been important in stacking studies of polynucleotides and in the protein field this activity parallels current interest in the techniques of cryobiology

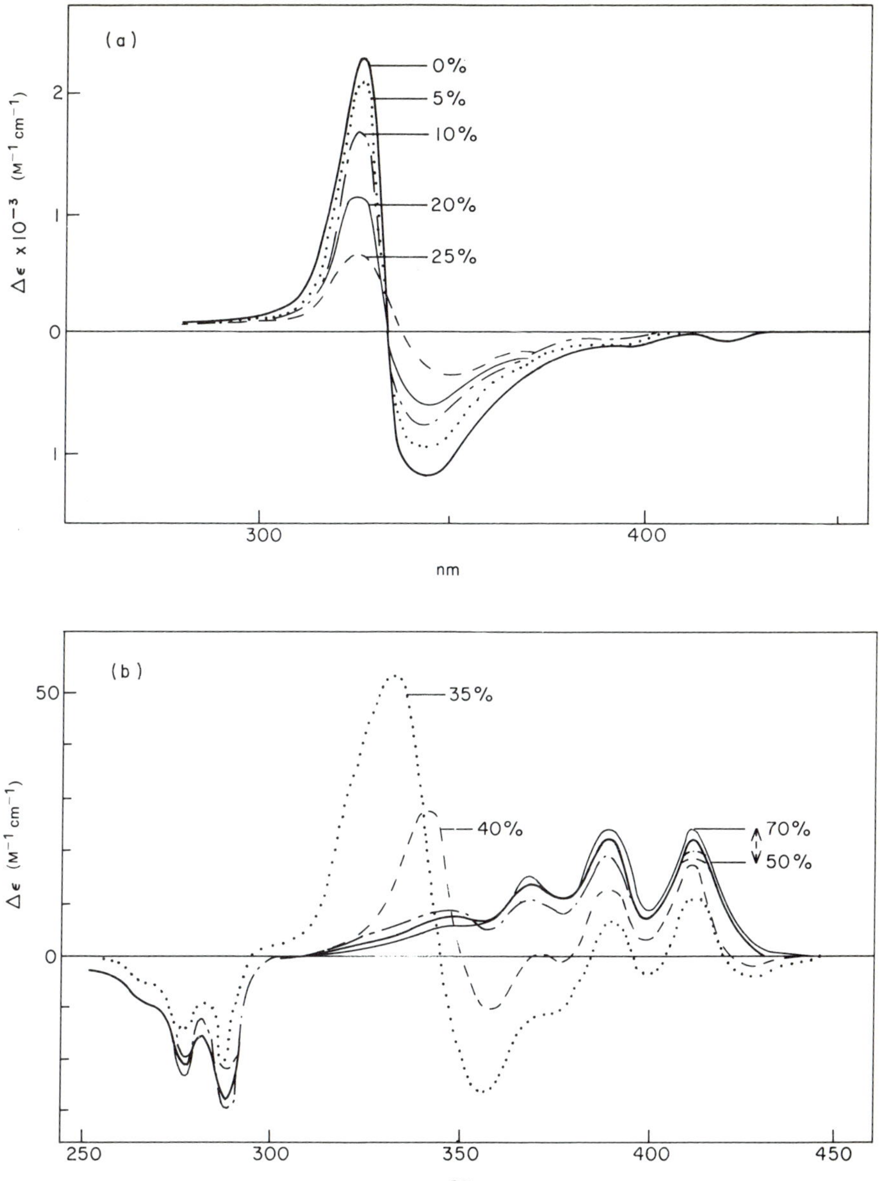

FIG. 5.38 Circular dichroism of micellar aggregates of Amphotericin B in aqueous ethanol: (a) 0, 5, 10, 20, 25% ethanol; (b) 35, 40 and 50–70% ethanol; (Data of Rinnert, H., Thriton, C., Dupont, G. and Lematre, J. (1977) *Biopolymers* **16,** 2419–2427).

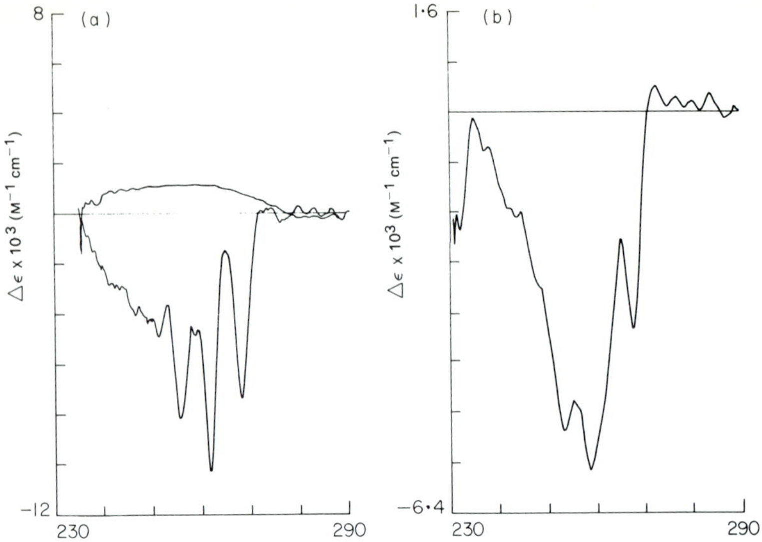

FIG. 5.39 Enhancement of weak CD by multiple scanning of isotopically substituted meso-hydrobenzoins (a) PhCH(OH)CD(OH)Ph (four scans) (b) PhCH(^{16}OH)CH(^{18}OH) Ph (24 scans, averaged). The g-value is $\sim 2\times10^{-5}$ for (b). (Data of Cullis, T. M. Lowe, G. and Bayley, P. (1978). *Chem. Commun.* 850–852.)

Magnetic CD (MCD)

Extensive work with MCD has been reviewed elsewhere. The relationship to natural CD (as discussed here) is as follows. The electronic transitions of symmetrical chromophores are split in the presence of a magnetic field so that all chromophores exhibit MCD. Where natural CD also exists, the two effects superimpose. Magnetic CD is not notably sensitive to conformation but yields important information of a more chemical nature about the types and symmetries of the electronic transition of a chromophore. However, certain analytical biochemical applications have been proposed, notably the determination of the tryptophan content of proteins and the assay of porphyrin derivatives in urine. Both of these depend on the unusually intense MCD properties of these chromophores.

Fluorescence detected CD and circular polarization of luminescence

Two techniques have been developed recently which use the emission properties of a chiral molecule. In *fluorescence detected CD* (FDCD), the differential intensity of fluorescence emitted after differential absorption of RH and LH-CPL is measured. The method complements CD absorption studies but clearly detects only those molecules which fluoresce (see Chapter 3). An excitation spectrum will be obtained for the fluorescence, weighted by the CD absorbance of the chromophore and its fluorescent quantum yield. In practice, applications have been restricted to relatively simple chromophores since, in more complex systems, energy transfer processes present considerable problems in interpretation. An elegant study of the Y-base of *t*RNA has been made by FDCD. As the sole fluorescent residue, it can be observed independently of all other bases.

The other somewhat related technique actually determines the circular polarization characteristics of the fluorescence emission of the chiral molecule. *Circular polarization of luminescence* (CPL) gives a measure of the chirality of the excited state which can then be compared with that of the ground state (from normal CD). A dissymmetry ratio for emission can be defined in terms of the difference in intensity of RH and LH CP fluorescence intensities $\Delta I = I_L - I_R$ as

$$g_{Em} = \frac{\Delta I}{I}$$

cf. $g = \Delta A/A$ for absorbance. If $g_{Em} = g$ the emitting state has the same asymmetry as the absorbing one. If not, there may have been a change of asymmetry during the excited state lifetime. Such interpretations have been made for tryptophan model compounds and for fluorescence indicators on chymotrypsin. The information clearly provides new insight into the mobility properties of the chromophore under excitation, and may also be relevant to interpretation of fluorescence data (e.g. polarization) as though it referred to ground-state geometries (see also Chapter 3). Further applications have been to transition metal complexes and lanthanide probes of protein structure, and the field was reviewed recently (see Bibliography, p. 234).

Rapid reaction techniques using CD detection

The availability of high-speed modulation techniques has opened up possibilities of measuring CD at millisecond time resolution and beyond. Both *stopped flow CD* (SFCD) and *temperature jump CD* (TJCD) have been developed with a resolution of approximately $\delta\Delta A \sim 1 \times 10^5$ at 1 ms time resolution. A full account of fast-reaction techniques is given in Chapter 2. Studies of binding of dyes to DNA, chromophoric ligands to enzymes, metal chelation by amino acids, metal–drug interactions, enzyme–coenzyme complexes and bilirubin binding to serum albumin have been performed. Where the different optical parameters (absorbance, fluorescence and CD) change with identical kinetic constants, a simple associative process is inferred. This was found with

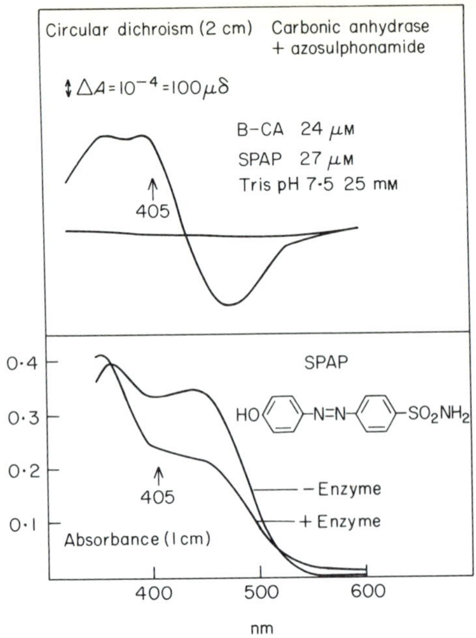
Circular dichroism (2 cm)
Carbonic anhydrase
+ azosulphonamide
$\Delta A = 10^{-4} = 100 \mu\delta$
B–CA 24 μM
SPAP 27 μM
Tris pH 7·5 25 mM
405
0·4
0·3
0·2
0·1
SPAP
HO–C6H4–N=N–C6H4–SO_2NH_2
405
– Enzyme
+ Enzyme
Absorbance (1 cm)
400
500
600
nm

(a) (b) (c) (d)

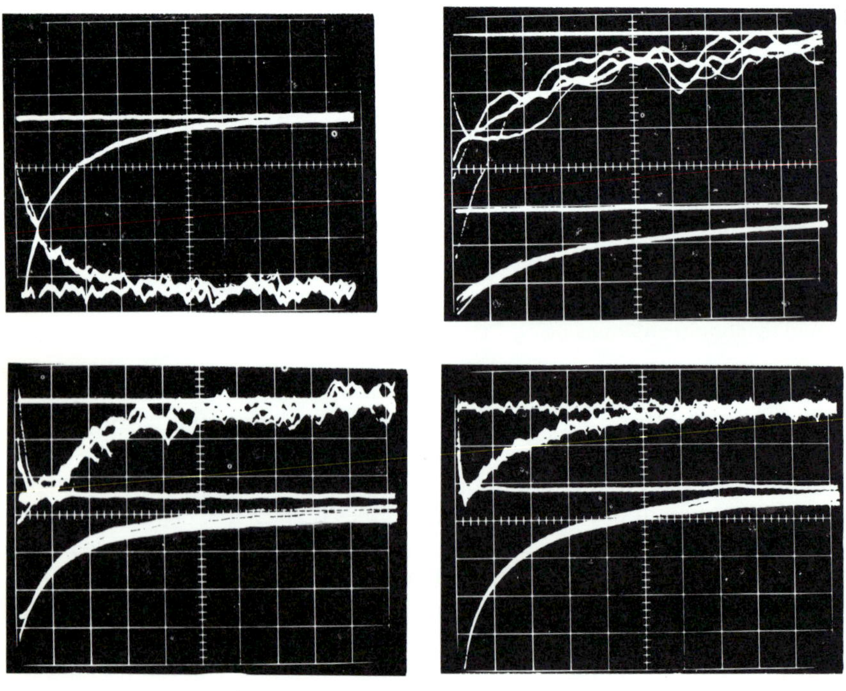

the binding of azosulphonamide inhibitors to isozymes of carbonic anhydrase (Fig 5.40) and the displacement of tetracycline from its complex with Mg^{2+} using excess Ca^{2+} (Fig 5.41). However ligand-induced conformational changes, denoted by

$$E+L \rightleftharpoons EL \rightleftharpoons EL^*$$

have been found in SFCD for the binding of L-glutamate to the binary complex of glutamate dehydrogenase with NADPH using simultaneous CD and fluorescence observations. A small difference is found in the observed rate constant for the two optical parameters; also the rate achieves a plateau at high [L-Glu].

In TJCD, with liver alcohol dehydrogenase, (LADH), and the inhibitor imidazolyl azobenzene, two relaxations are observed. The fast process is concentration dependent (Fig. 5.42(a)) and the slower one is concentration independent (isomerization step, Fig. 5.42(b)), as required for the above mechanism.

Circular dichroism is expected to be a most sensitive parameter for conformational changes and isomerization steps generally, and this has been illustrated by simulation techniques. These results assist the interpretation of complex mechanisms, particularly as found in relaxation techniques, and may help assign the CD properties, for example to complexes such as EL and EL*. Studies further into the UV are becoming practicable for SFCD and TJCD with more intense light sources. A range of biochemical problems is now becoming accessible to study, and questions can be posed about the kinetics of many of the conformational processes discussed in this chapter.

Conclusions

This chapter has illustrated the principles involved in the interpretation of CD properties of a number of biochemical systems. A wide range of spectra has been presented, in order to try to convey some familiarity with the data as it appears in the CD literature. The relationship between spectra and structure is not simple: it involves knowledge of the visible and UV absorption spectra

FIG.5.40 Stopped flow circular dichroism: the reaction of carbonic anhydrase isozymes with azophenyl sulphonamide inhibitors. Upper: equilibrium spectra, CD and absorption for inhibitor SPAP (see diagram for structure) ± bovine carbonic anhydrase (B–CA). Lower: stopped flow CD and transmission (Tx) recorded simultaneously. Sensitivity/division as indicated ($1\mu\delta = 10^{-6}$ CD absorbance units).
(a) CA (bovine) 48 μM; Inhibitor 54 μM; CD 50 $\mu\delta$ TX 1%; 2 shots; λ, 405 nm; Time 20 ms/div
(b) CA (human C) 48 μM; Inhibitor 48 μM; CD 167 $\mu\delta$ TX 2%; 4 shots, λ, 365 nm; Time 2 ms/div
(c) CA (bovine) 50 μM; Inhibitor 25 μM; CD 333 $\mu\delta$ TX 1%, 5 shots, λ, 365 nm; time 10 ms/div
(d) CA (bovine) 38 μM; Inhibitor 38 μM; CD 333 $\mu\delta$ TX 1%, 4 shots; λ, 365 nm; Time 100 ms/div
Curves (d) in presence of 48 μM perchlorate as competitive inhibitor. (cf. Bayley, P. M. and Anson, M. (1975). *Biochim. Biophys. Res. Commun.* **62,** 717–722).

(a)

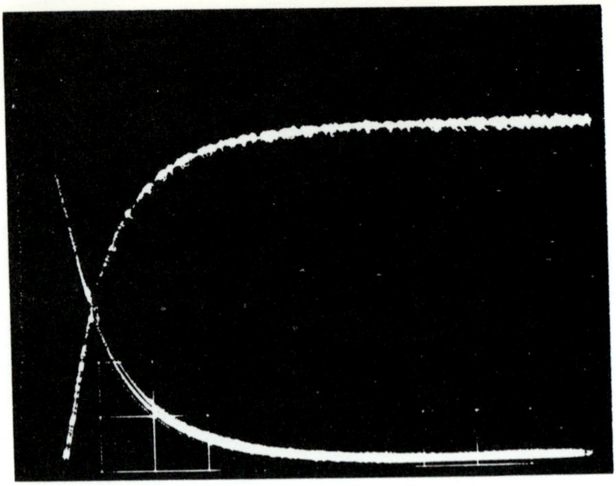

(b)

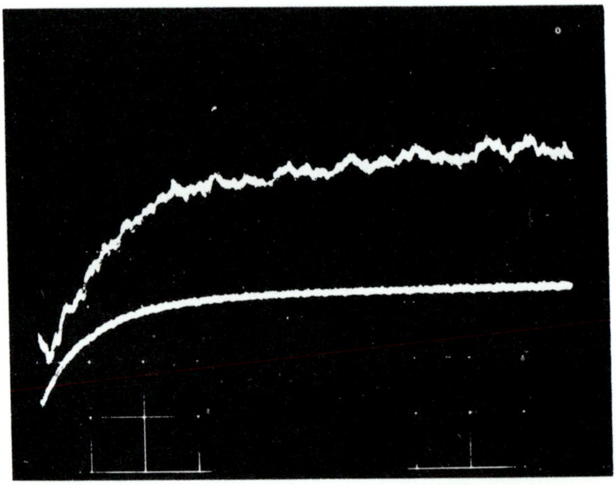

FIG. 5.41 Stopped flow traces from tetracycline using simultaneous monitoring of CD absorbance and fluorescence: λ=436 nm.
(a) CD (upper) and absorbance (lower). Tetracycline 72 μM + 19 mMMg^{2+} + 56 mMCa^{2+}; CD 200 $\mu\delta$ TX1%; Time 50 ms/div.
(b) Fluorescence (upper) and absorbance (lower). Tetracycline 72 μM + 19mMMg^{2+} + 56 mMCa^{2+}; TX 0·5%; Time 50 ms/div. (Bayley, P. M. and Martin, S. R.; unpublished).

and often depends on interpretations involving carefully chosen model compounds. The ability of theory to predict CD spectra is best expressed with periodic (i.e. helical) arrays of well-defined chromophores. At a more empirical level, the CD of the intrinsic chromophores of many molecules affords a sensitive indicator of native structure, and this is further extended by the use of

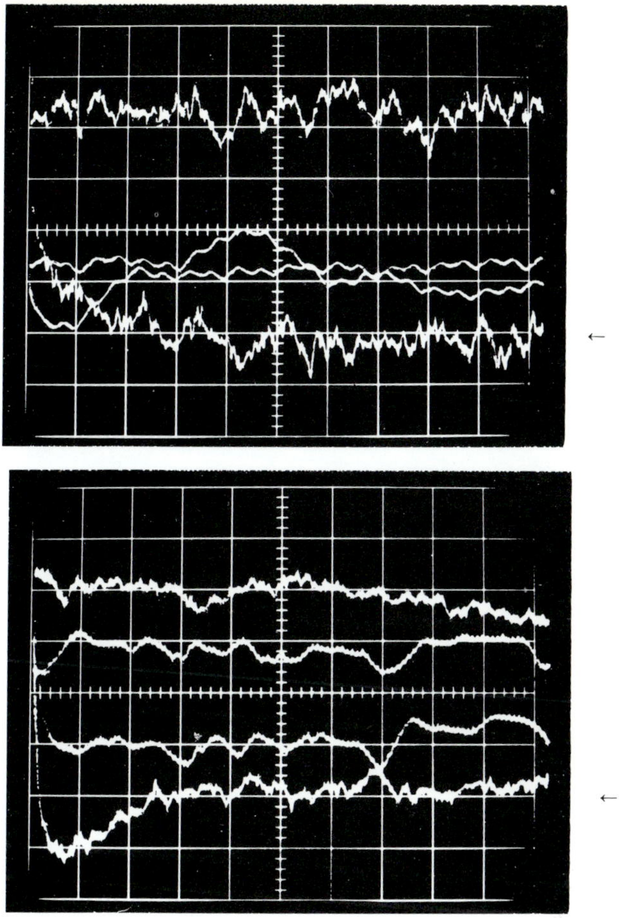

FIG. 5.42 Temperature-jump circular dichroism of the complex between LADH and the chromophoric inhibitor imidazolyl azobenzene, IAB. Each picture shows from top to bottom: (i) CD before temperature jump; (ii) and (iii) transmission before and after temperature jump (no transient is observed since 436 nm is an isosbestic point; (iv) the CD transient after temperature jump. In the upper picture this transient shows a rapid ligand-dependent relaxation with $t_{\frac{1}{2}}=30$ ms. In the lower picture (extended time scale) an additional slow ligand-dependent relaxation is observed with $t_{\frac{1}{2}} \approx 250$ ms. Top—a rapid, ligand-dependent relaxation with $t_{\frac{1}{2}} \sim 30$ ms; bottom—a slow ligand-independent relaxation with $t_{\frac{1}{2}} \sim 250$ ms. Both traces refer to CD; the TX observation is effectively at an isosbestic point for the system. (Bayley, P. M. and Martin, S. R.; unpublished).

natural extrinsic chromophores such as coenzymes and nucleotides as well as simple chromophoric dyes acting as molecular probes. Finally the direction of future applications is indicated by the extension of present techniques to extreme conditions, to large molecular assemblies and to the millisecond time domain for studying the kinetics of conformational changes.

Bibliography

In addition to specific references cited in Figure legends, the following reviews are helpful:

Adler A. J., Greenfield, N. J. and Fasman, G. D. (1973). CD and ORD of proteins and polypeptides. *In* "Methods in Enzymology", Vol. XXVII. (Eds. S. Fleischer and L. Packer), 675–735. Academic Press, London and New York.

Bayley P.M. (1973). The Analysis of Circular Dichroism of Biomolecules. *Prog. Biophys. Molec. Biol.* **27**, 1–76.

Bush C. A. and Brahms J. (1973). Conformations of nucleic acids, oligo- and polynucleotides by circular dichroism investigations. *In* "Physico-Chemical Properties of Nucleic Acids," Vol. 2, (Ed. J. Duchesne), 147–186.

Cathou, R. E. and Dorrington, K. J. (1975). "Sub-units and Biological Systems, Part C" (Eds S. N. Timasheff and G. D. Fasman), 91–224 Marcel Dekker, New York.

Ciardelli. F., and Salvadori. P. (Eds) (1973). "Fundamental Aspects and Recent Developments in ORD and CD" Heyden, London. Chapters by Tinoco, Brahms and Blout.

Sears. D. W., and Beychok. S. (1973). Physical principles and techniques of protein chemistry: Part C *In* "Circular Dichroism", (Ed. S. J. Leach), 446–593. Academic Press, London and New York.

Steinberg, I. Z. (1978). Circular Polarization of Luminescence: Biochemical and Biophysical applications. *A. Rev. Biophys, Bioeng.* **7**, 113–137.

Strickland. E. H. (1974) Aromatic Contributions to Circular Dichroism Spectra of Proteins. *Critical Rev. Biochem.* **2**, 113–175.

6 Nuclear Magnetic Resonance

R. Jones

Introduction to Magnetic Resonance

Magnetic resonance is the absorption of quanta of energy from a radiofrequency ($\sim 10^8$ Hz) or microwave frequency ($\sim 10^{10}$ Hz) electromagnetic field by a system containing unpaired nuclear or electron spins placed in a strong static magnetic field. The separation of energy levels and hence the frequency at which absorption occurs depends on the strength of the magnetic field. Magnetic resonance was first observed by physicists in the 1940s. It soon became a favourite tool of chemists for identifying and determining the structure of small molecules, but applications to more complex biochemical systems only began in the late 1950s. Since then the applications have rapidly proliferated with the realization that the techniques of magnetic resonance are peculiarly suited to studying certain types of biological and biochemical problem.

Magnetic resonance is one of the few techniques available at present for obtaining both structural and dynamic information about molecules at the atomic level; other spectroscopic techniques usually give information at the molecular level. X-Ray diffraction gives information at atomic resolution, but this is essentially static information. From the point of view of the biologist or biochemist, the additional attractions of magnetic resonance are that it is completely non-destructive, it can be used under near-physiological conditions, it is universally applicable (at least in principle) since all biological materials contain protons which give an NMR signal and it can often give highly specific information. The disadvantages are first, that sensitivity is very low compared with more conventional forms of spectroscopy because of the much smaller energy changes involved and secondly, that the wealth of information contained in magnetic resonance spectra gives rise to problems of interpretation. They explain the initial delay in applying the technique to complex biological systems, but these are steadily being overcome by new developments in instrumentation and methods. The development of applications of magnetic resonance to biology and biochemistry also owes much to the established techniques of X-ray crystallography, model-building and chemical modification, because of the need to test out new theory in well-characterized model systems. Magnetic resonance now has a sound theoretical basis and can thus become independent of other techniques to some extent, but for the complex systems of interest to the biochemist a variety of techniques will, as always, be necessary to solve problems.

NMR and EPR: relationship to other forms of spectroscopy and to each other

Magnetic interactions were detected as fine structure in atomic spectra (the *Zeeman splitting*) long before they were observed directly as magnetic resonance. The two techniques of nuclear magnetic resonance (NMR) and electron spin (paramagnetic) resonance (ESR, EPR) arise from the interaction of a nuclear spin with a radiofrequency field ($\sim 10^8$ Hz) and the interaction of an electron spin with a microwave field ($\sim 10^{10}$ Hz) respectively. The regions in which resonance occurs are near the low-frequency end of the electromagnetic spectrum relative to most other forms of spectroscopy (Fig. 1.3) indicating that the energy level differences ($\Delta E = h\nu$) are correspondingly small; the energy differences involved in a typical EPR experiment are about 10^5 times smaller than for UV spectra and in a typical NMR experiment, about 10^7 times smaller. A major difference between magnetic resonance and other forms of spectroscopy is that the energy levels arise from the application of a static magnetic field, and this leads to some overlap with other forms of spectroscopy utilizing the same wavelengths; e.g. the so-called "umbrella" transition of the NH_3 molecule occurs in the microwave region. It also means that we could in principle obtain a magnetic resonance spectrum either by varying the frequency (as in more conventional forms of spectroscopy) at constant magnetic field or by varying the magnetic field at constant frequency. For practical convenience, it is more usual to vary the magnetic field. However, there is another way of observing magnetic resonance which does not involve scanning the magnetic field or frequency in the conventional way. This involves short radiofrequency pulses and will be described later (see pp. 253–255).

Nuclear magnetic resonance experiments utilize magnetic field strengths of 10^4–10^5 gauss (1–10 tesla) which are about 10-fold greater than those for EPR (10^3–10^4 gauss, or 0·1–1·0 tesla). The energy levels in a given magnetic field are determined by the magnetic moments associated with the spins, which for electrons are $\sim 10^3$ times greater than those associated with nuclei, leading to the correspondingly greater energy differences and the higher frequencies of resonance in EPR experiments. The two techniques of NMR and EPR are sometimes treated separately, mainly because the instrumentation involved and the practical applications are somewhat different. However, from a theoretical point of view there is much overlap and many studies are specifically designed to utilize the interactions between NMR and EPR phenomena. For example, electron spins often lead to line-broadening in NMR spectra, and nuclear spins often lead to line-splitting in EPR spectra. Many of the examples given later illustrate the usefulness of these interactions in providing information on molecular structure.

Scope and problems of NMR and EPR spectroscopy

What are the limitations and problems in obtaining useful information from magnetic resonance, and how do they differ in NMR and EPR? These can be discussed under the headings: (i) Scope of application, (ii) Sensitivity, (iii) Resolution, (iv) Assignment and (v) Interpretation. The *scope of application* for NMR in biological systems is greater than for EPR because all biological molecules contain ^1H nuclei, all biological systems contain H_2O, and many contain ^{31}P, all of which give rise to NMR signals. Other nuclei which can be studied in biological systems to greater or lesser advantage include ^{13}C, ^{14}N, ^{19}F, ^{23}Na and ^{35}Cl. By contrast, EPR can only be observed in systems containing unpaired electrons and is thus limited to a study of systems containing free radicals or paramagnetic metals. Fortunately, many of the systems of greatest biochemical interest are of this type. In addition the scope of EPR can be greatly extended by the introduction of an exogenous paramagnetic centre which again may be a free radical ("spin-label") or metal.

The problem of *sensitivity* is a major one in all biological magnetic resonance studies and has probably restricted the biochemical application of the techniques more than any other factor. A single NMR experiment may demand quantities of, e.g. a purified enzyme, which would suffice for 10 or even 100 experiments of a more conventional type. Electron paramagnetic resonance is inherently more sensitive than NMR because of the larger energy changes involved. In a typical case satisfactory EPR spectra can be obtained on solutions of about 10 μM in the species containing unpaired spins, which can be readily obtained in many cases. For ^1H NMR, concentrations of about 10 mM are required to obtain satisfactory spectra without specialized instrumental techniques; these levels are unobtainable in solution for many biological macromolecules of interest. For nuclei other than ^{1}H the sensitivity is always less (often much less) and specialized instrumentation is imperative. Sensitivity problems are often by-passed by the so-called *indirect methods*, which use NMR signals of water or some ligand present at a high concentration to obtain information about the biological system.

The problems of *spectral resolution* and *assignment* of resonances are of more concern in NMR studies, where a multiplicity of lines is commonly observed (especially in ^{1}H studies) than in EPR studies where only one or a few paramagnetic centres are normally present. Nuclear magnetic resonance studies are broadly of two kinds. "High-resolution" NMR involves studying all the resonances of a complex biological system directly, with all the attendant problems of resolving lines and assigning them to particular atoms. The instrumental and experimental methods used to overcome these problems will be discussed later. Another type of study, which may be termed "low-resolution" NMR, involves observing only one or a few resonances from the solvent water or from ligands, thus overcoming the resolution and assignment problems as well as sensitivity problems. The corresponding disadvantages of this indirect approach are that the information obtained is severely restricted and there may be considerable interpretation problems. The problems of *interpreting* magnetic resonance spectra depend to a large extent on the purpose of the experiment; for example, where the technique is required simply to detect a small molecule or to measure its binding to a large molecule, a detailed interpretation of spectral properties may be unnecessary. For other applications such as the determination of molecular structure, more use must be made of magnetic resonance theory, with a good deal of reliance in many cases also on theoretical models, comparative studies, or results from X-ray diffraction.

Biological applications of magnetic resonance

Magnetic resonance can be used to study both intact tissues and chemically well-defined systems such as small biomolecules, purified enzymes, and model membranes. The most spectacular application to date has been the use of high-resolution NMR, particularly ^{1}H NMR, in providing insight into the three-dimensional structure and dynamics of nucleotides, peptides, antibodies, the catalytic centres of enzymes, *t*RNA, model membranes and biological membranes. In most cases a good deal was already known of the structure from X-ray diffraction and chemical techniques, but NMR has complemented crystallography in a remarkable way and done much to dispel lingering doubts that crystal structures might not be relevant to biological molecules in solution. It has also provided new information on molecular dynamics which is equally relevant to the function of biological molecules, e.g. enzyme catalysis.

Other applications of magnetic resonance, though not necessarily providing detailed structural information, have been equally exciting from the biological or biochemical point of view. These include the study of conformation changes in allosteric proteins and membranes by both NMR and (using the spin-labelling technique) by EPR, especially in response to the binding of substrates, effector molecules, or drugs. Electron paramagnetic resonance has been used to good effect in studying the action of metalloproteins and a variety of biological oxidation–reduction processes such as oxidative phosphorylation and photosynthesis. Finally, whole-tissue studies by magnetic resonance are beginning to come into their own after a period of being overshadowed by the "hard" structural studies. This is the result of a general awareness that detailed studies on isolated molecules are incomplete without corresponding knowledge of the environment *in vivo* and the realization that NMR and EPR, by their non-destructive and highly selective character, are particularly suitable analytical tools for investigating this. Examples of whole-tissue studies are the detection of metabolites in muscle by ^{31}P NMR and the use of ^{1}H NMR and free-radical EPR to study malignant tissue. The examples given at the end of this chapter and the following chapter cover most of the areas of application listed above.

Theoretical Principles of Nuclear Magnetic Resonance

To interact with a static magnetic field and produce the energy states which give rise to magnetic resonance, a nucleus must possess a *magnetic moment* ($\boldsymbol{\mu}_N$) analogous to that of a compass needle which tends to orientate itself in a magnetic field. Only nuclei possessing both *charge* and *spin angular momentum* have a magnetic moment, given by

$$\mu_N = \gamma_I \hbar \, [I(I+1)]^{\frac{1}{2}} \tag{6.1}$$

where $\hbar$ is Planck's constant/2π, I is the *nuclear spin angular momentum quantum number*, and γ_I is the ratio of the nuclear magnetic moment to the spin angular momentum, or *magnetogyric ratio*. The value γ_I essentially determines the resonance conditions for each nucleus. The term $\gamma_I \hbar$ is often replaced for convenience by the term $g_N \beta_N$ which enables all nuclear magnetic moments to be expressed in terms of a basic unit of magnetic moment, the nuclear magneton (β_N) and a dimensionless constant (g_N), the *nuclear g-factor*.

The energy of interaction of this nuclear magnet (or dipole) with a static magnetic field, $\boldsymbol{B}_0$, is given by the scalar product of the vectors $\boldsymbol{\mu}_N$ and $\boldsymbol{B}_0$, i.e.

$$E = -\mu_N B_0 \cos\theta \tag{6.2}$$

where θ is the angle between the two vectors. The minus sign results from the fact that (as with a compass needle) the energy is a minimum when the magnetic moment is orientated parallel to the magnetic field.

Quantum theory requires that this energy is quantized so that the component of $\boldsymbol{\mu}_N$ along $\boldsymbol{B}_0$ can only have discrete values given by $\gamma_I\hbar I$, $\gamma_I\hbar(I-1)$, $\gamma_I\hbar(I-2)\ldots\gamma_I\hbar(-I)$. The difference in energy between the adjacent levels is thus

$$\Delta E=\gamma_I\hbar\boldsymbol{B}_0=g_N\beta_N\boldsymbol{B}_0 \tag{6.3}$$

As in any form of spectroscopy, we can cause transitions between these levels by applying the appropriate quantum of electromagnetic energy $\Delta E=h\nu$. In magnetic resonance, the possible transitions are limited by *selection rules* to transitions between adjacent energy levels, i.e. $\Delta I_z=\pm 1$ where I_z is the projection of the nuclear spin I along the magnetic field direction. The resonant frequency is thus given by

$$\nu=\frac{g_N\beta_N}{h}\boldsymbol{B}_0 \tag{6.4}$$

Alternatively, it is often convenient to describe magnetic resonance in terms of the precession of the magnetic moment about the applied field direction (see below and p. 249). This precessional motion has an angular velocity given by

$$\omega_I=2\pi\nu=\gamma_I\boldsymbol{B}_0 \tag{6.5}$$

Classical versus quantum mechanical approach

Magnetic resonance is essentially a quantum mechanical phenomenon, and this is the only way in which the behaviour of *individual nuclei* can be rigorously described. However, for descriptive purposes, or for treating the behaviour of a large population of atoms, as with relaxation processes on pp. 248–252, the approach of classical physics is often simpler. For instance, a nucleus of $I=\frac{1}{2}$ (e.g. ^{1}H) has two discrete energy levels according to quantum theory. These may be visualised as corresponding to two orientations of the nuclear magnet in the applied field as shown in Fig. 6.1. The nuclear magnetic moment is constrained in these orientations, but it is also subjected to a torque as a result of interaction with the applied field. It thus precesses about the field direction (as a spinning gyroscope will precess about a gravitational field) with a frequency ω_I (the *Larmor precession frequency*) given by eqn (6.5).

If we now apply a second, smaller magnetic field, $\boldsymbol{B}_1$, perpendicular to $\boldsymbol{B}_0$ (i.e. in the x–y plane) the nuclear magnet will attempt to precess about this direction in addition to the $\boldsymbol{B}_0$ direction. The vector $\boldsymbol{B}_1$ will be relatively ineffective because it is smaller and because the orientation of the magnet relative to it will be continuously changing as a result of precession about $\boldsymbol{B}_0$. Suppose now, however, that $\boldsymbol{B}_1$ is the magnetic field vector of a circularly polarized electromagnetic field rotating with angular velocity ω in the x–y

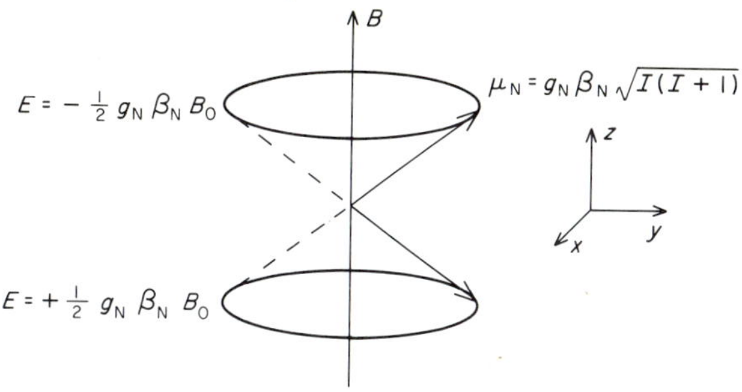

FIG. 6.1 Energy levels of a nuclear spin ($I=\frac{1}{2}$) precessing in a static magnetic field, $\boldsymbol{B}_0$. The higher-energy state has the z-component of nuclear magnetic moment antiparallel to the $\boldsymbol{B}_0$ direction.

plane, and we view the system in a frame of reference rotating with $\boldsymbol{B}_1$. When $\omega=\omega_I$ the nuclear magnet will appear stationary and maintain a constant orientation relative to $\boldsymbol{B}_1$. It can thus precess freely about $\boldsymbol{B}_1$ in addition to $\boldsymbol{B}_0$ and energy can be transferred from the field $\boldsymbol{B}_1$ to the nuclear magnet and vice versa. This is the *classical* description of magnetic resonance, where $\mathbf{B}_1$ is provided by the radiofrequency field. (It can equally well be plane polarized since it can then be resolved into two circularly polarized components rotating in opposite directions, only one of which is effective in inducing magnetic resonance.) However, according to quantum theory the nuclear magnet can only assume two orientations (for $I=\frac{1}{2}$) with respect to either $\boldsymbol{B}_0$ or $\boldsymbol{B}_1$. The radiofrequency field thus induces transitions between the two allowed orientations in the $\boldsymbol{B}_0$ field rather than causing continuous precession about $\boldsymbol{B}_1$.

Several important results follow from this treatment of magnetic resonance. Firstly, it is the magnetic field vector of the applied radiation, rather than the electric field vector as in most other forms of spectroscopy, that interacts with the system to give resonance. Secondly, because the nuclear magnets are constrained in their orientations relative to the rotating field $\boldsymbol{B}_1$, they tend to precess about $\boldsymbol{B}_0$ in phase with $\boldsymbol{B}_1$. Thirdly, magnetic resonance spectrometers only detect the component of magnetization induced in the sample in the plane of the applied electromagnetic field ($\boldsymbol{M}_{xy}$) and not $\boldsymbol{M}_z$. Magnetization ($\boldsymbol{M}$) is the bulk vector quantity corresponding to the magnetic moment of single nuclei. In the absence of an electromagnetic field the precessional motions of individual nuclei have random phases with respect to each other and thus in a bulk sample components of $\boldsymbol{M}$ in the x–y plane cancel out, i.e. $\boldsymbol{M}_{xy}=0$. The population of the two energy states, corresponding to the two spin orientations relative to $\boldsymbol{B}_0$, is not exactly equal (see below) giving a small component $\boldsymbol{M}_z$ along the $\boldsymbol{B}_0$ direction. Resonance causes some phase coherence of precessing spins and causes transitions which tend to equalize the energy level populations. Thus $\boldsymbol{M}_z < \boldsymbol{M}_z$ (equil) and $\boldsymbol{M}_{xy} > 0$ and the net effect of resonance is to rotate the magnetization vector $\boldsymbol{M}$ away from the z-direction.

The importance of relaxation processes in magnetic resonance

As with all forms of spectroscopy, NMR depends for its observation on the continuous absorption of energy from an electromagnetic field. This cannot occur without *relaxation processes* (see p. 248) which remove the energy from the spin system to other parts of the molecular structure (the "lattice") and transform it ultimately into thermal energy. The efficiency of the relaxation processes partly determines the rate at which the system will absorb energy and thus how easily magnetic resonance can be observed.

Consider again a nucleus of $I = \frac{1}{2}$ (e.g. ^{1}H). In the presence of a magnetic field $\boldsymbol{B}_0$ there are two energy levels with a separation given by eqn (6.3). At thermal equilibrium (no applied electromagnetic field) there is a slightly greater population ($N+$) in the lower energy level than in the higher energy level ($N-$) and the distribution is given by the Maxwell–Boltzmann law

$$\frac{N+}{N-} = \exp(\Delta E/kT) \sim 1 + \Delta E/kT \qquad (6.6)$$

In magnetic resonance, $\Delta E \ll kT$ and the excess population in the lower level is exceptionally small: about 1 in 10^5 for NMR and about 1 in 10^3 for EPR. When the system is irradiated at the appropriate frequency, there is an equal probability that both upward and downward transitions between the energy levels will be induced; the net absorption of energy thus depends on the small inequality of populations being maintained. As the intensity of irradiation increases, more and more transitions both ways are induced, but the inequality of populations is reduced and thus the *net absorption* of energy is reduced. At sufficiently high intensities the populations are equalized, no net absorption of energy occurs and the NMR signal disappears; this is the opposite extreme to thermal equilibrium and is termed *saturation*. It is normally necessary to avoid this condition in NMR experiments. It is less likely to occur where efficient relaxation processes are operating to restore thermal equilibrium.

Relaxation processes are thus essential to observe magnetic resonance in the first place, but conversely if they are too efficient they may cause magnetic resonance spectra to disappear through line-broadening (see pp. 247–250). However, under the right conditions their detailed study can provide a large amount of useful information on molecular structure and motion (see pp. 251–252).

The Characteristics of an NMR Spectrum

High-resolution NMR spectra may be obtained in more than one way as will be discussed on p. 253, but they are normally displayed as *continuous-wave* spectra, i.e., a plot of absorption of energy from a continuous radiofrequency field against magnetic field strength. For a particular nucleus, resonance can be obtained under any conditions of ν and $\boldsymbol{B}_0$ satisfying eqn (6.4), but $\boldsymbol{B}_0$ is normally varied at constant frequency for practical reasons. Figure 6.2 shows a representative ^{1}H NMR spectrum at moderate resolution (90 MHz) for a small biomolecule, 5′-AMP. Figure 6.3. shows representative ^{1}H NMR spectra for a small protein (ribonuclease) at moderate resolution (60 MHz) and also at high resolution (220 MHz). We now discuss the characteristics of NMR spectra in turn using these as examples.

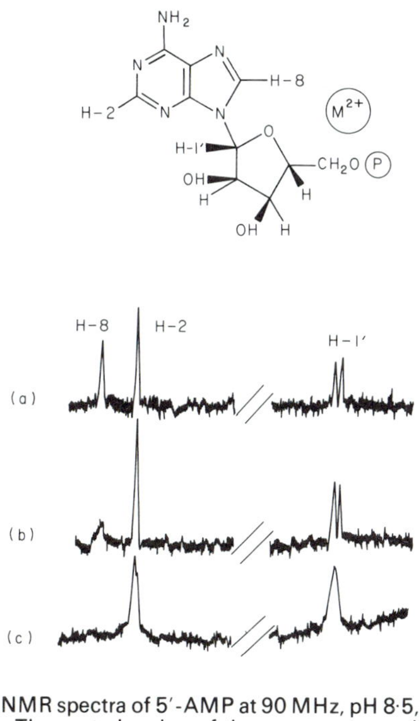

FIG. 6.2 Hydrogen-1 NMR spectra of 5′-AMP at 90 MHz, pH 8·5, 22°C in the presence and absence of Mn(II) ions. The central region of the spectrum containing the large solvent ^{1}H signal is omitted. (a) 18 mM AMP; (b) 18 mM AMP+9 μM Mn(II); (c) 18 mM AMP+45 μM Mn(II). (From Dwek, R. A. (1973). "Nuclear Magnetic Resonance in Biochemistry: Applications to Enzyme Systems". Clarendon Press, Oxford.)

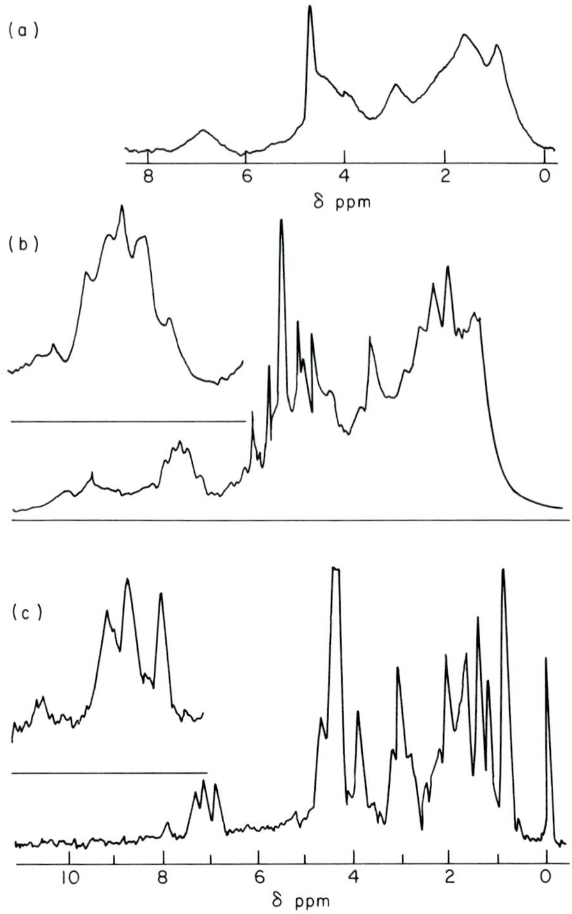

FIG. 6.3 Hydrogen-1 NMR spectra of ribonuclease, 11% in ^2H_2O. (a) 60 MHz, p^2H 7.5, 38°C; (b) 220 MHz, p^2H 6.8, 22°C (native form); (c) 220 MHz, p^2H 6·8, 72·5°C (thermally denatured). The insets to (b) and (c) are the expanded spectra from $\delta \approx 7$ ppm. (From Ferguson, R. C. and Phillips, W. D. (1967). *Science (N.Y.)* **157**, 257. Copyright 1967 by the American Association for the Advancement of Science.)

The nuclear g-factor, g_N

The first thing to note in the spectra is that only resonances from one type of nucleus (in this case ^1H) are displayed in any given NMR spectrum, and that all these resonances occur within a very small range of magnetic field: the total variation in field for the ^1H spectra of Figs 2 and 3 is only about 10 parts per million. The value of magnetic field for resonance for each type of nucleus is determined by g_N, the nuclear g-factor. Table 6.1 lists values of g_N for the magnetic nuclei of most interest along with the natural abundance of the isotope involved and the relative sensitivity. From these values of g_N and eqn (6.4), it is apparent that the magnetic fields for resonance for different nuclei are very widely separated compared to the spread of resonances from a single nucleus. As well as determining the resonance position, g_N affects the sensitivity of nuclei because of its effect on the energy level separation. This will be apparent from a consideration of eqns (6.3) and (6.6). The combined effects of g_N, natural abundance, and ubiquity of protons explain why ^1H NMR is so

TABLE 6.1 Properties of magnetic nuclei most commonly encountered in biological systems

Nucleus	I	g_N	Natural abundance	Relative sensitivity
^1H	$\frac{1}{2}$	5·585	100	100
^{13}C	$\frac{1}{2}$	1·405	1·1	0·018
^{19}F	$\frac{1}{2}$	5·257	100	83·3
^{31}P	$\frac{1}{2}$	2·263	100	6·63

commonly used in comparison with other nuclei. However, ^1H NMR has two major disadvantages. First, except in the smallest molecules (e.g. Fig. 6.2) many overlapping resonances are present (e.g. Fig. 6.3) limiting the complexity of molecules which can be studied. Secondly, for high resolution studies ^1H_2O must be eliminated as far as possible from solutions and be replaced by ^2H_2O because otherwise its large ^1H signal would obliterate most other signals. The effect of a slight trace of ^1H_2O is apparent in many ^1H NMR spectra even when great care has been taken to eliminate it from solutions (e.g. the large resonance at $\delta \sim 5$ ppm in Fig. 6.3). This is not surprising if it is remembered that the concentration of protons in water is 110 M, whereas even in a relatively concentrated solution (100 g l^{-1}) of a small protein (of molecular weight 10 000) the concentration of a particular proton is only 10 mM.

The chemical shift, δ

As we saw above, not all resonances from a particular type of nucleus occur at exactly the same magnetic field. The variations are caused because adjacent electron clouds interact with the applied field to give small induced magnetic fields, and these induced fields alter the effective magnetic field felt by each nucleus. Because the variations are caused by *induced* local fields, they are proportional to the applied magnetic field, and thus these *chemical shifts* are expressed in parts per million (ppm) of the applied field rather than in absolute magnetic field units. The chemical shift is measured downfield from some reference which for protons is usually tetramethylsilane (TMS) or a related compound, and is thus defined as

$$\delta = \frac{\boldsymbol{B}_{\mathrm{ref}} - \boldsymbol{B}}{\boldsymbol{B}_{\mathrm{ref}}} \times 10^6 \tag{6.7}$$

Different proton groupings in a molecule, e.g. methyl protons, methylene protons, aromatic ring protons, have characteristic chemical shifts, which are often little changed when the groupings form part of a complex molecule (see p. 262, Example 1). Thus chemical shifts are an important feature in assigning resonances in complex spectra. The study of chemical shifts is of particular value in the following situations: (i) where abnormal shifts occur due to the magnetic fields associated with neighbouring aromatic rings (*ring-current* shifts) (e.g. haemoglobin, p. 266 Example 4) (ii) where specific probes (e.g. some lanthanide metals) can be introduced to perturb the shifts (e.g. nucleotides, p. 264 Example 2, and lysozyme, p. 264 Example 3) and (iii) where nuclei such as ^{19}F which normally give large shifts are being studied.

Line-splittings: the spin–spin coupling constant, *J*

Many NMR lines are split into doublets or more complex patterns as a result of magnetic interactions between a particular nuclear spin and other spins which may be those of completely equivalent nuclei, non-equivalent nuclei of the same type, or nuclei of a different type. The simplest case is that of a molecule AX where both nuclei have $I = \frac{1}{2}$. The A nucleus senses a slightly different magnetic field according to whether the spin state of X is parallel or antiparallel to the applied magnetic field. There is an approximately equal probability for the two orientations and thus the resonance of A is split into two components of equal intensity, and similarly for the resonance of X. Figure 6.2. shows a simple example of this type where the H-1′ resonance of AMP is split into two due to interaction with the proton on the adjacent C-atom of the ribose ring. Unlike the chemical shift, line splittings are the result of a direct magnetic spin–spin interaction and are independent of the applied field, so that they can be expressed in absolute magnetic field or frequency units (usually in Hz). The interaction is, however, transmitted via the electron clouds surrounding the nuclei and the splitting depends on the separation of the nuclei and the intervening chemical bonds and their orientation. Thus a study of coupling constants (*J*) can often provide useful information on bond angles and molecular conformation (p. 263 Example 2).

If a nucleus is coupled to n equivalent spins, there will be $(n+1)$ lines with intensities given by a binomial distribution, e.g. 1:2:1 for 2 equivalent spins, to allow for all possible combinations of spin state (Fig. 6.4). For each additional coupling all the pre-existing lines are split again and thus where

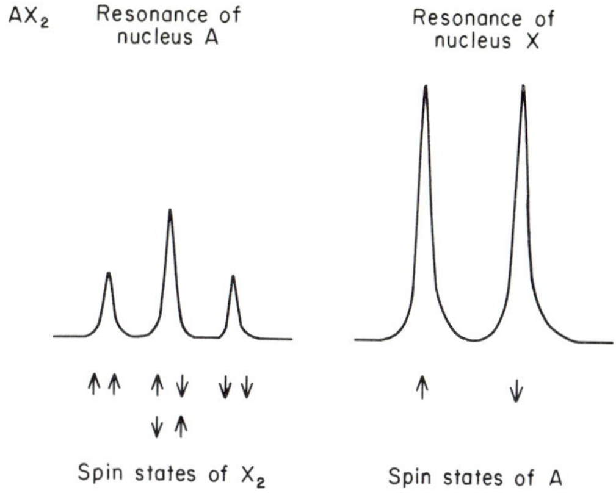

FIG. 6.4 Schematic representation of NMR line-splittings for a simple molecule AX_2 where $I=\frac{1}{2}$ for both nuclei and the two X-atoms are equivalent.

non-equivalent nuclei or many equivalent nuclei interact the spectrum can become complex. Methods are available (see p. 259) for specifically "decoupling" spin–spin interactions, thus simplifying spectra and assisting assignment.

Area under the peak: number of nuclei

The area under each peak of an NMR spectrum is directly proportional to the energy absorbed from the radiofrequency field by each chemically distinct atom and hence to the concentration of that atom. It is necessary that (a) saturation is avoided, (b) instrumental conditions are kept constant, (c) the temperature is constant and (d) line-splittings are allowed for, e.g. each H-1′ peak in Fig. 6.2 is equivalent to half a proton. With these checks, NMR spectra can be made quantitative and can be used to measure concentrations, interconversions of different spectral forms and reaction kinetics much like other spectroscopic techniques.

The linewidth: relaxation rates

All lines in magnetic resonance spectra have a finite width, apart from any instrumental broadening (see p. 253) which may occur. This is a consequence of a particular type of relaxation process called spin–spin relaxation characterized by a *relaxation time*, T_2. There is a reciprocal relationship between T_2 and linewidth given by the equation

$$T_2=\frac{1}{\pi\Delta\nu_{\frac{1}{2}}} \qquad (6.8)$$

where $\Delta\nu_{\frac{1}{2}}$ (measured in Hz) is the spectral linewidth at half-peak height. A typical narrow, fully resolved line in a ^{1}H NMR spectrum has a Lorentzian line-shape with T_2 in the range 10^{-2} s to 1 s and $\Delta\nu$ of the order 10 Hz. This reciprocal relationship is an important theme in magnetic resonance and we return to it again later (see pp. 249–253). In general, processes which lead to more efficient nuclear relaxation also lead to spectral line-broadening. This may simply be a nuisance where high resolution is required (Fig. 6.3) but in other applications (Fig. 6.2) a specific line-broadening effect can be used to calculate interatomic distances or obtain information on molecular motion.

Nuclear Relaxation Processes

Nuclear relaxation is concerned with ways in which the radiofrequency energy absorbed by a nuclear spin system is transferred either to other spin systems or to other parts of the environment in which the spin exists ("the lattice"). It can most appropriately be discussed using the classical approach to magnetic resonance and the concept of pulsed NMR which is much utilized in modern instrumentation.

The two relaxation times T_1 and T_2

We saw earlier (pp. 240–242) how we can represent a population of identical nuclear spins ($I=\frac{1}{2}$) at thermal equilibrium by a small net magnetization ($\boldsymbol{M}$) lying in the $\boldsymbol{B}_0$ or z-direction (Fig. 6.5(a),(b)) and that the effect of resonance is to produce a component of $\boldsymbol{M}$ in the x–y plane and effectively rotate $\boldsymbol{M}$ away from the z-direction. In a conventional magnetic resonance experiment, measuring the continuous absorption of energy from a low-power radiofrequency field, a steady state is set up with $\boldsymbol{M}$ near to its equilibrium orientation along $\boldsymbol{B}_0$. If, however, the radiation is given as a high-intensity pulse, $\boldsymbol{M}$ can be rotated through any given angle by a suitable choice of intensity and duration provided that the processes which re-establish thermal equilibrium are slow compared to the pulse duration. Thus $\boldsymbol{M}$ can be turned through exactly 180° (Fig. 6.5(c)). After the pulse, $\boldsymbol{M}_z$ will return through zero to its original direction to re-establish thermal equilibrium in an exponential manner (Fig. 6.5(d)), characterized by a relaxation time T_1, the *longitudinal* or *spin-lattice relaxation time*. (It should be stressed that this classical description in terms of rotating magnetization is only possible because we have a large population of spins: quantum mechanics does not allow this behaviour for a single precessing nuclear magnet.)

If attention is now confined to the individual precessing spins it will be

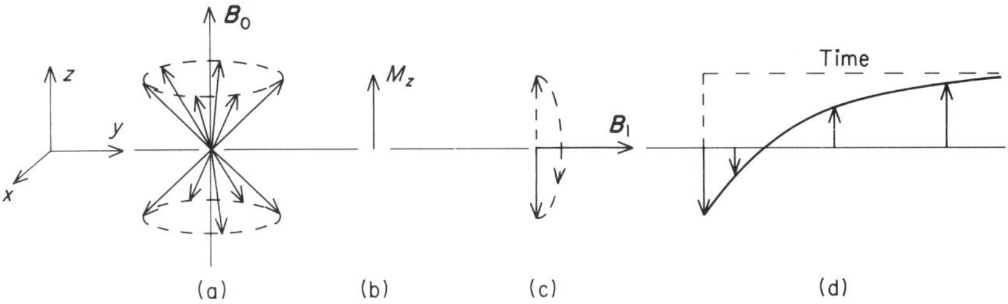

FIG. 6.5 Spin–lattice relaxation (T_1) following a 180° pulse as viewed in a coordinate system rotating in the x-y plane. (a) A population of nuclear spins ($I=\frac{1}{2}$) at thermal equilibrium precessing about the static magnetic field direction, $\boldsymbol{B}_0$. (b) The resultant magnetization $\boldsymbol{M}$ in the $\boldsymbol{B}_0$ direction. (c) Rotation of $\boldsymbol{M}$ through 180° by a radiofrequency pulse with its magnetic field vector $\boldsymbol{B}_1$ perpendicular to $\boldsymbol{B}_0$ and rotating with the coordinate system. (d) Exponential return of the magnetization to thermal equilibrium with the radiofrequency field switched off.

apparent that there is another type of relaxation process, at first sight independent of T_1. By a suitable pulse, $\boldsymbol{M}$ can be turned through exactly 90° instead of 180° so that it now lies in the x–y plane. When the pulse is switched off the spins (which now show phase coherence) begin to precess freely about the z-direction as before. If all the spins were in an identical magnetic field and could precess independently of each other they would have the same Larmor frequency and would always remain coherent: $\boldsymbol{M}_{xy}$ would remain constant. This is not the case; individual spins gradually get out of phase and $\boldsymbol{M}_{xy}$ decays to zero. The decay is again exponential and characterized by the relaxation time T_2 (see p. 247) the *transverse* or *spin–spin relaxation time*. We now see the basis of the relationship of T_2 to linewidth since a rapid decay of $\boldsymbol{M}_{xy}$ to zero and a short T_2 value imply a large range of Larmor frequencies and thus a large range of magnetic field values for resonance at constant frequency.

The decay of $\boldsymbol{M}_{xy}$ to zero in a uniform magnetic field after the radiofrequency field is switched off is called a *free induction decay* and can be followed directly in an NMR spectrometer as an induced radiofrequency signal in a receiver coil (unlike the conventional way of observing magnetic resonance as the absorption of energy from a continuously applied radiofrequency field). The free induction decay could in principle be used to measure T_2 directly for a single-line spectrum, but it is more commonly used as a rapid means of obtaining complex NMR spectra, using Fourier transform methods (p. 254). More elaborate pulse methods are used to obtain accurate estimates of T_1 and T_2. They are obligatory for T_1 since $\boldsymbol{M}_z$ cannot be measured directly in an NMR spectrometer.

Molecular relaxation mechanisms

In many forms of spectroscopy, efficient relaxation from an excited energy state occurs by collision with other molecules or by spontaneous emission. In magnetic resonance these effects are insignificant in causing relaxation; nuclear spins in particular are well shielded from their chemical environment. What is the physical basis of the two relaxation processes we have defined above?

Nuclear relaxation is mainly caused by "magnetic noise", or time-dependent fluctuations in the local magnetic field. The noise in turn is mainly due to neighbouring magnetic dipoles undergoing relaxation processes or rotational motions, or to exchange of the nucleus between different chemical environments. Each process has a characteristic time-scale and thus the noise has a well-defined frequency distribution. Any components of the noise close to the Larmor frequency ω_I will tend to induce transitions between energy levels just as an applied radiofrequency field induces magnetic resonance, whereas components of the noise greatly above or below ω_I will be ineffective. The combined molecular motions responsible for the noise can be expressed in terms of a *correlation time*, τ_c, which indicates how effective relaxation processes will be at a particular frequency. The effect of inducing transitions between energy states is to restore thermal equilibrium between them, so that magnetic noise is the major factor contributing to the spin-lattice relaxation rate, i. e. $1/T_1$. The transitions also reduce the lifetimes of energy states and cause a loss of phase coherence between the precessing spins. Heisenberg's Uncertainty Principle indicates that when the lifetime of an excited energy state is reduced by relaxation processes, the uncertainty in measuring its energy is correspondingly increased ($\Delta E \times \Delta t \sim h$) leading to spectral line broadening. Hence, any process which increases spin-lattice relaxation rates ($1/T_1$) also contributes to line-broadening and T_2 relaxation (uncertainty broadening or lifetime broadening). Since such processes tend to restore thermal equilibrium they can alleviate saturation (see p. 242).

Now consider what happens in the limit where molecular motion is very slow. There are still local variations in magnetic field throughout the sample due to other magnetic dipoles, but they do not now vary with time (or at least, vary only slowly compared with the precession frequency ω_I), and thus there is no mechanism for inducing transitions between energy states which consequently have a long lifetime. However, the static field variations cause the spins to have slightly different precession frequencies and so provide a further means for the loss of phase coherence and for line-broadening. This means that there is a second type of relaxation process which only contributes to spin–spin relaxation rates ($1/T_2$) and not to spin-lattice relaxation rates ($1/T_1$). It does nothing to alleviate saturation since unlike spin-lattice relaxation it is not concerned with net energy level changes. It may be the dominant form of relaxation in the solid state and at low temperatures ($T_1 \gg T_2$) but is often relatively unimportant in liquids having a good deal of thermal motion ($T_1 \approx T_2$).

Thus in most systems of interest to the biologist or biochemist relaxation is dominated by *dipole–dipole interactions* modulated by molecular motion, and relaxation processes for both T_1 and T_2 can usually be described by equations of the form

$$\frac{1}{T}=\frac{K}{r^6}\mathrm{f}(\tau_c) \tag{6.9}$$

where K is a constant for any two interacting spins depending on their respective magnetic moments and r is the distance between them. The dependence on K and r^6 follows from classical physics. The function $\mathrm{f}(\tau_c)$ is a complex function specifying how the efficiency of relaxation varies with τ_c, the correlation time and on the Larmor frequencies of the spins involved (and hence on the spectrometer frequency).

Paramagnetic relaxation

An important and frequently encountered case is where a paramagnetic centre is used to increase the nuclear relaxation rate. Electron magnetic moments are greater than nuclear magnetic moments by a factor of $\sim 10^3$, and thus nuclear relaxation rates can be greatly increased by dipolar interactions with a neighbouring paramagnetic centre. The ^1H resonance of pure water has an extremely long relaxation time ($T_1 \sim 2$ s) because only the weak nuclear dipolar interactions are present. If a small concentration (~ 100 μM) of magnetic Mn(II) is now added, the observed relaxation time decreases to about 0·5 s. Manganese(II) in water is surrounded by a hydration sphere of six water molecules which are rapidly exchanging with the bulk solvent. The correlation time for the dipolar interaction between the electron spin and the proton spin is given by

$$\frac{1}{\tau_c}=\frac{1}{\tau_R}+\frac{1}{\tau_s}+\frac{1}{\tau_M} \tag{6.10}$$

where τ_R is the rotational tumbling time of the hydrated Mn(II) ion, τ_s is the Mn(II) electron spin relaxation time, and τ_M is the water exchange rate. The time τ_R is very short ($\sim 10^{-11}$ s) whereas τ_s and τ_M are longer ($\sim 10^{-8}$ s). It is the fastest motion (τ_R) which dominates τ_c, and thus $\tau_c \approx \tau_R \approx 10^{-11}$ s. As suggested above, relaxation is most efficient when the motions have the same timescale as the nuclear resonance, i.e. when $\omega_I \tau_c \approx 1$. For typical NMR frequencies $\omega_I \approx 10^8 \mathrm{s}^{-1}$, and hence, although Mn(II) has some effect on relaxation times, the molecular motion is not ideal for efficient relaxation.

If we now add to the solution an enzyme or other biological macromolecule which can bind Mn(II), the relaxation time is further decreased. This is essentially due to the fact that τ_R is now increased to the region of 10^{-8} s,

$\tau_c \approx 1/\omega_I$, $f(\tau_c)$ in eqn (6.9) increases, and T_1 decreases. This is termed *proton relaxation enhancement*. It will be obvious from the above example that relaxation enhancement probes can give internuclear distances (e.g. p. 263, Example 2 and Fig. 6.2; p. 264, Example 3) binding information (p. 269 Example 6) and molecular motion parameters.

Exchange phenomena

We saw above a several-fold increase in ^1H relaxation rate of water (55 M) as a result of adding 10^{-4} M Mn(II). At any instant only six water molecules are sufficiently close to an Mn(II) ion to be relaxed by it: this is only about one in 10^5 molecules. How is it possible to observe an effect from such a small population? The reason is that the exchange rate of water between the bound site and bulk phase is rapid compared to the relaxation rate in the bound site ($\tau_M \ll T$) and we see a weighted average of the NMR relaxation rates in the two sites; this is called *fast exchange*. The fast exchange condition is a very useful one in NMR because it allows information to be obtained about ligand-binding sites on large molecules (p. 269 Example 6) which would otherwise be inaccessible because (i) the signal from the bound site would be too small to observe, (ii) the signal would be too broadened by relaxation to observe.

As τ_M increases a point may be reached where fast exchange no longer occurs. Two separate signals are then present, and only that from the bulk may be observable (Fig. 6.6(a)); this is termed *slow exchange*. The ideas of fast and slow exchange can be applied to molecules exchanging between different chemically shifted environments (Fig. 6.6(b)) as well as between environments different in linewidth and relaxation rate. All exchange phenomena are temperature dependent, and thus a study of NMR spectra or relaxation rates at different temperatures may give useful kinetic information (p. 264 Example 3).

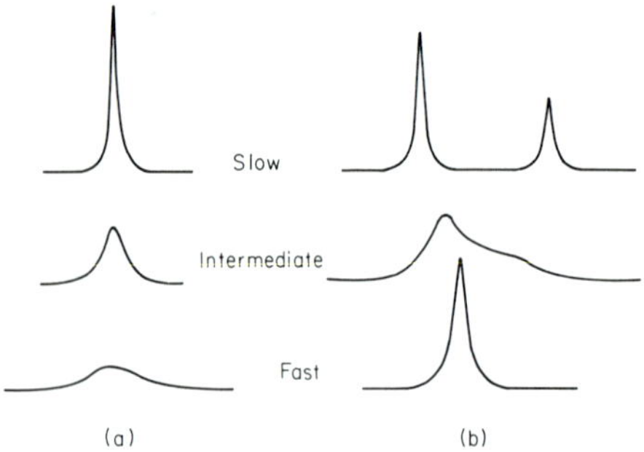

FIG. 6.6 Schematic representation of NMR signals under conditions of slow, intermediate and fast exchange between (a) environments of different relaxation rate and (b) different chemically shifted environments.

NMR Instrumentation and Methods

Nuclear magnetic resonance technology is mainly directed towards (i) improving resolution by the provision of high, stable, and homogeneous magnetic fields, (ii) improving sensitivity by the design of electronic circuitry and the use of signal-accumulating and computing techniques and (iii) providing versatility to accommodate the various specialized methods.

The two ways of doing NMR: continuous wave and pulse methods

The traditional way of observing an NMR spectrum is to scan the magnetic field slowly while a steady radiofrequency field is applied, and record the absorption by each resonance in turn. Low power is used and the system is always close to thermal equilibrium (i.e. $\boldsymbol{M}_z \approx \boldsymbol{M}_z$ (equilibrium), and $\boldsymbol{M}_{xy}$ is small (p. 241). This *continuous wave* method is used with instruments of moderate frequency (up to about 60 MHz) on samples with relatively simple NMR spectra where concentration is not a problem, e.g. for identifying small organic compounds. A single spectrum is obtained in a few minutes. For more demanding applications (which include most biological applications) the sensitivity must be improved. This can be done by increasing the magnetic field (eqn 6.3), decreasing the temperature (eqn 6.6) or averaging over many spectra to improve the signal/noise ratio. The scope for the first two possibilities is limited, and the continuous wave method is not suited to signal-averaging because of the time requirement. *Pulsed* methods are used instead.

These methods depend on measuring the *free induction decay* following a 90° pulse, the technique described earlier for measuring T_2 (p. 249). We saw how the rate of decay of $\boldsymbol{M}_{xy}$, and hence of the signal induced in a receiver coil in a uniform magnetic field with the exciting radiation switched off, depends on the rate at which the precessing spins get out of phase with each other, i.e. the free induction decay is the result of interference between the precessing spins. The rate of loss of phase coherence depends on the distribution of Larmor precession frequencies due to local field variations and on the lifetime of the spin states and these are the factors that also govern the linewidth in a continuous-wave spectrum (p. 250). For the case of an infinitely narrow NMR line (theoretically impossible since it would require both a perfectly uniform local field and an infinite lifetime) the spins would always remain in phase and no decay would be seen (Fig. 6.7 (a)); broader lines give more rapid decay (Fig. 6.7 (b)). If now the precession frequency of the spins (ω_I) is made slightly different from the detection frequency of the instrument ("off resonance") by altering the magnetic field, further interference or "beating" occurs between

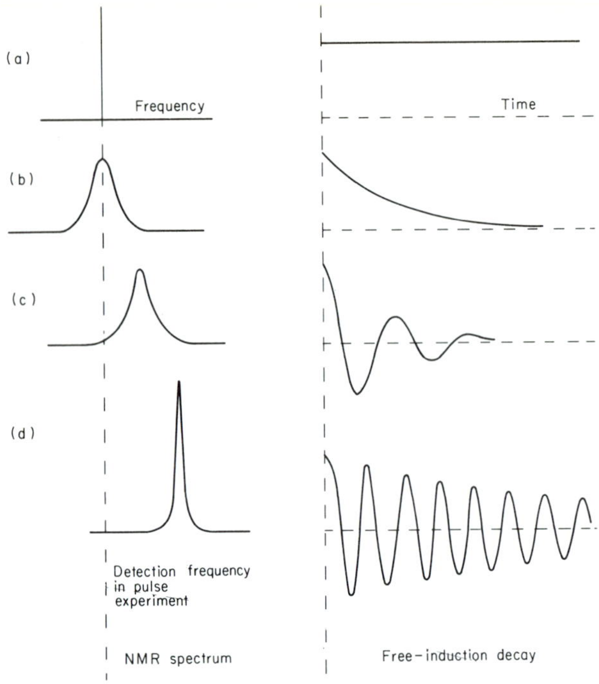

FIG. 6.7 Relationship between free induction decay and NMR spectrum (a) for an infinitely narrow line at resonance, (b) for a broadened line at resonance, (c) for the same line as (b) with the resonant frequency slightly different to the detection frequency, (d) for a narrower line further "off-resonance".

the two frequencies (Fig. 6.7(c)). As the separation of frequencies is increased the beats get more frequent (Fig. 6.7(d)); similar effects occur between two chemically shifted resonances in a spectrum. There is a reciprocal relationship between features in the free induction decay ("time domain") and features in the NMR spectrum ("frequency domain").

This concept can be extended to spectra of any degree of complexity with different chemical shifts and linewidths; the waveform of the free induction decay completely determines the continuous-wave spectrum. The mathematical relationship which exists is similar to that between the optical interference pattern of a row of slits and the object or between a complex sound waveform and the frequency spectrum into which it can be resolved and is a *Fourier transform*. In Fourier transform NMR spectroscopy, the information content of the whole spectrum is obtained in a few seconds from the free induction decay pattern. Many of these can then be stored and added together to improve signal/noise, then transformed by a computing procedure and displayed as a conventional spectrum.

In summary, Fourier transform NMR involves exciting all the resonances of the spectrum simultaneously, allowing them to interfere, summing a large number of these "interference patterns" and then transforming them back into the original object (spectrum).

The elements of an NMR spectrometer

A block diagram of a "typical" NMR spectrometer is shown in Fig. 6.8. For practical convenience the spectrometer is divided into several more or less spatially separate units (i) the *magnet* and *probe*, i.e the sample and coils for irradiating it and detecting resonance, (ii) the *magnet power supply*, (iii) the *main console* which contains the radiofrequency source, signal receiver, amplifier, oscilloscope, and the chart recorder and (iv) *computers* and other ancillary equipment.

The properties of the magnet are of central importance. Field stability is important to prevent drift of spectra. Sensitivity is a function of magnetic field strength (eqns 6.3 and 6.6). Resolution is a function of both field homogeneity and field strength. This is because, if the magnetic field is inhomogeneous over the sample, identical nuclei in different parts of the sample will resonate at slightly different field settings, giving rise to *inhomogeneity broadening*. In addition, because the chemical shift is proportional to the magnetic field while spin–spin coupling constants and linewidths are not, increasing the field strength (and frequency) "spreads out" chemically shifted lines, leading to an improvement of resolution in complex spectra. This is well illustrated in a comparison of Fig. 6.3 (a) and (b), and explains the emphasis placed on ever-increasing frequencies and magnetic field homogeneity in the design of NMR spectrometers.

Nuclear magnetic resonance magnets may be permanent magnets (up to about 90 MHz operating frequency), electromagnets (at higher frequencies) or superconducting magnets (200–400 MHz). In all cases elaborate cooling or thermostatting arrangements are necessary. Field homogeneity of the degree demanded by high resolution NMR (~ 1 in 10^9 over the sample) cannot be built into even the most carefully constructed magnet and is thus normally provided by a process of compensation using subsidiary electromagnets ("shimming"). Subsidiary magnets are also required to provide the field sweep when a spectrum is recorded, and are directly linked to the x-movement on the chart recorder. Field stability is usually provided by an automatic feedback ("locking") system which compares a reference radiofrequency signal with some strong resonance (^1H or ^2H) which can act as an indicator of field drift and which applies a suitable correction to the field.

The radiofrequency source has to be controlled with a comparable precision to that of the magnetic field. The signal is usually based on the frequency

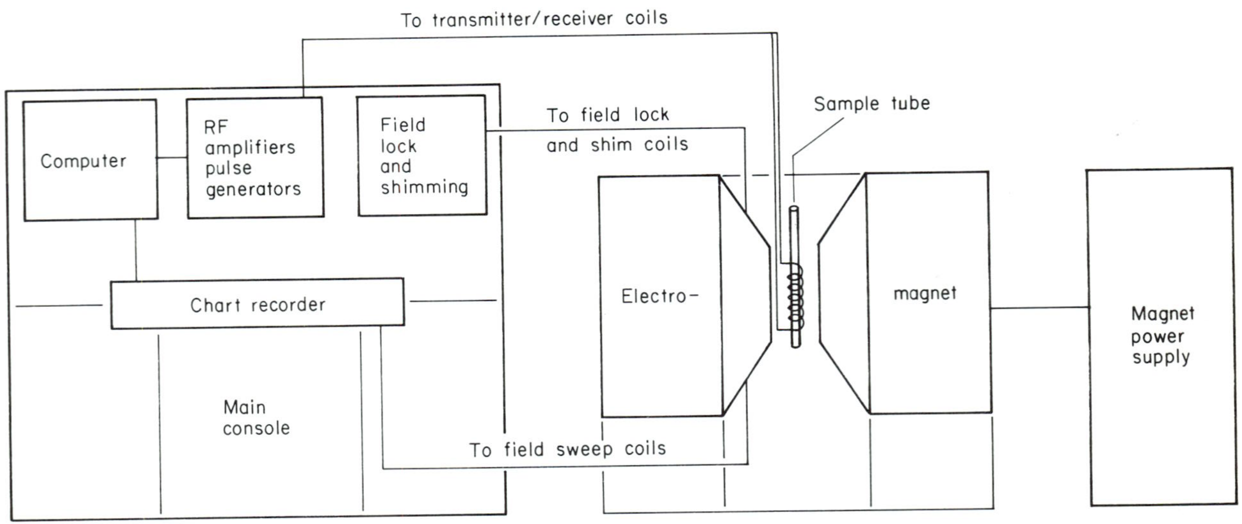

FIG. 6.8 The main elements of an NMR spectrometer.

of a crystal oscillator and is amplified before passing to a transmitter coil around the sample in the magnetic field. In some cases the absorption of radiation by the sample at resonance is detected by a separate receiver coil incorporating a preamplifier for detecting the small signals. In other cases one coil may serve both purposes and a bridge arrangement for detecting very small *changes* at resonance is incorporated. Pulsed NMR spectrometers usually have a single-coil probe since, unlike continuous-wave operation, irradiation and detection are temporally separated events. The signal is then amplified, filtered, rectified and applied to an oscilloscope (which is mainly used in setting-up procedures) or a chart recorder. For operation in the Fourier transform mode signals are stored, summed and transformed by computer into a spectrum.

Sample preparation

Most biological NMR studies are made on aqueous ($^{1}H_2O$ or $^{2}H_2O$) solutions although small molecules are sometimes studied in organic solvents and dispersions (e.g. of membranes) and intact tissues can also be studied. Sample volumes are kept small because of the problems of field inhomogeneity (0·1–0·4 ml for ^{1}H studies but rather more than this for ^{13}C or ^{31}P studies where sensitivity is a problem). For high-resolution ^{1}H studies, H_2O must be replaced by $^{2}H_2O$ and exhaustive dialysis or the use of lyophilized samples is necessary. The effect of the replacement on biochemical systems is usually small, but resonances are usually lost from those protons such as > NH protons in a peptide which exchange rapidly with the solvent, and in addition a small correction has to be applied to pH values. A variety of other requirements must also be fulfilled, particularly in samples for high-resolution studies. These include (i) elimination of trace paramagnetic impurities which can drastically affect linewidths, (ii) inclusion of a chemical shift reference, which may be internal or external, (iii) a means of spinning the sample in the probe to average out small residual magnetic field inhomogeneities and (iv) a means of controlling temperature which affects the properties of NMR spectra as well as the biological system itself. Requirements (iii) and (iv) are routinely met on most NMR spectrometers.

The limiting factor in many biological NMR studies is, as we saw earlier, concentration. The observation of single ^{1}H resonances from small molecules without signal accumulation is not usually a problem since the required molarity ($\sim$ 10 mM) can easily be obtained. To observe a single proton in a macromolecule also requires a similar molarity whereas, as we have seen (p. 245), for even a small protein (molec. wt $\sim$ 10 000) this represents a solution of 100 g l^{-1}. At best, this concentration may lead to problems of aggregation or high viscosity (which causes line broadening). At worst it may be physically impossible to obtain or be prohibitively expensive in material. Thus, signal-averaging and other devices to improve sensitivity are of the utmost importance and greatly increase the scope for studying dilute solutions or nuclei which are inherently less sensitive (^{13}C, ^{19}F, ^{31}P). Even so, they may be difficult to apply to unstable samples or to time-dependent studies because of the time requirement. Despite all these problems, NMR spectroscopy has considerable advantages over many other techniques in that optically opaque or intact systems may often be studied under nearly physiological conditions.

Specialized methods

In view of rapid advances, the description of particular methods as routine or specialized is necessarily arbitrary. The use of pulse methods to measure relaxation times, Fourier transform NMR, and the use of paramagnetic probes have already been mentioned. These would have been described as specialized a decade or so ago, but are now routine. *Paramagnetic probes* can, if they are incorporated specifically into a biological molecule, be used to broaden selectively or shift resonances from neighbouring atoms and thus simplify complex spectra. If the perturbed spectrum is subtracted from the original (by computer) a *paramagnetic difference spectrum* showing only the perturbed resonances is obtained (p. 264 Example 3 and Fig. 6.9; p. 268 Example 5). A particular case of the use of paramagnetic probes is *water proton relaxation enhancement* mentioned earlier (p. 252) and much used because of its practical simplicity to determine the binding of paramagnetic metals to macromolecules, particularly enzymes which require a divalent metal for activity (p. 269 Example 6 and Fig. 6.10). Paramagnetic probes are thus more than a device for assisting in the interpretation of high-resolution spectra, but can give quantitative information on ligand binding, structure and molecular dynamics (p. 263, Example 2 and Fig. 6.2).

Nuclear magnetic resonance difference spectroscopy can also be useful where the perturbation is a pH change or other environmental change or a specific chemical modification of the molecule. The *pH difference spectra* allow the pK values for ionizing groups such as carboxyl or imidazole groups to be determined *in the molecule* (See p. 268, Example 5) since the ionization usually results in a chemical shift change and only the shifted resonances appear in the difference spectrum. *Isotope exchange*, e.g. the $N^1H \rightleftharpoons N^2H$ exchange in the amide group of polypeptides, can be studied. Normally the amide ^1H resonance is lost in ^2H_2O due to exchange, but it can be selectively observed in difference spectra before and after a change of solvent from ^1H_2O to ^2H_2O. The re-equilibration of the proton with the solvent may be slow, allowing the kinetics of the process to be studied. *Specific isotopic substitution* has also been used in conjunction with difference spectroscopy where other methods for resolving and assigning resonances have failed. In a few cases it has been possible to prepare enzymes with specific isotopically substituted amino acids and thus selectively study their resonances and their role in ligand or substrate binding.

An important group of special methods are the *double irradiation* methods which are used to simplify spectra, to enhance the intensity of specific resonances, and to provide structural information. The relatively common technique of *spin–spin decoupling* serves as a typical example. Consider an NMR line from a nucleus A ($I=\frac{1}{2}$) which can interact with another nuclear spin (X) of $I=\frac{1}{2}$. If the lifetime of the two spin states of X is sufficiently long, then the resonance A senses these independently and A is split into a doublet as we saw before. If X is now strongly irradiated at its resonant frequency, rapid transitions are induced, the lifetime is reduced, and A detects only an average magnetic field: the doublet collapses. Only a few resonances in a complex NMR spectrum will generally be affected at a given irradiation frequency because spin–spin couplings are restricted in range. A different type of double irradiation technique involves keeping the measuring frequency constant while the irradiating frequency is scanned. This leads to a greatly simplified spectrum because changes in absorption at the measuring frequency only occur when the irradiating frequency passes through a resonance which is coupled to the measured resonance (an approximate analogy is the case of fluorescence excitation spectra, which may be much simpler than the corresponding absorption spectra—see pp. 74–79). The technique is called *internuclear double resonance* (INDOR). Other forms of double irradiation are electron–nuclear double resonance (ENDOR) (p. 301) and nuclear Overhauser enhancement (NOE).

Nuclear magnetic resonance has also specialized in the direction of studying nuclei other than ^{1}H. Carbon-13, ^{19}F and ^{31}P NMR all suffer from a lower sensitivity than ^{1}H NMR (Table 6.1) but have specific advantages over ^{1}H NMR in certain applications. Phosphorus-31 NMR is particularly useful for studying nucleotides and phosphorylated metabolic intermediates, and gives spectra which are often simpler and easier to interpret than ^{1}H spectra (p. 272, Examples 8 and 9; Fig. 6.13). Carbon-13 NMR is potentially very powerful for the smaller biomolecules, with large chemical shifts and small spin–spin couplings (p. 272, Example 8 and Fig. 6.12), but the sensitivity is extremely low without artificial isotopic enrichment. Fluorine-19 NMR spectra also have large chemical shifts which have proved very useful in detecting interactions between specifically fluorinated enzymes and substrates. Finally, for those magnetic nuclei with $I>\frac{1}{2}$ (including ^{2}H, ^{14}N, ^{17}O, ^{23}Na, ^{25}Mg, ^{35}Cl and ^{39}K) a different type of relaxation mechanism to that already discussed assumes importance because the nuclei possess an *electric quadrupole moment*.

This relaxation is sensitive to the environment and provides opportunities for studying the binding of ions such as Na^+ and Cl^- to macromolecules.

The Types of Information Available from NMR in Biological Systems

These range from relatively simple information (which may nevertheless be inaccessible to other techniques) to relatively complex structural and dynamic information requiring a detailed knowledge of NMR theory. The information content of NMR has been arbitrarily divided into five broad categories but any one study usually gives a variety of information.

Detection and identification

Nuclear magnetic resonance is an established tool of the chemist for "fingerprinting" or determining the identity of a newly synthesized compound. The spectral characteristics of line intensity, δ, and J are clearly interpretable in small molecules in terms of the number of each kind of chemical group, proximity, bond-angles, and isomers present. This can be extended to biomolecules such as nucleotides, hormones, peptides, sugars and metabolic intermediates, which can be detected *in vivo* if they are sufficiently mobile to give narrow lines and are present at a sufficiently high concentration, thus providing an analytical technique of unique value (see p. 275, Example 9).

Monitoring changes

Nuclear magnetic resonance can be used empirically without knowing the mechanism of spectral changes to determine concentrations, measure time-courses or investigate equilibria just like any other spectroscopic technique. Ideally, the spectral changes should correspond to some biochemically meaningful change, e.g. the binding of the substrate to an enzyme (see p. 269, Example 6) or of a drug to a membrane, they should be of suitable magnitude, and the number of components involved in the system should be defined since it is often difficult to treat more than a two-state system.

Probing the general environment

Nuclear magnetic resonance spectra are sensitive to molecular motion because of exchange phenomena and because, in general, spin–spin relaxation is favoured by slow rotational motion (p. 250). Under ordinary solution conditions, rapid molecular motion usually results in narrow NMR lines, and slow molecular motion results in broad NMR lines. Nuclear magnetic resonance spectra are also perturbed by environmental polarity, pH, the proximity of aromatic rings, paramagnetic ions, and other effects. These effects have been used to study conformation changes in proteins (see below Examples 1, 3, 4, 5) and membranes and can often be used to probe very localized variations in the internal environment of macromolecules or detect specific interactions between groups in a macromolecule.

Kinetic and dynamic information

Studies of molecular motion can be made quantitative in suitable cases. Exchange phenomena and relaxation theory give rate information on a range of molecular motions including chemical exchange (see p. 264, Example 3) rotational tumbling, and "flipping" motions between discrete molecular conformations. The temperature-dependence of the rate processes can then sometimes be used to determine thermodynamic characteristics of molecular reactions and conformation changes (i.e. free energy, enthalpy and entropy changes).

Structural information: distances and orientations

In favourable cases, NMR can give structural information *in solution* comparable in precision to that obtained in the solid state by X-ray crystallography. Paramagnetic relaxation probes incorporated in a molecule give "through-space" information on interatomic distances (Examples 2, 3, 5, below) Paramagnetic shift probes and intrinsic ring-current shifts give distance and directional information. Spin–spin coupling constants depend on interactions transmitted through chemical bonds and give bond angles (p. 263, Example 2). The most rigorous structural studies have inevitably been made on small biomolecules such as nucleotides and oligopeptides, but many useful results have also been obtained on the active centres of larger molecules such as enzymes and antibodies.

Applications of NMR to Biological Systems

The intention below has been to provide a general survey of some of the more important areas of application and of the types of problem that NMR can help to solve, and to show the different approaches which have to be used with different biological systems (treated roughly in order of the complexity involved). As a first introduction to the subject, where the emphasis must be on brevity and readability, it has been thought unnecessary to attempt to be comprehensive or to provide sufficient detail on individual examples to enable the reader to carry out experiments and verify the conclusions himself. The present text could not possibly do justice in this respect and the interested reader who may at some point wish to carry out experiments himself, will necessarily have to consult some of the excellent specialized texts available of which the reading list at the end of this chapter provides a small selection.

Example 1. General features of polypeptide and protein NMR spectra: native and random-coil spectra

The most successful direct NMR studies on proteins to date have been with small polypeptides or with small, monomeric, extracellular enzymes such as lysozyme or ribonuclease. With a few notable exceptions, the problems of sensitivity, resolution and assignment set a practical limit of about 20 000 on the molecular weight of proteins that can be directly studied by high-resolution NMR even with special techniques such as signal accumulation; the difficulties of obtaining sufficiently high concentrations and resolution problems increase with molecular weight. The resolution problems arise not only from the increased number of overlapping resonances but also from actual increases in line width due to spin–spin relaxation which is more efficient in general in rigid and slowly rotating molecules (p. 250). In addition, the smaller proteins often have a known crystallographic structure, which greatly assists spectral interpretation.

Figure 6.3 shows ^{1}H NMR spectra of ribonuclease at 220 MHz in its native state (b) and in its thermally denatured state (c) which may be taken as typical of small proteins. In the denatured state (c) the resonances are well-resolved and the total spectrum is approximately a weighted summation of the spectra from all the constituent amino-acids present as a random-coil polypeptide. The spectrum is dominated in the high-field region by ^{1}H resonances from aliphatic side chains (CH_3 at $\delta \approx 1$ and CH_2 at $\delta \approx 2$–4 parts per million) and in the low-field region by aromatic side-chain resonances (Phe and Tyr at $\delta \approx 7$, His at $\delta \approx 8$). In the native state (b) the spectrum is much less well resolved and many resonances are shifted. The differences are caused by a variety of intramolecular interactions in the secondary and tertiary structure such as spin–spin broadening, ring-current shifts and H-bonding. The shifts mean that individual amino-acids of the same type in a protein can often be distinguished.

The *primary assignment* of resonances in protein spectra, i.e. to a particular kind of amino acid, is done by a variety of methods mainly (i) comparing δ and

J-values with those from small peptides and model compounds, (ii) spectral simulation, (iii) observing effects of temperature, which detects H-bonding, and pH, which detects ionization of histidines, (iv) spin–spin decoupling, (v) solvent exchange, ^1H_2O to ^2H_2O, which detects amide protons, (vi) ligand-binding effects and (vii) selective chemical modifications of residues. A complete primary assignment of the resonances has now been achieved in many peptides and nucleotides but not as yet in any protein of the order of complexity of ribonuclease. *Secondary assignment*, i.e. to specific residues in the sequence, also utilizes these methods, but is possible in far fewer cases.

The denaturation of ribonuclease by a variety of agents is reversible, and this has enabled the pathway of the folding process to be studied in detail by NMR.

Example 2. Small peptides and other biomolecules

A cyclic structure or repeating structure places constraints on possible conformations and H-bonding schemes in a number of small peptides of biological importance and, together with the small number of resonances involved, this has enabled complete solution structures to be determined entirely by NMR in favourable cases. Nuclear magnetic resonance can thus attempt to answer questions such as: what is it about the conformation of a peptide hormone or antibiotic in solution which causes such potent biological activity and why does a slight modification such as a change in one amino acid sometimes cause a complete loss of activity?

The case of valinomycin, which transports the K^+ ion across membranes by forming a 1 : 1 complex with it, is a typical illustration of this. Valinomycin is a cyclic depsipeptide with a triply repeated structure:

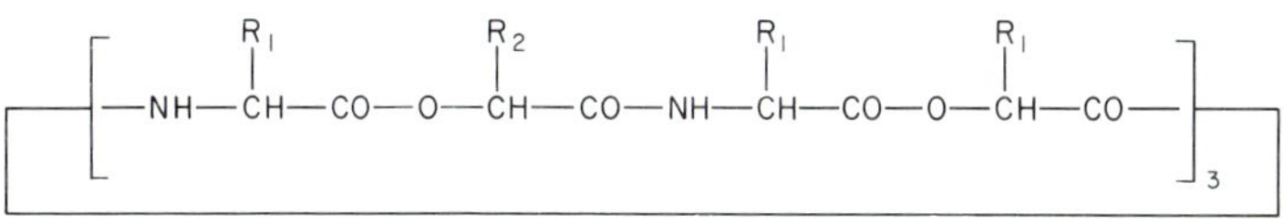

NMR, in conjunction with the techniques of IR and ORD, showed that the specificity for K^+ depended on the conformation in solution which in turn depended on the polarity of the medium. In a hydrophobic solvent, the six NH-resonances in ^{1}H NMR spectra were more or less equivalent whereas in a polar solvent, three of these were chemically shifted. This indicated that in the hydrophobic solvent all six NH and CO groups were internally H-bonded to each other whereas in the polar solvent three were H-bonded to solvent molecules. The studies also showed that adjacent amide and ester groups were in a *trans* conformation. Because of the simple repeating structure, the solution conformation was narrowed down to one of two possibilities, but only

NMR could distinguish between them. This was done by a study of spin–spin coupling constants (p. 246) in the amide linkage, which also provided further information on the side-chain orientations, thus completely defining the conformation in solution. Further NMR studies on the 1 : 1 complex with K^+ showed why the binding was so critically dependent on conformation: the binding is caused by interaction of K^+ with the six ester carbonyl groups, and the central binding site is protected from disruption by the solvent by the arrangement of hydrophobic side-chain groups.

The complete conformation of some nucleotides in solution has also been determined by NMR, by using lanthanide ions, which have a specific binding site, as paramagnetic relaxation probes (giving internuclear distances) or as paramagnetic shift probes (giving distance and directional information). A very simple illustration of the use of a relaxation probe (in this case Mn(II)) to determine *relative* distances of protons from a bound paramagnetic centre is shown in Fig. 6.2. In this experiment, very low concentrations of Mn(II) ions are added to a large concentration of AMP (18 mM) and the binding equilibrium is such that at these concentrations almost all the metal added is complexed to AMP. Thus the fraction of AMP bound to Mn(II) can easily be calculated (e.g. 1 part in 2000 for the middle curve). The bound and free forms of AMP are in fast exchange (p. 252) so that at any Mn(II) concentration, the linewidth for each proton is a weighted average of the linewidths in the free and bound forms of AMP. It is obvious that the contribution to the linewidth due to Mn(II) will increase linearly with the Mn(II) concentration under these conditions and the linewidth in the bound state (or T_2) can be calculated for each proton. Use of equations such as eqn (6.9) can then give the absolute values of the interatomic distances, but for this, τ_c (and hence details of the molecular motion) must be known. However since the relaxation mechanism is the same for each proton we can calculate *relative* distances quite easily since (line broadening) $\propto r^{-6}$.

In the example given, although H-8 is only about 30% closer to the bound Mn(II) than H-2 and H-1′, the line-broadening is about an order of magnitude greater. This example illustrates the extreme distance-sensitivity of relaxation effects and also why chemical exchange is such a useful phenomenon since even with only 1 in 400 sites bound by Mn(II), the H-8 resonance is already lost by broadening.

Example 3. High-resolution NMR studies of lysozyme: an enzyme of known three-dimensional structure

Lysozyme was the first enzyme to have its complete three-dimensional structure determined by X-ray crystallography. The mechanism of substrate binding and catalysis is now known in considerable detail and several excellent accounts of this have been given. What more can NMR studies add, and what methods were used?

Lysozyme cleaves the glycosidic bond of its substrates by a general acid catalysis involving residues Glu 35 and Asp 52. Lanthanide metals were found to bind specifically between the negatively charged carboxyl groups of these residues and were used as broadening (relaxation) probes (e.g. Gd(III)) or shift probes (e.g. La(III)) for neighbouring residues. Using paramagnetic difference spectroscopy (p. 258) ^{1}H resonances from these residues were specifically observed, as shown in Fig. 6.9 for the high-field (methyl) region of the ^{1}H NMR spectrum of lysozyme. Figure 6.9(a) is the unperturbed spectrum, and Fig. 6.9(b) is a Gd(III) difference spectrum. The secondary assignment of resonances (i.e. to specific amino acids in the sequence) has now been done for about 20 of the 139 residues of lysozyme, including most of these perturbed residues near the active site. This enabled an excellent correlation to be made between crystallographically determined Gd(III)–proton internuclear distances and (line-broadening)$^{-1/6}$ (eqn 6.9) for the residues close enough to the active site (< 2 nm) to be studied. Similar correlations between crystallographic structure and NMR theory were made for paramagnetic shifts and ring-current shifts in selected resonances. Thus NMR has shown, as no other technique could, that the crystal structure of lysozyme is essentially preserved in solution, removing any doubts that this structure might not be relevant to biological function.

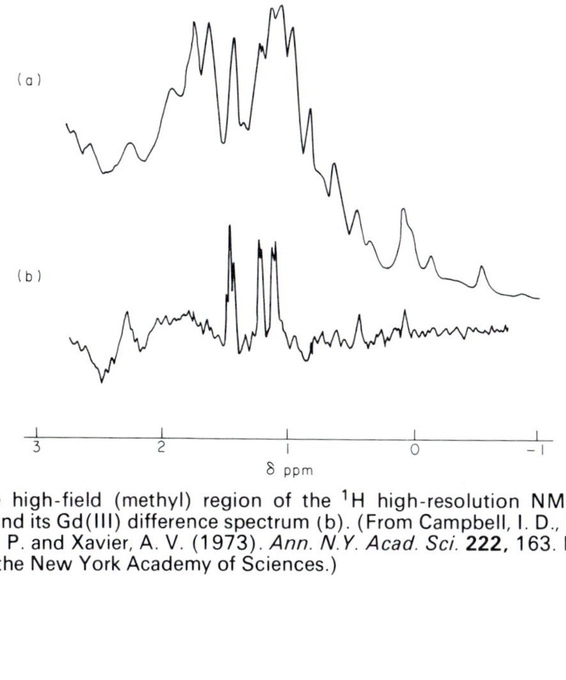

FIG. 6.9 The high-field (methyl) region of the ^{1}H high-resolution NMR spectrum of lysozyme (a) and its Gd(III) difference spectrum (b). (From Campbell, I. D., Dobson, C. M., Williams, R. J. P. and Xavier, A. V. (1973). *Ann. N.Y. Acad. Sci.* **222**, 163. Reproduced by permission of the New York Academy of Sciences.)

Having shown that the crystal and solution structures of lysozyme were essentially the same, NMR went on to highlight the small differences that do occur. Recent NMR studies appear to show that changes in the solution conformation in response to different treatments (e.g. inhibitor binding, temperature increase, or chemical modification) are generally more extensive than changes in the crystal. Thus, when Trp 108—a residue involved in substrate binding—is specifically modified, changes in resonances from several regions of the protein are detected in solution, but in the crystal the observed changes are localized to the vicinity of Trp 108/Glu 35. Such differences could reflect a real crystal constraint or they could simply mean that NMR is more sensitive than crystallography in detecting the same small changes since it is difficult to quantify NMR spectral changes in terms of the size of conformation change responsible. In either case, NMR is clearly a powerful tool for studying conformation changes in solution.

Finally, NMR has given dynamic information on conformation changes in lysozyme, unlike the time-averaged information from crystallography. An example of this is a conformation change induced in some inhibitors following binding to lysozyme, according to the scheme

$$E+I \rightleftharpoons EI_1 \rightleftharpoons EI_2$$

which is accompanied by a chemical shift change in the ligand resonances. The temperature-dependence of the ligand resonances was studied and found to be similar to that in Fig. 6.6(b); i.e. the detection of separate resonances from the two conformations depended on whether the two states were in fast or slow exchange. The study thus gave data for the rate of isomerization of the two states. The particular significance of this example is that similar isomerizations are often rate-limiting in enzyme catalysis; if I is a substrate analogue, valuable insight into the reaction mechanism may be obtained.

Example 4. High-resolution NMR studies of haemoglobin: an allosteric protein of known structure

Haemoglobin was the first allosteric protein to have its complete tertiary structure determined by X-ray crystallography. Its tetrameric structure, $(\alpha\beta)_2$, the structure of the haem group, and the sigmoidal O_2-binding characteristics will be familiar to most biochemists. It is the "prototype" allosteric system and has proved particularly amenable to study by spectroscopic methods, but the last word on the old controversy: "Does it conform to the Monod two-state allosteric model or the Koshland sequential allosteric model?" has yet to be said.

Haemoglobin is well suited to study by NMR. The Fe-atom acts as an intrinsic paramagnetic shift probe and the tetrapyrrole ring produces ring-current shifts. As a result several resonances from protons of interest near the active centre are outside the usual range of chemical shifts ($\delta = 0$ to $+10$ parts per million). A group of low-field resonances ($\delta = +10$ to $+25$) have been assigned to protons from the haem group and the His residue involved in Fe-binding, mainly as a result of studies with chemically modified haems. Another group of resonances at high field ($\delta = 0$ to -5) has been assigned mainly to ring-current shifted CH_3-groups in the protein. The haem group can exist in different oxidation and ligation states which can be distinguished by absorption spectroscopy and also by NMR, e.g. deoxy (Fe(II)), oxy (Fe(II)O_2), "met" (Fe(III)H_2O) and cyano (Fe(III)CN) (see p. 39). In this system, only Fe(III) is paramagnetic since Fe(II) occurs in its low(zero)-spin form (Chapter 7). Hence Fe(III) is the only state in which the shifted resonances can be observed. The paramagnetic Fe(III) can in turn exist in high- or low-spin forms according to the ligands present. Thus in methaemoglobin the high-spin form occurs (Chapter 7, Example 1) but in the following example the paramagnetic species is Fe(III)CN where the CN ligand causes the Fe(III) to adopt the low-spin state.

A hybridization study was designed with the object of answering the question: what is the effect of ligand binding on one subunit on the conformation of the other subunits? In this study the shifted resonances from α-subunits in the cyano- (Fe(III)CN) state were used to detect changes resulting from ligand binding at β-subunits in the Fe(II) state which lacked the shifted resonances. Cyanide was used as the ligand because its binding is essentially irreversible and unaffected by the ligation state of the other subunits. The oxidation and spin state of the α-haems thus remained constant throughout the experiments, and any changes in the NMR spectrum of the α-subunits could be unambiguously interpreted as being due to conformation changes associated with ligand binding at β-subunits. Spectra were observed from

(i) α(Fe(III)CN) subunits in isolation
(ii) α(Fe(III)CN)$_2$ hybridized with β(Fe(II) O_2)$_2$
(iii) α(Fe(III)CN)$_2$ hybridized with β(Fe(II))$_2$

The difference between spectra from (i) and (ii) was small indicating that association with other liganded subunits had little effect on the conformation of isolated liganded subunits. In (iii) a new type of spectrum from the α-subunits was seen as a result of deoxygenating the β-subunits. It consisted of a superposition of two spectra, one of which was the same as (ii). Furthermore, only these two basic types of spectrum could be obtained when the same experiment was repeated with α(Fe(II)) and β(Fe(III)). The conclusion drawn from these studies was that the whole tetramer had to exist in one of two conformational states regardless of the oxygenation states of individual subunits, i.e. there were no hybrids containing both conformations, and thus the haemoglobin molecule appeared to obey the allosteric model of Monod *et al.* The same series of experiments also showed that adding other allosteric effectors (such as phosphates and H^+) only caused an interconversion of the two states without producing further states (as expected on the Monod model) and incidentally, showed that the rate of exchange of conformational states must be slow to allow the observation of separate spectra (see p. 252).

Example 5. Immunoglobulin structure

Where the three-dimensional structure of a biomolecule is not fully known a combined approach using several techniques is usually the most profitable. This is illustrated by recent studies on the interactions between the mouse myeloma protein, MOPC 315, and its dinitrophenyl haptens. A myeloma protein was used rather than a conventionally raised antibody because it was available in larger amounts and (unlike the antibodies raised conventionally against a single antigen) was homogeneous. Proteolytically cleaved fragments consisting of the combined variable domains of the heavy and light chains were mostly used for NMR studies. X-Ray diffraction indicates that within these *variable* domains there is much homology of tertiary structure between different immunoglobulins: the high degree of specificity arises from three *hypervariable* regions within each of the heavy and light chains which together form the hapten binding site. The approach used was as follows.

1. The primary sequence of MOPC 315 was compared with sequences of two immunoglobulins of known crystal structure, with particular attention to differences in the hypervariable regions. A preliminary model was built incorporating the MOPC 315 hypervariable loops in the general framework of the other structures.

2. A series of spin-labelled dinitrophenyl haptens of graded lengths was used to probe the geometry and polarity of the combining site; their EPR spectra indicated that the site was a hydrophobic cleft with considerable rigidity and the overall dimensions $1{\cdot}1 \times 0{\cdot}9 \times 0{\cdot}6$ nm (see Chapter 7).

3. High-resolution NMR spectra were obtained (a) in the absence of haptens, (b) in the presence of diamagnetic haptens, and (c) in the presence of spin-labelled haptens. Paramagnetic difference spectroscopy, (b)−(c), showed which resonances were from residues near to the combining site. Similar studies showed which resonances were perturbed on binding diamagnetic haptens, (b)−(a). Many of these were the same, indicating that the conformation changes on binding haptens were localized near the binding site. Assignment of resonances to residues in or near the combining site was done by a variety of methods, e.g. pH difference spectroscopy for histidines, ring-current shift studies for aromatic residues (p. 246 and p. 258).

4. The results from the various studies were combined and a refined

structure was proposed, showing many of the interactions responsible for binding specificity. This type of study may thus help to answer some of the fundamental questions of immunology. How does the great diversity of specificity among antibodies arise from superficially similar structures? How does the binding trigger off further events in the immune response? Is the antibody molecule conformationally flexible, so that events such as complement fixation are triggered off by antigen binding via conformation changes or is the molecule relatively rigid, producing its response in a manner completely unlike allosteric enzymes?

Example 6. Studies on large enzymes: the indirect approach

Some of the systems of great potential interest, including most regulatory enzymes, are too large to give well-resolved NMR spectra. In these cases the only approach possible is to study NMR signals from small ligands and water molecules undergoing fast exchange between the bulk solution and some binding site of interest and hope that useful information about the binding site can be deduced. Nevertheless, many useful NMR studies have been made in this way, particularly with the kinases or phosphoryl-transfer enzymes and with metalloenzymes.

The kinases all require a divalent metal (normally Mg(II)) for activity, which functions by forming a complex with ATP. This complex can bind to the enzyme in one of several coordination schemes depending on the enzyme studied. In most cases Mn(II) can replace Mg(II) without loss of activity and thus act as a paramagnetic relaxation probe. Figure 6.10 shows the water

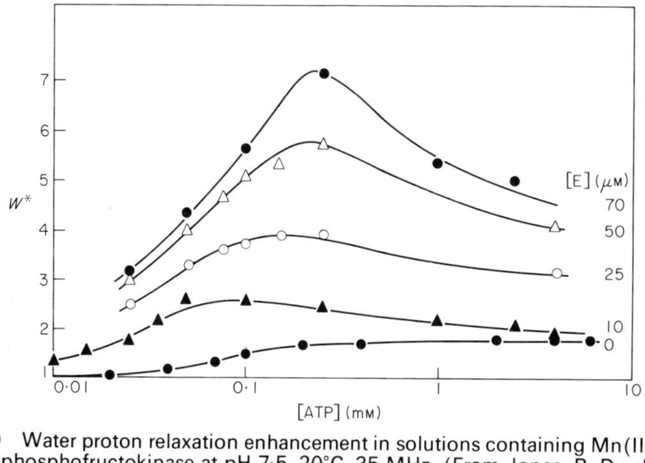

FIG. 6.10 Water proton relaxation enhancement in solutions containing Mn(II) (100 μM), ATP and phosphofructokinase at pH 7·5, 20°C, 35 MHz. (From Jones, R. Dwek, R. A. and Walker, I. O. (1972). *Eur. J. Biochem.* **28**, 74.)

proton relaxation enhancement in solutions containing Mn(II), ATP, and rabbit-muscle phosphofructokinase, a regulatory enzyme of glycolysis which catalyses the transfer of the terminal phosphate group of ATP to fructose-6-phosphate. The enhancement, W^*, is defined as the ratio of paramagnetic contributions to the relaxation rate in the presence and absence of the macromolecule, i.e.

$$W^* = \frac{\dfrac{1}{T_{1^*}} - \dfrac{1}{T_{1^*}(0)}}{\dfrac{1}{T_1} - \dfrac{1}{T_1(0)}} \tag{6.11}$$

where the asterisk denotes solutions containing the macromolecule and the $T_1(0)$'s denote control solutions containing no metal. The mechanism of relaxation enhancement by macromolecules was discussed earlier (p. 251). In this particular example, it is of interest that very little enhancement occurred unless ATP was added (Fig. 6.10), indicating that the substrate was required in order to coordinate the metal to the enzyme, i.e. the coordination scheme was E—S—M to a first approximation. The decrease in enhancement at high levels of ATP was due to competition between free ATP and Mn–ATP for the binding sites.

By analysing the enhancement data at different concentrations of Mn(II), ATP and enzyme it was possible to determine dissociation constants for the equilibria $M + S \rightleftharpoons MS$, $MS + E \rightleftharpoons MSE$, and $E + S \rightleftharpoons ES$, since each species containing M had its own characteristic enhancement and the observed enhancement was an average, weighted according to the equilibrium concentration of each species, i.e.

$$W^* = \frac{[M]}{[M_T]} + \frac{[MS]}{[M_T]} \cdot W_{MS} + \frac{[MSE]}{[M_T]} \cdot W_{MSE} \tag{6.12}$$

where $[M_T]$ represents the total metal concentration. This represents an empirical use of the technique. However, by using the fast exchange condition for the water molecules between the bulk solvent and each of the metal environments (free M, MS, and MSE) it was possible to obtain further information on the water relaxation rate in each bound site which could then be used in the theoretical equations (of which eqn (6.9) is a simplified version) to investigate in more detail the structure and molecular motion in the binding site.

In many cases, individual substrates in fast exchange between the bulk solution and a binding site can be studied by high resolution NMR. If a paramagnetic centre is also present, its relaxing effects on individual substrate resonances can be studied, and thus substrate-mapping studies *on the enzyme* are sometimes possible. In principle, mapping studies can be made quite detailed as shown schematically in Fig. 6.11, by utilizing the interactions between water, ligands, and different paramagnetic centres, but information is always restricted to small regions of the macromolecule. This type of approach is well illustrated by studies on phosphorylase and creatine kinase.

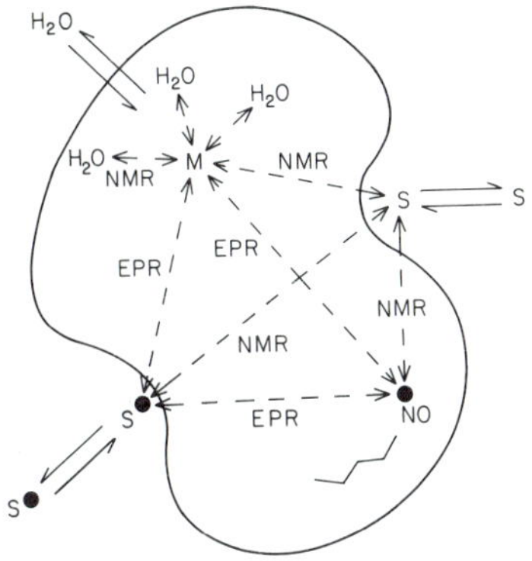

FIG. 6.11 Some of the interactions between binding sites on a large molecule which can be studied by magnetic resonance using paramagnetic probes.

Example 7. Long-stranded biomolecules

Chromatin, collagen, and muscle proteins have been studied by high-resolution ^1H NMR. The studies illustrate how only limited, but nevertheless often very interesting, information can be obtained in highly polymerized systems. *Chromatin* is a complex of DNA with a small group of basic proteins called histones and with other non-histone proteins and is commonly prepared from calf thymus. *In vitro* it behaves as a gel which undergoes condensation followed by re-expansion as the concentration of NaCl is increased. The histones are thought to be involved in conformation changes in chromatin thoughout the cell cycle which may in turn be involved in the regulation of cell division, and they progressively dissociate from chromatin as the NaCl concentration increases. Histone H-1 is the first to dissociate, and its dissociation has been studied by NMR in the range 0–0·5 M NaCl. In low salt, almost no NMR signal could be seen, and similarly with H-1-depleted chromatin in 0·5 M NaCl, but both isolated H-1, and chromatin in the presence of fully dissociated H-1 in 0·5 M NaCl, gave a similar well-resolved spectrum. The conclusion drawn from this was that the resonances observed came entirely from dissociated H-1 and that neither fully bound H-1 in low salt nor any of the other histones still bound to chromatin in 0·5 M NaCl were sufficiently mobile to give a resolved spectrum. In intermediate salt concentrations (0·1 M) where the chromatin is condensed, *part* of the NMR spectrum of H-1, due mainly to the apolar amino acids alanine and proline, appeared. At somewhat higher concentrations (0·25–0·5 M NaCl) where re-expansion of the gel and dissociation of H-1 occur, a further large spectral change occurred with the appearance of resonances from lysines. Histone H-1 has an unusual amino acid sequence in that the central part is rich in apolar residues whereas one end of the sequence is highly polar with many lysine residues. The conclusion was that dissociation of H-1 occurs in two stages: in condensed chromatin in 0·1 M NaCl, the central apolar region becomes detached from the main complex and thus somewhat mobile, while the lysine-rich end is still rigidly bound. In 0·25–0·5 M NaCl complete dissociation occurs with the release of the lysine-rich end. The physiological significance of these studies is that, since histone H-1 appears to be directly responsible for the condensation of chromatin *in vitro* and has also been implicated in the contraction of chromosomes before mitosis *in vivo*, the two changes may be directly related.

In *collagen* the fibres are insoluble under normal solution conditions as a result of close packing of polypeptide chains arranged in triple helices. At first sight this might seem to be unsuitable material for NMR studies because of excessive line-broadening in the rigid structure, but when the ^{1}H resonance from the *water of hydration* was observed its was found to be split into a doublet by spin-spin interactions, leading to the discovery that the water of hydration was present as H-bonded chains.

Example 8. Biological membranes and model membranes

Nuclear magnetic resonance is well suited to studying membrane structure and the number of applications in this field is second only to the applications in protein structure. Magnetic resonance is helping to answer the following types of question. How does the mobility and flexibility of the alkyl hydrocarbon chain vary along its length? How are different groups in the membrane structure orientated? Does the structure allow molecules in it such as proteins to diffuse in the plane of the bilayer? How does the membrane environment alter the properties of a bound protein and vice-versa? What are the effects of temperature, cholesterol, drugs and tumour-formation on membrane properties?

Model membranes are generally prepared from phospholipids, e.g. dipalmitoyl lecithin. They can be studied by ^{1}H, ^{13}C, or ^{31}P NMR and may be prepared in various ways, i.e. as unsonicated dispersions, in which the bilayers have random orientations and exist as large sheets which spontaneously form multilamellar structures, as orientated multilayers, or as sonicated vesicles which are relatively small structures of a single bilayer. Hydrogen-1 NMR spectra of *unsonicated dispersions* can be obtained in a few minutes but show very few resolved features, mainly the resonance from $-\overset{+}{N}(CH_3)_3$ in the choline head group. This is because the overall Brownian (tumbling) motion of the membrane is very slow and the relative rigidity of the structure favours relaxation processes and line-broadening. The resolved methyl resonances indicate that the group has rapid internal motion independent of the rest of the molecule. To perform ^{13}C NMR on the same material, signal accumulation for about four hours is required but the resulting spectra are better-resolved with several resonances.

With *sonicated vesicles* there is an improvement in resolution, and again ^{13}C NMR is superior. The limited improvement is brought about by the faster rotational motion of the vesicles, but the most dramatic effects on spectral resolution for both ^{1}H and ^{13}C are brought about by an increase in temperature (e.g. Fig. 6.12). The lines in these spectra were assigned by selective proton decoupling. There is a sharp transition at about 40°C; above this temperature the alkyl chains have, in effect, melted, although the polar head groups are still rigid (with the exception of the internal motion of CH_3-groups noted earlier) and the overall integrity of the vesicle is maintained. The linewidths, i.e. T_2-relaxation, for the alkyl ^{1}H and ^{13}C resonances are now dominated by rapid internal motion rather than the slower motion of the whole vesicle. By suitable pulse methods it was possible to simultaneously measure T_1 values for all the resolved ^{13}C resonances and this showed a clear "flexibility gradient" with the longest T_1 values and hence the least efficient relaxation and most rapid internal motion, near the apolar centre of the bilayer.

Phosphorus-31 NMR studies of model phospholipid membranes give a different type of information. The single resonance present shows chemical shift *anisotropy*, i.e. the measured value depends on the orientation of the molecule in the magnetic field. (Nuclear magnetic resonance spectral parameters frequently show a small degree of anisotropy, but this is usually averaged out by the rapid isotropic rotation normally present under the conditions for observing high-resolution spectra. This is in contrast to EPR (Chapter 7) where anisotropy is frequently larger and is often not averaged out by molecular motion. For this reason studies utilizing spectral anisotropy are much commoner in EPR than in NMR.) In *sonicated vesicles*, rotation is isotropic and fast enough to produce averaging, and a single narrow ^{31}P signal is seen. In the *unsonicated dispersion* overall tumbling motion is slow and orientations of the bilayer are random, but the polar head group giving rise to the resonance has internal motion which is anisotropic. The resulting spectrum shows a continuous spread due to an overlap of spectra from all possible orientations of the bilayer in the magnetic field. In the *orientated bilayer*, the spectra obtained depend on the orientation and the orientation-dependence can be used to define the anisotropic motion of the polar head group relative to the bilayer; it is found to rotate about an axis perpendicular to the plane of the bilayer.

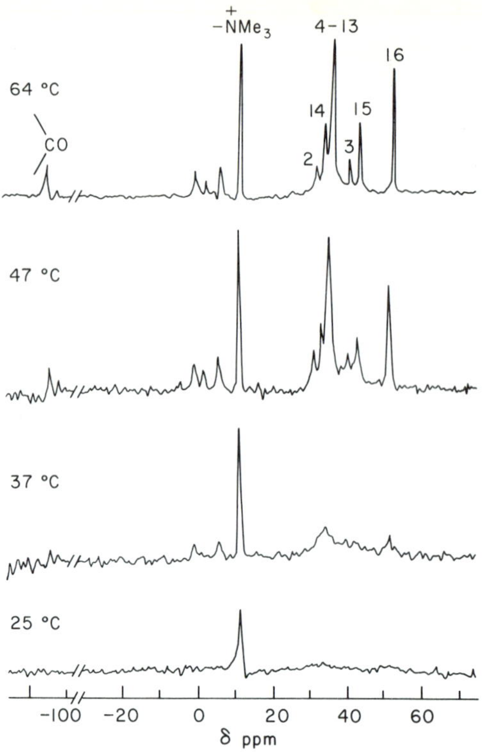

FIG. 6.12 Carbon-13 NMR spectra of sonicated vesicles of dipalmitoyllecithin at different temperatures. (From Y. K. Levine, N. J. M. Birdsall, A. G. Lee, and J. C. Metcalfe (1972). *Biochemistry* **11**, 1416.)

The most interesting developments in NMR of membranes from the biological point of view have been the extension of phospholipid model membrane studies to include the effects on structure and mobility of incorporating other components into the membrane, especially different saturated and unsaturated lipids, sterols, drugs and anaesthetics, and isolated membrane proteins. It has been found, for instance, that the lipid composition and cholesterol content of a membrane act together in a rather complex way to alter the mobility of the bilayer. In addition, a variety of intact biological membranes have now been studied in the form of sonicated vesicles, but high-resolution spectra are more difficult to obtain than in model membranes. The $-\overset{+}{N}(CH_3)_3$ ^1H resonances are not generally seen and this is attributed to interaction of the polar head group with membrane proteins. Comparative studies have suggested that in some tumour cells, the cell membrane proteins are more mobile than in normal cells. Phosphorus-31 NMR and spin-labelling studies failed to reveal any differences in membrane fluidity as a whole, but it was found that interactions between the polar head group and membrane proteins appeared to be weakened in the transformed cells.

Example 9. Metabolic studies on intact tissues

This is a recent development in NMR which promises to provide unique information of physiological interest. In suitable tissues, concentrations of individual metabolites are large enough to be studied by ^1H or ^{31}P NMR. If the metabolites are small, mobile molecules with well-characterized spectra, the signals can often be resolved, identified and interpreted even in complex biological systems. Information can be obtained on (i) concentrations, (ii) compartmentation, (iii) complexing of metabolites with other ions or molecules, (iv) pH of the intracellular environment and (v) time-courses of metabolic changes.

In the earliest studies, metabolic changes were monitored in excised mammalian skeletal muscles and hearts without taking special measures to maintain sample viability. Phosphorus-31 NMR signals were detected from ATP, P_i, phosphocreatine and sugar phosphates. The signal from P_i could be used as an intracellular pH indicator because P_i has an ionization with a pK in the physiological range which is accompanied by a chemical shift change. The intracellular pH was found to be 7.2, decreasing with time and after activity. Changing the extracellular pH caused the P_i ^{31}P signal to split into two signals, only one of which was pH-sensitive; this was taken to mean that inorganic phosphate exists in two cellular compartments only one of which is readily accessible to the solvent. The time-dependence following excision demonstrated the ATP-buffering function of the creatine kinase reaction in muscle; levels of ATP and sugar phosphate were maintained nearly constant for a considerable period at the expense of creatine phosphate, but when the reserves of this were used up the ATP level fell sharply. The intracellular ATP:ADP ratio, important for metabolic regulation, could not be determined directly from NMR spectra, but was deduced from the creatine kinase equilibrium:

$$K_{eq} = \frac{[\mathrm{ATP}]\,[\mathrm{creatine}]}{[\mathrm{ADP}]\,[\mathrm{creatine\text{-}P}]\,[\mathrm{H}^+]}$$

Here, K_{eq} is known and the other concentrations can be directly measured by NMR. The metabolic studies by NMR confirmed results obtained from the traditional method of freeze-clamping tissue followed by chemical analysis.

More recently, specially constructed sample tubes have been used to maintain muscles in a physiologically viable state for many hours during which the effects of electrical stimulation, tetanization, recovery, poisoning and anaerobiosis have been studied. Because of the low sensitivity of ^{31}P NMR, a major problem was that of accumulating sufficient spectra without the physiological state of the muscle changing significantly. In studies where frog gastrocnemius muscles were stimulated repeatedly for one second every minute, the problem was avoided by poisoning the muscles with cyanide and N_2 to prevent oxidative recovery; thus the metabolic effects of stimulation could be observed over a long period and are shown in Fig. 6.13. The main effects are a decline in phosphocreatine levels and a corresponding increase in P_i and sugar phosphates. These experiments allow a correlation to be made between the metabolic state of the muscles and their tension response at any given time.

The adrenal medulla is another tissue which appears to be very suitable for ^{1}H and ^{31}P NMR studies. In this case, spectra are dominated by resonances from storage structures in the internal fluid of the chromaffin granules, opening up the possibility for investigating these structures *in vivo* and the mechanism of adrenaline storage and release.

Example 10. Larger objects and whole organisms

Hydrogen-1 NMR can be used as a simple non-destructive analytical tool to determine the water content of biological materials such as seeds and food products. A more useful technique than simply measuring the line intensity is to measure T_1 which can provide information on the relative amounts of free and bound water. Samples containing a large proportion of free water will have longer T_1 values than samples where the water is largely bound and hence subject to efficient relaxation. It is now known that: (i) T_1 varies considerably between different tissues as a result of differences in water content, e.g. between bone, muscle, adipose tissue and lung tissue, (ii) T_1 varies in different physiological states of the same tissue, e.g. it increases in lung oedema, and varies between normal, ischaemic and infarcted cardiac muscle and (iii) T_1 increases in malignant tumours compared to normal tissue.

The possibilities for using NMR as a diagnostic tool in medicine are considerably increased if it can be shown that the changes are statistically significant and can be localized *in vivo* without the need for excising tissue. Both conditions are fulfilled by a new development termed *zeugmatography* which enables spatially resolved images of organs or whole bodies in some NMR parameter such as T_1 to be obtained. Normal tissues can be used as an internal control for abnormal tissues in the same image. The principle of the method is as follows. If the magnetic field intensity over the sample can be varied in some known way, e.g. in a linear gradient, it is possible by a pulse of suitable frequency to excite only the spins lying in a particular region of the sample, e.g. in a single plane, and thus determine T_1. The irradiated region is

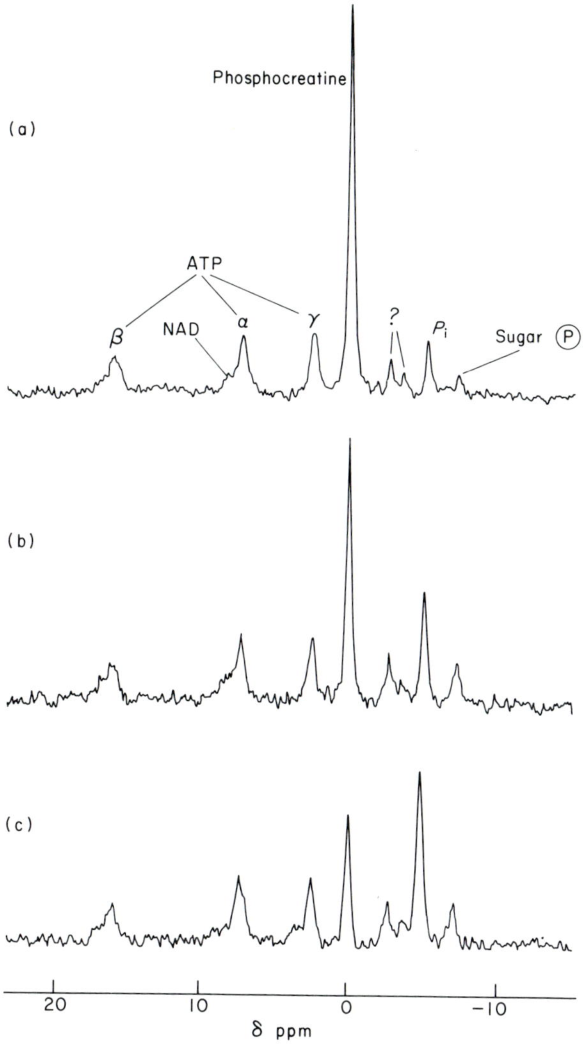

FIG. 6.13. Phosphorus-31 NMR spectra at 129 MHz of metabolites in poisoned frog gastrocnemius muscles at 4°C. Muscles were poisoned for 1 h with nitrogen and cyanide, then stimulated for 1 s at 1 min intervals. (a) Before stimulation; (b) spectra accumulated between 5–10 min after start of stimulation; (c) spectra accumulated between 15–20 min. (From M. J. Dawson, D. G. Gadian and D. R. Wilkie (1978). *Nature* (*London*) **274**, 861)

varied in a controlled way and computing procedures used to provide an image (in practice a T_1-contour map of a cross-section of the sample). The instrumental problems are considerable and mainly involve the provision of a suitably uniform magnetic field over a volume large enough to contain the sample, but whole-body studies, in which the gradual development of a tumour is followed, are now possible.

General References

Carrington, A. and McLachlan, A. D. (1967). "Introduction to Magnetic Resonance". Harper International Edition, Harper and Row, New York and London.

Dwek, R. A. (1973). Nuclear Magnetic Resonance in Biochemistry. "Applications to Enzyme Systems". Clarendon Press, Oxford.

Dwek, R. A., Campbell, I. D., Richards, R. E. and Williams, R. J. P. (Editors). (1977). "Nuclear Magnetic Resonance in Biology". Academic Press, New York and London.

Farrar, T. C. and Becker, E. D. (1971). "Pulse and Fourier Transform NMR". Academic Press, New York and London.

James, T. L. (1975). "Nuclear Magnetic Resonance in Biochemistry: Principles and Applications". Academic Press, New York and London.

Knowles, P. F., Marsh, D. and Rattle, H. W. E. (1976). Magnetic Resonance of Biomolecules. An Introduction to the Theory and Practice of NMR and ESR in Biological Systems. Wiley–Interscience, New York.

Roberts, G. C. K. and Jardetsky, O. (1970). Nuclear magnetic resonance spectroscopy of amino acids, Peptides and Proteins. *Adv. Protein Chem.* **24,** 448.

7 Electron Paramagnetic (Spin) Resonance

R. Jones

Introduction

Electron paramagnetic resonance involves the absorption of energy from a microwave electromagnetic field ($\sim 10^{10}$ Hz) by a system containing unpaired electron spins when placed in a strong static magnetic field. The pairing of electrons in orbitals is normally an energetically favoured process in the formation of ionic or covalent bonds and the observation of EPR is restricted to the minority of systems which are exceptions to this, i.e. free radicals and paramagnetic metals. Electron paramagnetic resonance is thus less universal in its applications than NMR.

Nuclear magnetic resonance and EPR are complementary techniques and cannot be treated fully in isolation; this is evident from the similarity of their theoretical base and from the large number of practical situations where the techniques have been used in conjunction. Nevertheless, each has its own specialized instrumentation arising mainly from the different technologies involved in the generation and transmission of microwave and radiofrequency radiation and each technique comes into its own in different types of application. In the last chapter we saw how the main strengths of NMR were in studying structure and conformation in relatively small molecules with some outstanding exceptions where the technique had been applied to large macromolecules and intact tissues. The main strengths of EPR are that it can give highly specific information on paramagnetic centres which are directly involved in catalysing biological oxidation–reduction reactions, it can give valuable information on transient free radicals which occur in many biological systems and it can often be used in complex systems or intact tissues. In addition, with the use of stable synthetic free radicals ("spin labels"), EPR can now be applied to a wide range of biological systems which lack an endogenous paramagnetic centre.

Electron paramagnetic resonance has greater sensitivity than NMR and fewer problems of resolution and assignment since in most studies only one, or at most, a few, paramagnetic centres are involved. This explains the greater suitability of EPR for studying intact biological systems and transient effects. Nevertheless, despite the relatively small number of signals observed in an EPR spectrum, the technique is in some ways inherently more complex than NMR. There are two main reasons for this. First, many of the paramagnetic centres of interest have more than one unpaired electron(electron spin quantum number $S > \frac{1}{2}$) giving rise to a multiplicity of energy levels and therefore of possible transitions giving rise to resonance. This is in contrast to the magnetic nuclei commonly encountered in biological systems (^{1}H, ^{13}C, ^{19}F, ^{31}P) which all have $I = \frac{1}{2}$ and only one possible transition. Secondly, the electron interacts much more closely with its chemical environment (i.e. other electrons in orbitals) than does the nucleus, and may also be delocalized over different atoms. This means that electron spin relaxation times in a chemical environment are usually very short, and other spectral parameters usually deviate considerably from the values found in a free electron. In addition, because most electronic orbitals have directional character, the interactions are directional and the spectra are *anisotropic*, i.e. the measured values of EPR spectral parameters are usually strongly dependent on the orientation of the molecule in the applied magnetic field. The result of these effects is that the EPR spectrum from a single paramagnetic centre may be a complex overlap of several lines showing broadening and anisotropy.

The extra information contained in all but the simplest EPR spectra is usually put to good use in biological or biochemical studies. The simplest EPR spectra are those from the transient free radicals generated by irradiation of biological tissues and during a variety of biochemical reactions. The EPR spectra individually do not usually contain a great deal of information but are used in an empirical way to monitor the changes in complex systems. The biological processes which have been studied in this way include photosynthesis, the mitochondrial respiratory chain, and radiation damage of nucleic acids. More complex EPR spectra are seen when nitroxide spin-labels are incorporated into macromolecules and where spectral anisotropy is used to provide information on molecular motion, orientation and environmental polarity. This approach has proved extremely useful in studying the conformation changes of proteins and membranes. Finally, the most complex EPR spectra are those obtained from transition-metal ions where frequently $S > \frac{1}{2}$

and there is a good deal of spectral anisotropy. Interpretation problems may be considerable, but in favourable cases structural information (e.g. the orientation of haem groups in haemoglobin) or information on the chemistry of the metal relevant to catalysis (e.g. in haem-proteins, iron–sulphur-proteins, or copper-proteins) may be obtained. Thus although EPR appears at first sight to be limited in scope compared to NMR, we see on closer acquaintance that it is well suited to studying many of the more interesting biochemical systems. Most of the systems mentioned above are illustrated in the examples given at the end of this chapter.

Theoretical Principles: Spectral Characteristics

The basic resonance condition for EPR is analogous to that for NMR (eqn 6.4), i.e.

$$h\nu = \Delta E = g\beta \boldsymbol{B}_0 \tag{7.1}$$

where g is a dimensionless number equivalent to the nuclear g-value (g_N) which determines the resonance position and β is the unit of electron magnetic moment, the Bohr magneton. The value of $\beta = 1836\ \beta_N$ (the nuclear magneton) and $g \approx g_N$, hence electron magnetic moments are greater than nuclear magnetic moments by a factor of $\sim 10^3$. Electron paramagnetic resonance experiments are carried out at about ten times smaller magnetic fields than NMR experiments, so that EPR frequencies ($\sim 10^{10}$ Hz) are ~ 100 times greater than NMR frequencies.

The timescale of EPR

The theoretical treatment and significance of relaxation processes and exchange processes in EPR are much the same as for NMR (see pp. 239–242 and pp. 248–252) in principle, but in practice there are large differences between the two techniques which make relaxation and chemical exchange much less useful subjects of study in EPR. First, the unpaired electron is more exposed to its chemical environment ("the lattice") than is a nuclear spin, leading to more efficient relaxation mechanisms. In free radicals, *electron delocalization* is very common and in transition metal ions *spin–orbit coupling* is an important mechanism for the loss of energy from the spin system. Electron paramagnetic resonance relaxation times are thus often too short (10^{-7}–10^{-8} s) to be accurately measurable using the pulsed techniques common in NMR and are less studied. Secondly, EPR linewidths and splittings correspond to a much shorter timescale than NMR linewidths and splittings.

To illustrate this, consider a typical spin-label EPR line-splitting of about 30 gauss in a total field of 3000 gauss, i.e. about 1 in 100 for a resonant frequency of about 10^{10} Hz, and typical NMR shifts and splittings of about 1 in 10^6 for a frequency of about 10^8 Hz. The former corresponds to a frequency of about 10^8 Hz whereas the latter corresponds to about 10^2 Hz. Nuclear magnetic resonance approaches the short timescale of EPR only in special cases such as paramagnetic relaxation. Thus, a process may be in fast exchange as seen by NMR and slow exchange as seen by EPR and a rotational motion which is fast enough to average out anisotropy and give sharp lines in high resolution NMR may be "frozen" by the shorter "exposure time" of EPR, which detects a mixture of all the possible molecular orientations. This is well illustrated by the different pictures of molecular mobility in membranes obtained from NMR and from EPR studies. Each spectroscopic technique is sensitive to a particular timescale of motion. The information lost in high-resolution NMR by motional averaging is much utilized in EPR studies of molecular motion and orientation.

The EPR spectrum

Electron paramagnetic resonance spectra are generally obtained by a procedure similar to the continuous-wave method of obtaining NMR spectra, in which the absorption of energy from a continuous electromagnetic field is plotted against magnetic field strength. However, for instrumental reasons, the *first differential* of this absorption spectrum is normally measured and recorded in EPR spectra (see p. 299). Figures 7.1 and 7.2 illustrate the two main types of EPR spectrum encountered in biological studies: those from free radicals and from paramagnetic metals respectively. The corresponding integrated (absorption) curves are also shown. Figure 7.1 is the EPR spectrum from a nitroxide spin-label in dilute aqueous solution and Fig. 7.2 is representative of spectra observed from a number of copper-proteins. Let us now consider the main characteristics of EPR spectra taking these two very different kinds as examples.

The g-factor

Electron paramagnetic resonance spectra are characterized by similar parameters to NMR spectra (see pp. 243–248) except that EPR has no equivalent to the NMR chemical shift parameter δ. Instead, a single parameter, g, is used to specify the wide variations of magnetic field strength (at constant frequency) for resonance in different paramagnetic species. For an electron in free space, $g = 2{\cdot}0023$. Nearly all free radicals including the nitroxide spin-label (Fig. 7.1) have g-values very close to this. For transition-metal ions,

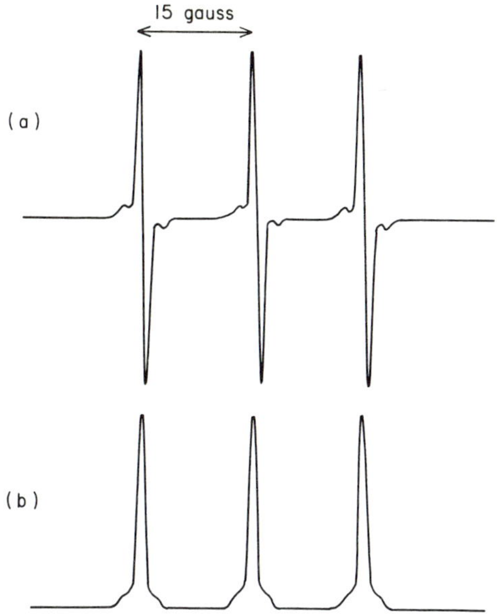

FIG. 7.1 The EPR spectrum of a nitroxide spin-label in dilute aqueous solution. (a). The three principal lines are due to hyperfine interactions of the unpaired electron with the three ^{14}N nuclear spin states ($I=1$) and the satellite peaks are due to interactions with the ^{13}C nuclear spin. The corresponding absorption spectrum (b) is also shown.

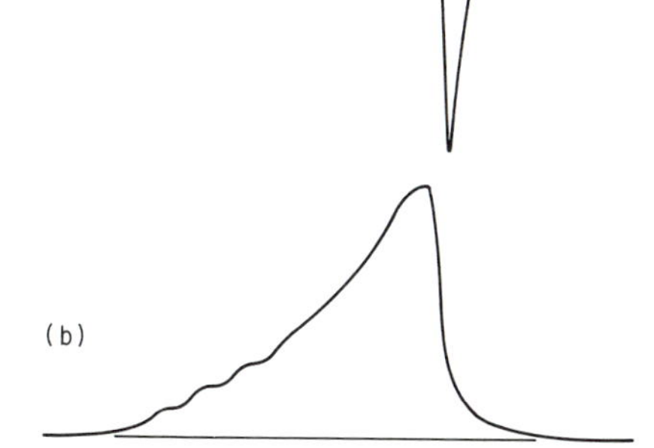

FIG. 7.2 Representative EPR spectrum of a Cu-protein in frozen solution (a), and the corresponding absorption spectrum (b). The overall shape of the spectrum is dominated by g-value anisotropy, and the subsidiary peaks are due to Cu nuclear hyperfine interactions ($I=\frac{3}{2}$).

g-values may deviate very considerably from 2 (Table 7.2) and in addition may be highly *anisotropic* (p. 287). The peculiar broad and asymmetric EPR line-shape seen in Fig. 7.2 is largely due to g-value anisotropy; the resonance position is dependent on the orientation of the molecule in the applied magnetic field, and the spectrum is a superposition of resonances from all orientations. It will be obvious from a glance at spectra such as Figs 7.1 and 7.2 that the range of magnetic field covered in recording EPR spectra is very much larger than for NMR spectra from a particular type of nucleus, particularly for paramagnetic metals. g-Value studies are of little importance in free radicals where very little variation occurs, but they can provide valuable information on the ligands around a paramagnetic metal and on their orientation (p. 303, Examples 1 and 2). A free radical with sharp EPR lines and $g \sim 2{\cdot}0$ is generally used as a g-value marker for more complex spectra. The most commonly used is diphenyl picryl hydrazyl (DPPH), $g = 2{\cdot}0036$.

Fine structure

Several degrees of fine structure may occur in EPR spectra. One type occurs only in paramagnetic systems with $S > \frac{1}{2}$ and a multiplicity of energy levels resulting from the ligand field. Transitions among these levels may give rise to resonances at several magnetic field values, but this type of fine structure is not usually very important in biological EPR spectra. *Hyperfine structure* is a very commonly encountered form of line-splitting in EPR spectra and is due to the interaction of the electron spin (S) with a spin on the same nucleus (I). In general, $(2I+1)$ lines of equal intensity are produced because each orientation of the nuclear spin in the applied magnetic field has an equal probability. For free radicals the nuclei most likely to give hyperfine splitting are ^{1}H ($I = \frac{1}{2}$) and ^{14}N ($I = 1$). The three equally spaced lines of equal height in Fig. 7.1 are due to the interaction of the nitroxide unpaired electron (which is mainly localized in the nitrogen 2p orbital) with the three allowed orientations of the ^{14}N nuclear spin. In Fig. 7.2 the subsidiary peaks are due to the interaction of the unpaired electron with the Cu nuclear spin ($I = 3/2$). In Mn^{2+} EPR spectra ($I = 5/2$) the dominant feature is six equally spaced lines of equal intensity due to the nuclear hyperfine interaction.

Superhyperfine structure is the term sometimes used for the narrower splittings arising from the interactions of the electron spin with spins of nuclei other than the one on which the electron normally resides. Most electrons are delocalized to a greater or lesser extent and so the distinction between this and hyperfine structure is somewhat arbitrary. Examples of line-splittings of this type are the small satellite lines seen in Fig. 7.1 due to the ^{13}C nuclear spin (natural abundance $1{\cdot}1\%$) of C atoms adjacent to the nitroxide group, and the narrow splittings sometimes detected in spectra like Fig. 7.2 due to the delocalization of unpaired electrons from transition metals onto ^{14}N ligands. As with nuclear hyperfine structure, when an electron interacts with n equivalent nuclei of $I=\frac{1}{2}$ (e.g. ^{1}H) there are $n+1$ equally spaced lines with intensities given by a binomial distribution. Some simple examples of hyperfine and superhyperfine structure are given in Table 7.1; note the difference in the three-line spectra produced when $I=1$, $n=1$ and $I=\frac{1}{2}$, $n=2$.

Nearly all of the useful information in EPR studies of free radicals and spin-labels is derived from a study of hyperfine structure. In free radicals (p. 307, Examples 3 and 7) the splitting pattern observed can often provide clues as to the extent of delocalization of the unpaired electron and the equivalence of different groups of protons in the structure and thus sometimes distinguish

TABLE 7.1 Some common hyperfine splitting patterns seen in EPR spectra

	Nucleus	I	Number of equivalent nuclei (n)	Intensity ratio
$>\dot{N}O$	^{14}N	1	1	1:1:1
Mn^{2+}	^{55}Mn	$\frac{5}{2}$	1	1:1:1:1:1:1
$>CH—\dot{C}<$	^{1}H	$\frac{1}{2}$	1	1:1
$—CH_2—\dot{C}<$	^{1}H	$\frac{1}{2}$	2	1:2:1
$CH_3—\dot{C}<$	^{1}H	$\frac{1}{2}$	3	1:3:3:1

between different isomeric structures. The magnitude of the splitting, or *hyperfine splitting constant, A* (in Hz or gauss), is sensitive to the spin density of the electron on any given atom. In nitroxide spin-labels the information obtained from hyperfine structure is rather different. Here although the unpaired electron is essentially localized on the ^{14}N atom, there are small increases in splitting constant as the polarity of the medium increases, and so one of the principal uses of spin-labels is to probe different hydrophobic and hydrophilic molecular environments. In addition, the hyperfine interactions of spin-labels are anisotropic and this feature is much used to study molecular motion (p. 289, and pp. 309–315, Examples 4 and 5).

The most useful EPR parameters are the g-values and the line-splittings, by contrast with NMR where the chemical shift, line splittings and relaxation times provide most of the useful information.

Linewidths and relaxation times

All EPR lines have a finite width (in Hz or gauss) related to the transverse relaxation time T_2 as in NMR. However the molecular mechanisms which dominate relaxation differ considerably between NMR and EPR and between the solid and liquid state. In Chapter 6 we saw that nuclear spin-lattice (T_1) relaxation in solution is dominated by magnetic noise and is extremely inefficient except under a small range of conditions where the molecular motion corresponds to the nuclear frequency, whereas for T_2 additional relaxation mechanisms are present in the limit of slow motion (p. 250). The result of this is that T_2 increases continuously with temperature giving narrower lines at high temperatures, whereas T_1 passes through a minimum. In EPR, much more efficient relaxation mechanisms are invariably present, giving short relaxation times and (where single homogeneous lines can be observed) broad linewidths compared to NMR. The linewidths in Fig. 7.1 are essentially due to relaxation and are about 10^7 Hz; by contrast, high-resolution NMR lines may only be about 10 Hz wide. Relaxation is most efficient of all in some transition-metal ions, e.g. Fe(III), (p. 303, Example 1) and here linewidths may be so great at normal temperatures that signals are almost unobservable. The reason for this is that spin–orbit coupling (p. 295) provides a very direct spin–lattice relaxation mechanism not paralleled in NMR, and this relaxation process also causes line broadening by the uncertainty principle (p. 3). This spin–lattice interaction decreases as the thermal motions of the lattice are reduced, and hence in this case it is possible to obtain narrower lines by decreasing the temperature.

In many EPR spectra, linewidths cannot be studied directly because of a broad overlap due to spectral anisotropy and other effects, and pulse methods are not readily applicable. Consequently the approach often used is to measure other spectral parameters such as anisotropic g-values as accurately as possible, then computer-simulate the spectrum by making reasonable assumptions about linewidth and shape. It may be important to know EPR relaxation times when the effect of the paramagnetic centre on relaxation in another paramagnetic centre or in a neighbouring nucleus is being studied (p. 251).

Area under the integrated curve

The area under a single EPR absorption curve is proportional to the rate of absorption of energy from the microwave field, which in turn depends on the total population of unpaired spins, the separation of the energy levels involved, and the net transition probability due to the difference in the energy level populations. Thus, as with NMR, if suitable precautions are taken (constant instrumental conditions, constant temperature, no saturation of the signal) studies may be made quantitative. In some simple cases the use of peak-to-peak heights of derivative spectra as a direct measure of concentration may be justified.

Special Features of EPR Spectra

Spectral anisotropy

The simple resonance condition expressed in eqn (7.1) applies strictly only to an isolated electron and is of little practical interest to a biologist or biochemist. Equation (7.1) and, to a lesser extent, the NMR equations (see pp.239–240) must be modified in two ways to take account of the spectral anisotropy and hyperfine structure which arise from the existence in molecular systems of electronic orbitals and other spins. First, in addition to the main energy term for the interaction of an electronic spin S or nuclear spin I with an applied magnetic field $\boldsymbol{B}_0$, we have to include terms for the minor interactions such as that between S and I which are responsible for EPR hyperfine structure. Secondly, we require a mathematical form for specifying anisotropic interactions, i.e. the fact that the interaction between two vectors may depend on the orientation of both of them relative to a set of coordinates fixed in the molecule and not simply on their orientation relative to each other as assumed in eqn (6.2). We will illustrate this further by reference to g-value anisotropy in EPR, giving at the same time some indication of its physical origin.

The energy of a nuclear or electronic spin in an applied magnetic field is classically given by the scalar product of the magnetic moment and applied field vectors, or the component of magnetic moment along the field direction multiplied by the field strength (e.g. eqn (6.2)). This component, however, is quantized (p. 240), so that the energy of interaction becomes (for the nuclear spin)

$$E = -\gamma \hbar\, I_z\, \boldsymbol{B}_0 = -g_N\, \beta_N\, I_z\, \boldsymbol{B}_0 \tag{7.2}$$

where I_z is the component of nuclear spin along the field direction. The equation for EPR is analogous except that β and β_N have opposite signs, and hence

$$E = g\, \beta\, \boldsymbol{S}_z\, \boldsymbol{B}_0 \tag{7.3}$$

For resonance transitions $\Delta S_z = \pm 1$, leading to eqn (7.1). However, in some systems the electron spin angular momentum is coupled to orbital angular momentum (see later in this section). The result of this can be viewed as a *new* effective spin vector quantized along the applied field direction or alternatively, as a *new* magnetic field vector along which the electron spin is effectively quantized. The spin and applied field no longer interact directly and the energy of interaction becomes a more complex function written as

$$E = \beta\, \boldsymbol{S}\, \boldsymbol{g}\, \boldsymbol{B}_0 \tag{7.4}$$

where $\boldsymbol{S}$ and $\boldsymbol{B}_0$ are again vectors but $\boldsymbol{g}$ is now a *second rank tensor*. It is a 3 × 3 matrix containing nine terms (g_{xx}, g_{xy}, g_{xz}, etc.) each of which gives the scalar interaction of one component of $\boldsymbol{S}$ with one component of $\boldsymbol{B}_0$ in an arbitrary set of coordinates fixed in the molecule. The expression is thus a scalar product like eqn (6.2) except that it now consists of nine terms instead of one. It can also be regarded as the scalar product of a new vector $\boldsymbol{S}\cdot\boldsymbol{g}$ with $\boldsymbol{B}_0$ or as the scalar product of $\boldsymbol{S}$ with a new vector $\boldsymbol{g}\cdot\boldsymbol{B}_0$. The mathematical form for representing A-value anisotropy (i.e. for the interaction of $\boldsymbol{S}$ with $\boldsymbol{I}$) is entirely analogous. The complete energy term for magnetic resonance is then a sum of such expressions including ones for the spin–spin interactions and is called a *spin Hamiltonian*. The term originates from quantum mechanics. Spin Hamiltonians and tensors thus provide a completely general and mathematical representation of the resonance condition but their use is not really essential at an elementary level since (as we see below for anisotropy) a simpler representation is often possible in practice. However, these relatively unfamiliar terms have been introduced since they are common in magnetic resonance texts.

By choosing the appropriate coordinate system relative to the molecular structure, the anisotropy of g and A can be completely specified by three *principal values*, e.g. g_{xx}, g_{yy}, g_{zz}. These are then simply the values observed with the magnetic field direction aligned along the three chosen molecular axes (the *principal axes*). If all three g-values are equal, the interaction is *isotropic*. This condition is approximately obeyed in free radicals and spin-labels ($g \approx 2{\cdot}00$). If $g_{xx} \neq g_{yy} \neq g_{zz}$ we have complete *anisotropy*. If $g_{xx} = g_{yy} \neq g_{zz}$, we have *axial symmetry* where by convention the z-axis is the symmetry axis. This situation is found in many transition metal–ligand complexes such as the haem group. The g-value measured when $\boldsymbol{B}$ is at an angle θ to the symmetry axis (Fig. 7.3(a)) is given by

$$g_\theta^2 = g_\parallel^2 \cos^2\theta + g_\perp^2 \sin^2\theta \tag{7.5}$$

where $g_\parallel$ (parallel to the symmetry axis) $= g_{zz}$ and $g_\perp$ (perpendicular to the symmetry axis) $= g_{xx} = g_{yy}$.

The same considerations apply to the hyperfine splitting tensor $\boldsymbol{A}$, which has three principal values A_{xx}, A_{yy}, A_{zz}. Some degree of anisotropy in the hyperfine splitting is usual in all EPR spectra. This is because spin–spin interactions, although essentially magnetic, are transmitted through electron clouds (p. 246). Thus any anisotropy in electronic orbitals (present in all systems apart from simple s orbitals) causes A-value anisotropy since interactions will be stronger along some molecular axes than along others. In free radicals the orbitals occupied by the unpaired electron usually have s- or σ-character and the electron is often delocalized, resulting in very little directional character and almost isotropic A-values. In nitroxide spin-labels there is approximately axial symmetry with $A_{zz} > A_{xx}$, A_{yy}. The anisotropy arises because the unpaired electron is localized in the nitrogen $2p_z$ orbital as shown

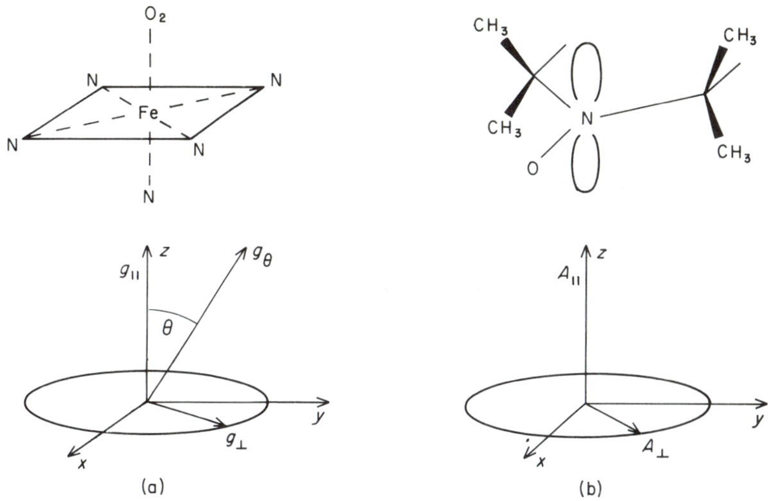

FIG. 7.3 Axial symmetry of the g-value in the haem group (a) and axial symmetry of the hyperfine splitting constant, A, in a nitroxide spin-label (b).

in Fig. 7.3(b). An equation analogous to eqn (7.5) holds for the case where A has axial symmetry.

Spectral anisotropy and molecular motion

From the point of view of anisotropy in EPR spectra we have to consider three principal situations: (i) completely ordered molecules with no molecular motion, e.g. a single crystal, (ii) randomly ordered molecules with no molecular motion or very slow motion, e.g. polycrystalline solid samples or slowly tumbling macromolecules in solution and (iii) molecules undergoing rapid tumbling motion in solution. For the case of g-value anisotropy in metalloproteins the second situation is usual. The peculiar line shape of Fig. 7.2 is typical of EPR spectra of many metalloproteins both in solution and in frozen samples. The tumbling motion is slow compared to the g-value anisotropy (in Hz) and the spectrum is a superposition of g-values corresponding to all possible orientations of the molecule in the magnetic field. There is a continuous range of absorption between $g_{\parallel}$ and $g_{\perp}$. In practice, spectra may extend beyond these extremes due to line-broadening and hyperfine splitting. The probability that the molecular symmetry axis will be perpendicular to the magnetic field direction is greater than the probability that it will be parallel to the field and hence the absorption increases towards the position corresponding to $g_{\perp}$.

For some metalloproteins single crystals large enough for EPR studies have been prepared. In these fortunate cases it is possible to measure the g-value directly as a function of the crystal orientation and obtain the principal values of the g-tensor, leading to information about molecular orientations and metal–ligand interactions (see p. 303, Example 1).

A-Value anisotropy has been most utilized in molecular motion studies with nitroxide spin-labels, which have been studied in all three situations above. The simplest case to consider is that of complete motional averaging of the anisotropy in free solution (Fig. 7.1). Here the rotational motion of the spin-label is fast ($\approx 10^{11}$ s^{-1}) compared to the hyperfine splitting (maximum value $A_{zz}=32$ gauss, or about 10^8 Hz). The observed hyperfine splitting is then an average of the three principal values and is called the *isotropic hyperfine splitting constant,*

$$a_0=\tfrac{1}{3}(A_{xx}+A_{yy}+A_{zz}) \tag{7.6}$$

In the nitroxide spin-labels, $A_{xx}=A_{yy}=6$ gauss, and $A_{zz}=32$ gauss, thus $a_0\approx 15$ gauss. Studies on orientated nitroxide spin-labels have been made in single host crystals and the three principal A-values have been observed directly by rotating the crystal in the magnetic field (Fig. 7.4(a),(b),(c)).

The case of randomly orientated immobilized spin-labels is rather more complex and gives rise to an unusual EPR spectrum (Fig. 7.5(d)). The absorption curve for this is also shown (Fig. 7.5(e)). The shape of these curves can be explained qualitatively as follows. A single randomly orientated immobile spin label will have a sharp three-line spectrum with a splitting somewhere between the extremes shown in Fig. 7.4 according to its orientation in the magnetic field. In a large population of such labels the middle lines will all coincide (assuming there is no g-value anisotropy) and thus the resultant

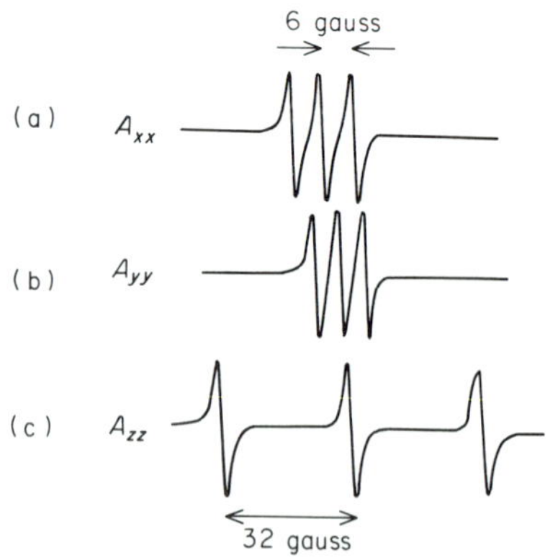

FIG. 7.4 (a), (b), (c) The three principal A values (A_{xx}, A_{yy}, A_{zz}) measured in a nitroxide orientated in a single host crystal with the magnetic field parallel to the x-, y-, and z-molecular axes respectively.

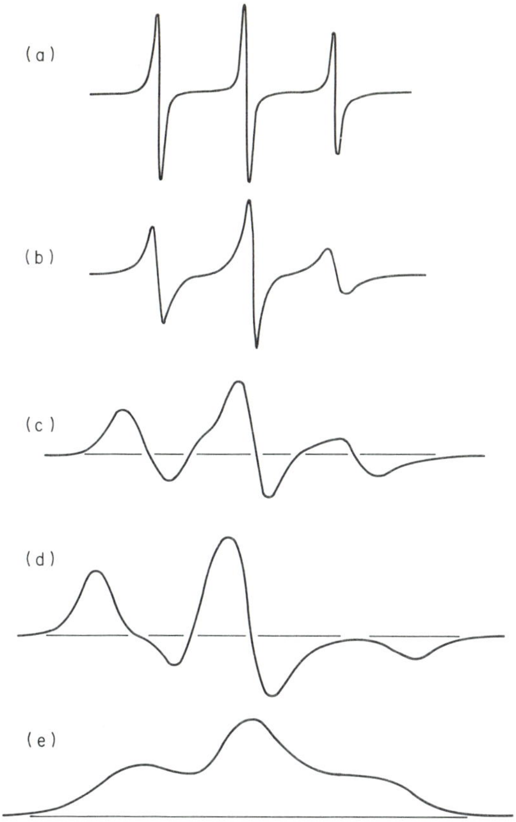

FIG. 7.5 Electron paramagnetic resonance spectra of nitroxide spin-labels: (a) fully mobile ($\tau_R \approx 0{\cdot}1$ ns); (b) weakly immobilized ($\tau_R \approx 1$ ns); (c) moderately immobilized ($\tau_R \approx 10$ ns); (d) strongly immobilized ($\tau_R \approx 100$ ns). The corresponding absorption curve for (d) is included (curve (e)).

spectrum has a large central peak. However, the outer two lines will be scattered over a range of magnetic fields and they do not superimpose but give a broad continuum. Clearly, the intensity of the spectrum must decrease towards the extremes since there will be fewer labels with the symmetry axis parallel to the field (giving the A_{zz} splitting) than with the axis perpendicular to the field (giving the A_{xx} and A_{yy} splittings). The extremes of the spectrum thus correspond approximately to A_{zz} (32 gauss relative to the central line) but since immobilization may be accompanied by relaxation broadening of lines, some absorption usually occurs outside these limits. For spin labels which are only "weakly immobilized", spectra intermediate between Fig. 7.5(d) and that of the mobile spin label (Fig. 7.5(a)) are obtained (Fig. 7.5 (b),(c)). The changes shown in Fig. 7.5 can be observed in free spin-labels dissolved in solutions of increasing viscosity, and rotational correlation times (τ_R) can be calculated by Stokes' law. Conversely, the shape of the EPR spectra for partially immobilized spin-labels attached to macromolecules can be used to estimate τ_R. This may represent the motion of the whole macromolecule to which the spin-label is attached rigidly or may represent independent rotational motion of the attached spin-label (p. 309, Example 4).

In some macromolecular systems, particularly membranes with intercalated lipid spin-labels, the motion of the partially immobilized nitroxide may itself be anisotropic, e.g. there may be rapid motion of the spin-label perpendicular to a given molecular axis but very little motion parallel to the axis. For lipid spin-labels it has been found that this molecular axis is the normal to the membrane surface (Fig. 7.6). From EPR spectra it may be possible to estimate hyperfine splitting constants for the magnetic field parallel to the normal ($A_\parallel$) and for the magnetic field in the plane of the membrane ($A_\perp$). These parameters can then be compared with A_{zz} and A_{xx} for the isolated nitroxide spin-label and the difference between the two sets of parameters then provides an estimate of the deviation of the spin-label from rigid order within the membrane. The *order parameter* is given by

$$S=\frac{A_\parallel - A_\perp}{A_{zz}-A_{xx}} \tag{7.7}$$

For complete order (lipid spin-label rigid and perpendicular to the membrane surface but still free to rotate rapidly about its long axis), $S=1$. This is illustrated by the outer parts of Fig. 7.6. For isotropic motion (no constraints by the membrane), $S=0$. An intermediate case (S between 0 and 1) is illustrated in the central part of Fig. 7.6 where some deviation from complete

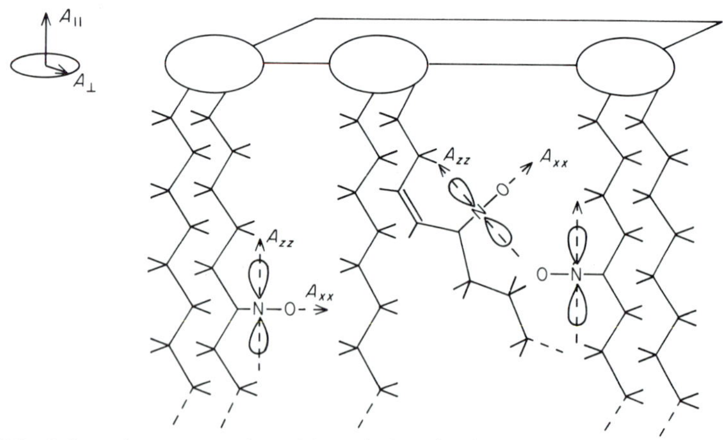

FIG. 7.6 Schematic representation of the ordering of a phospholipid nitroxide spin-label in a membrane. The hyperfine splitting constants in the membrane frame of reference are $A_\parallel$ and $A_\perp$ and the splitting constants in the nitroxide frame of reference are A_{zz} and A_{xx}.

order is introduced by a *cis*-double bond causing a tilting of the nitroxide relative to the bilayer surface. Example 5 (p. 312) illustrates further the use of order parameters.

Multi-spin centres in a ligand field: transition metal ions

Up until now we have considered mainly free radicals with $S=\frac{1}{2}$ and have only mentioned in passing the complexities which occur in transition metal ions. In free radicals, there are two spin states (spin-multiplicity $2S+1=2$) of equal energy, i.e. degenerate, in the absence of an applied magnetic field. The degeneracy is removed by applying a strong magnetic field to give two energy levels with a separation determined by the g-value ($\sim 2{\cdot}00$) and there is only one magnetic field value for resonance (Fig. 7.8(a)). In transition metal ions at least three additional factors have to be allowed for: (i) more than one unpaired electron may be present ($S>\frac{1}{2}$); (ii) the energy levels available to the unpaired electrons may not be degenerate in the absence of an applied field but may be split by the intrinsic field of the ligands and thus depend on their arrangement. Because of (i) and (ii) there may be more than one possible transition giving rise to resonance (Fig. 7.8(c)). (iii) The orbital angular momentum associated with electrons in d orbitals may be coupled to the spin angular momentum to give g-values which deviate from 2 and are anisotropic, and the effects are again dependent on the arrangement of ligands.

Table 7.2 shows the electronic configuration of the paramagnetic metals most often encountered in biological systems. The unpaired spins arise from the energy states of the five 3d orbitals available to valence electrons, and these in turn depend on the coordination scheme of the ligands around the metal. Figure 7.7 illustrates the very common *octahedral coordination scheme* of six

TABLE 7.2 The electronic configurations and g-values of common paramagnetic metal ions as usually found in biological systems

Metal ion		Electronic configuration					S	g-value
Mn(II) $3d^5$		↑	↑	↑	↑	↑	$\frac{5}{2}$	2·0
Fe(III) $3d^5$	high-spin	↑	↑	↑	↑	↑	$\frac{5}{2}$	wide variations according to spin state and ligands
	low-spin	↑↓	↑↓	↑			$\frac{1}{2}$	
Fe(II) $3d^6$	high-spin	↑↓	↑	↑	↑	↑	2	
	low-spin	↑↓	↑↓	↑↓			0	
Cu(II) $3d^9$		↑↓	↑↓	↑↓	↑↓	↑	$\frac{1}{2}$	2·0–2·3
Cu(I) $3d^{10}$		↑↓	↑↓	↑↓	↑↓	↑↓	0	—

ligands, with the corresponding energy levels. This picture is oversimplified but shows how the coordination scheme minimizes repulsive effects between orbitals and is thus energetically favoured. It also shows why the d_{z^2} and $d_{x^2-y^2}$ orbitals have the highest energies; they point directly towards the octahedral positions bearing the ligands (usually negatively charged). The energy level diagram shown in Fig. 7.7 would be altered by changing the coordination scheme and is sensitive to the exact form of the ligand orbitals (commonly nitrogen p orbitals which have directional character). The arrangement of electrons in orbitals shown in Table 7.1 depends on whether the energy (Δ) required to promote an electron to the higher set of orbitals is more than or less than the energy involved in pairing. If it is less than the pairing energy, electrons occupy separate orbitals and a "high-spin" system results. Many transition metals can occur in "*high-spin*" or "*low-spin*" forms, and the form

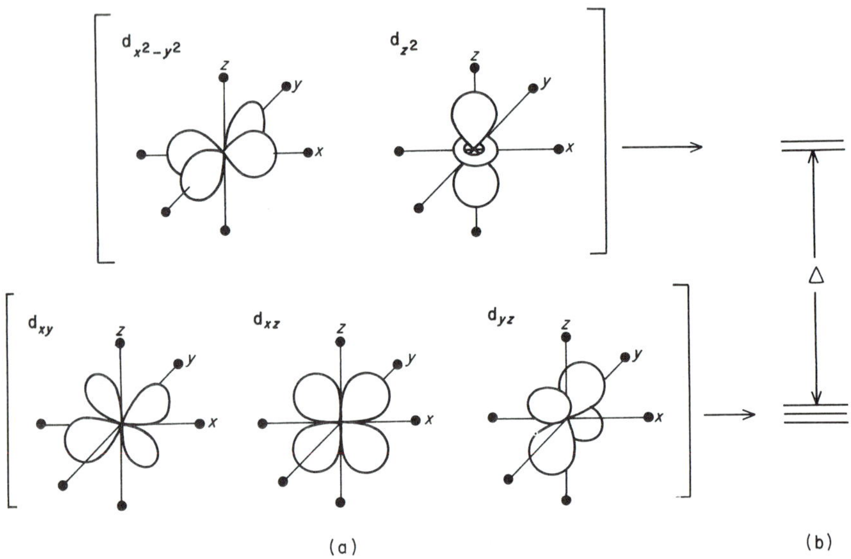

FIG. 7.7 (a) The five d orbitals in transition-metal ions seen in relation to six ligands (represented by point charges) in octahedral coordination, and (b) the corresponding energy levels.

preferred is very sensitive to the ligand field because of its effect on energy levels. Thus Fe(III) in the haem group occurs as its low-spin form in the presence of a CN^- ligand in the sixth coordination position because the higher set of orbitals is energetically unfavourable partly due to repulsive effects between ligand and orbitals (p. 266, Example 4) whereas if the ligand is less charged a high-spin form may result (e.g. H_2O in metmyoglobin—p. 303, Example 1).

For a paramagnetic metal with $S=\frac{1}{2}$, (e.g. Cu^{2+}), the situation is similar to

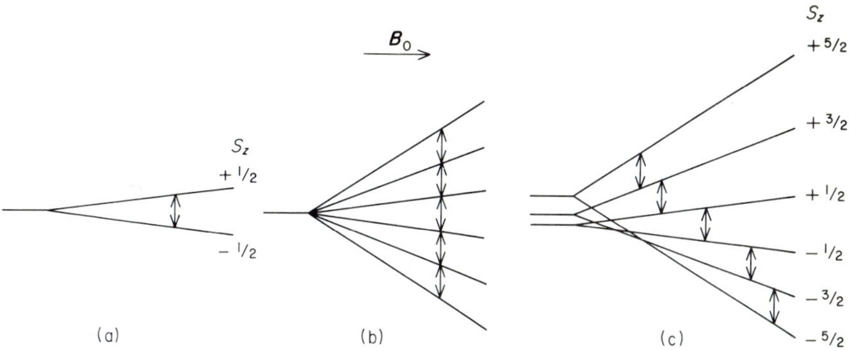

FIG. 7.8 Energy level diagrams for unpaired electronic spins in an applied magnetic field for (a) $S=\frac{1}{2}$, (b) $S=\frac{5}{2}$ (e.g. Mn(II)), (c) $S=\frac{5}{2}$ with zero-field splitting in the presence of a crystal field. The allowed transitions are also illustrated.

that for free radicals where there is only one resonance position (Fig. 7.8(a)) but the possibility of g-value anisotropy has also to be considered. For metals with $S>\frac{1}{2}$, more complex situations may occur. For Mn(II) and for high-spin Fe(II), $S=\frac{5}{2}$ and the spin-multiplicity is $2S+1=6$, corresponding to spin projections (S_z) along the magnetic field direction of S, S-1, S-2. . . $-S$. If all six spin states are of equal energy in the absence of an applied field, the application of a magnetic field will cause the energy-level separations shown in Fig. 7.8(b), but because a selection rule $\Delta S_z = \pm 1$ applies, there is still only one magnetic field value for all allowed transitions. The spin states may not, however, all be of equal energy in the absence of an applied field but be split into three doublets, $S_z = \pm\frac{1}{2}$, $\pm\frac{3}{2}$, $\pm\frac{5}{2}$ by the crystal field ("zero-field splitting"). The result of this is that, with the same selection rule, there are now five magnetic field values for resonance and five EPR lines (Fig. 7.8(c)). In practice, these splittings in Mn(II) and Fe(III) are not well resolved and are not usually an important feature of the EPR spectra in biological systems.

g-Values and relaxation times in paramagnetic metals

We have already discussed the g-value as a spectral parameter which may be anisotropic and used it to illustrate the concept of a tensor earlier in this section. The deviations of the g-value of transition metals from the free electron value $g_e = 2{\cdot}0023$ depend on the system possessing *orbital angular momentum* (quantum number, L) which is coupled to the spin angular momentum. If the coupling is efficient, a single resultant spin (J) precesses about the magnetic field direction with an increased g-factor. For s atomic orbitals and σ molecular orbitals, $L=0$ and $g=2$. For d orbitals in general, $L\neq 0$ and $g\neq 2$. In practice, spin–orbit coupling in a ligand field is less efficient than in a free metal ion and the observed g-value is given by equations of the form

$$g = g_e\,(1 + \lambda/\Delta) \tag{7.8}$$

where λ is called the *spin–orbit coupling constant* and Δ is the energy-level separation referred to earlier. In a strong crystal field, as when the ligands have strong negative charges, the splitting of energies between the d orbitals will be large, $\Delta \gg \lambda$, and $g \sim g_e$. This is termed *decoupling*. A study of g-values thus gives information about the separation of energy levels in liganded metal ions and hence on the arrangement of ligands. In addition the anisotropy of g depends on energy levels associated with particular d orbitals with particular directional properties and gives further information on the ligand system. Large deviations of g from 2 and large anisotropy usually occur together.

We can illustrate this by taking again Mn(II) ($S=\frac{5}{2}$), Cu(II) ($S=\frac{1}{2}$) and Fe(III) ($S=\frac{5}{2}$) as examples. Manganese(II) has all five d orbitals half-filled and the net electron density is spherically symmetrical—the so-called S-state. In this case $L=0$, $g=2{\cdot}0$. Spin–orbit interactions are a major relaxation mechanism in EPR and so Mn(II) has a long relaxation time ($\sim 10^{-8}$ s), allowing it to be used as a relaxation enhancement probe (see p. 251), and relatively narrow EPR lines (though broader than for free radicals). The spectra are dominated by nuclear hyperfine interactions (Table 1.1). In Cu(II) the single spin is associated with some orbital angular momentum ($L \neq 0$) but in the usual ligand fields the energy-level separations are large, the electron is mainly localized in a single orbital and S and L cannot interact strongly. The value of g deviates from 2 to some extent and shows anisotropy ($g=2{\cdot}0$–$2{\cdot}3$) and relaxation times are shorter than for Mn. EPR spectra are dominated by g-value anisotropy and hyperfine interactions with the Cu(II) nucleus (Fig. 7.2). In Fe(III) the situation is very complex because the splitting (Δ) allows both low-spin and high-spin forms to occur. With a strong negatively charged ligand (e.g. CN^-), Δ will be large, leading to a low-spin form with relatively weak spin–orbit coupling and g-values close to 2. With less strongly charged ligands, Δ is smaller and a high-spin form is produced with strong spin–orbit coupling leading to extremely short relaxation times and extremely anisotropic g values (2–6) (e.g. metmyoglobin, p. 303, Example 1).

EPR Instrumentation and Methods

The elements of an EPR spectrometer

Figure 7.9 illustrates the main units of a typical EPR spectrometer. The basic requirements are very similar to those of an NMR spectrometer, i.e. a

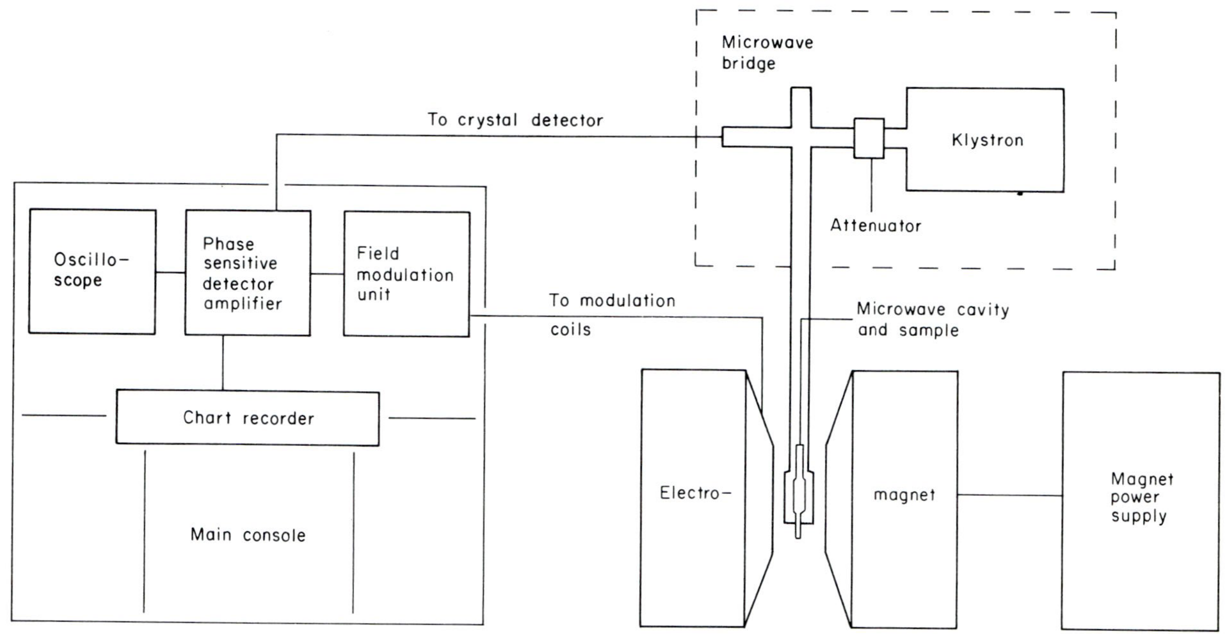

FIG. 7.9 The main elements of an EPR spectrometer.

magnet and magnet power supply, a source of electromagnetic radiation (in this case microwave radiation), a means of irradiating the sample in the magnetic field, a means of detecting the absorption at resonance and amplifying the signal by the usual electronic procedures, and an oscilloscope and chart recorder for displaying the signal.

In a typical EPR experiment, magnetic field strengths about 10 times smaller than for high-resolution NMR are used (about 3000 gauss for $g = 2$ and $\nu \sim 9 \times 10^9$ Hz). The resulting energy-level separations are, nevertheless, about 100 times greater than for NMR ($\nu \sim 10^8$ Hz) because electronic magnetic moments exceed nucear moments by a factor of about 10^3. In addition EPR spectra are spread over a much larger fraction of the total magnetic field (compare the g-value variation of 2–6 for some Fe(III) EPR spectra with the 10 parts per million range in many ^{1}H NMR spectra). This means that the technological problems associated with the production of very high magnetic fields, field homogeneity, and stability for NMR are much less acute for EPR spectrometers. The problems of signal detection and amplification are also less severe and the specialized pulse methods and computers used to overcome sensitivity problems in NMR are not generally necessary for EPR.

The main point of departure in EPR spectrometers is in the microwave power source and means for irradiating the sample. The power source is a *klystron* or specialized type of valve designed to oscillate at microwave frequencies. The klystron is connected to a length of *waveguide* terminating in a cavity containing the sample of dimensions such that a standing electromagnetic wave pattern is set up similar to the standing sound waves in a length of pipe. The wavelengths involved, and the waveguide dimensions, are a few cm. As with sound waves, matching the source frequency to the resonant frequency of the cavity is critical for maximizing the microwave power ("tuning"). The sample is placed in a specially designed flat cell (for aqueous samples) at a point in the cavity near the centre of the standing wave pattern where the microwave magnetic field vector is at a maximum. Different standing wave patterns ("modes") may be used, depending on the instrumental design. The absorption of power from the field by the sample at resonance is detected by a bridge arrangement of waveguides, one arm of which contains a crystal detector. The bridge is balanced so that microwave power only flows into this arm when the sample is at resonance. The sensitivity of this arrangement means that tuning the system (either by slightly varying the klystron frequency or by introducing dielectric material into the waveguides) is an important part of setting up an EPR spectrometer. Once tuning is achieved, there is normally some locking device to prevent drift of the klystron frequency. Electron paramagnetic resonance experiments are generally performed at one of several fixed frequencies termed X-band, K-band and

Q-band from their original use in radar. X-Band is the lowest frequency ($9{\cdot}4 \times 10^9$ Hz) and the most commonly used.

The other main difference from NMR spectrometers is that while the static magnetic field is being swept through resonance, it is modulated by a subsidiary alternating magnetic field (usually 100 kHz) and the response to this alternating field is then subjected to phase-sensitive detection and amplification before the signal is displayed. The result is that EPR spectra are the first differential of the absorption curve (Fig. 7.10). The sensitivity of detection

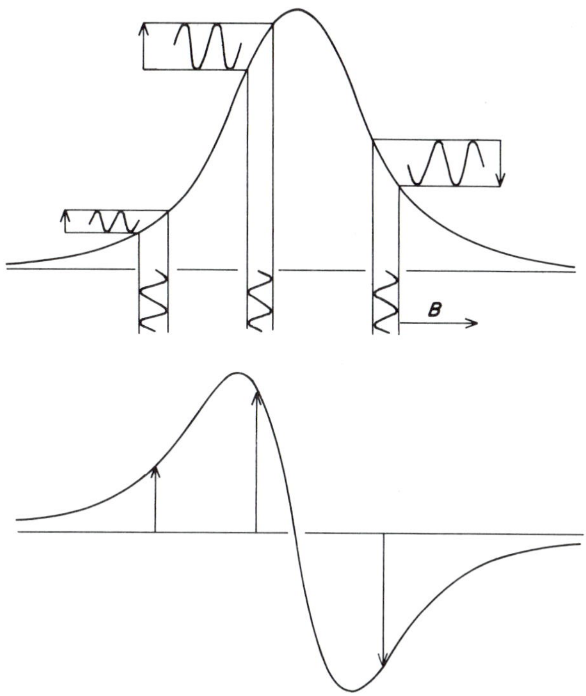

FIG. 7.10 The EPR derivative spectrum produced in response to magnetic field modulation and phase-sensitive detection.

increases as the amplitude of the modulation increases but beyond a point (determined by the linewidth relative to the modulation amplitude) unacceptable spectral distortion occurs. Other important sources of distortion in EPR spectra (as in NMR) are incorrect setting of phase controls, saturation due to excessive microwave power, and recording the spectrum too rapidly in comparison with the response time of the recorder.

Sample preparation

Biological EPR is most commonly performed on aqueous solutions near to room temperature or on frozen solutions, but polycrystalline solids, single crystals, membrane dispersions and intact tissues have also been studied. The great advantage of EPR is that there is little interference from the diamagnetic bulk of the sample. The simplest systems to study are spin-labels in aqueous solutions because they are stable and usually have fairly narrow lines, giving large signals for relatively low concentrations (1–10 μM). For spin-labels immobilized in biological systems (p. 309, Examples 4 and 5) somewhat larger concentrations are required. Aqueous samples are restricted in size by the critical positioning necessary and the effect of the high dielectric constant of water in causing power losses, and are usually contained in a special flat quartz cell (capacity 0·1–0·4 ml) which can be opened at both ends. The detection of transient free radicals in solution may not be a problem of sensitivity, but special arrangements may be necessary to generate, stabilize, and observe them within a very short time interval. These may include provision for (i) irradiating the sample with UV or γ-radiation, (pp. 315–319, Examples 6 and 7), (ii) continuous flow systems allowing radicals to be observed a short time after mixing reagents, (iii) rapid freezing to stop reactions a known time after mixing in a flow system so that EPR spectra of unstable radicals can be observed at leisure (p. 307, Example 3) and (iv) maintaining anaerobic conditions necessary to stabilize many free radicals by working with solutions under N_2. With frozen solutions, dielectric losses of power are reduced so that sample size, shape and position are less critical than with liquid samples.

The observation of satisfactory EPR spectra from paramagnetic metals in biological systems may be extremely difficult. The main problem is the low sensitivity and resolution due to the broad spectra, particularly with iron; in many studies liquid nitrogen temperatures or below are required (pp. 303–307, p. 315, Examples 1, 2 and 6). This alters the thermal equilibrium between energy-level populations in favour of absorption and improves resolution by reducing the relaxation rate (see p. 286). Even so, millimolar concentrations are required for most metals except Mn(II). Because metalloproteins are already "immobilized" on the EPR timescale in aqueous solution, freezing does not introduce further problems of mobility. Low temperatures are also usually required for whole tissue studies (p. 318, Example 7) in addition to a specially constructed sample cell. In single crystal studies (see p. 303, Example 1) provision for rotating the sample about different axes in the spectrometer is necessary and in all EPR studies variable-temperature facilities are useful and are available with most instruments.

Specialized methods

In addition to the somewhat specialized techniques for sample preparation and handling mentioned above there has been a variety of specialized instrumental developments in EPR. *High frequencies* can be used to improve spectral resolution and detect hyperfine interactions. As with NMR, the basis of this method is that the separation of regions of a spectrum differing only in g-value is proportional to the magnetic field, whereas hyperfine splittings and linewidths (measured directly in gauss or Hz) are independent of the field. *Double irradiation* methods are also very useful for studying hyperfine interactions in complex spectra. In *electron–nuclear double resonance* (ENDOR), the EPR spectrum is monitored at some fixed point corresponding to a nuclear hyperfine splitting while the sample is irradiated with a varying radio frequency. The EPR signal is saturated by applying high microwave power so that the energy-level populations are equalized and no signal is seen. When the radiofrequency field passes through resonance for the nucleus causing the EPR splitting, the energy-level populations are disturbed and EPR absorption occurs. This technique combines the sensitivity of EPR with the resolution of NMR and has considerable potential for studying nuclei (particularly protons) in the vicinity of a paramagnetic centre, since individual nuclei involved in hyperfine interactions can be unambiguously identified where the normal EPR spectrum is too complex to provide useful information (see p. 303, Example 1). A related technique is *electron–electron double resonance* (ELDOR) which resembles ENDOR and gives similar information on hyperfine interactions, except that the varied irradiating frequency is also an EPR frequency.

Relaxation rate studies in EPR do not assume the importance of relaxation studies in NMR, but are useful where more than one paramagnetic centre is present, as in many metalloproteins. Sometimes two paramagnetic centres can only be distinguished by EPR because they have different saturation behaviour and different relaxation rates. The EPR spectrum of one paramagnetic centre may be perturbed or quenched by a neighbouring centre and the quenching may give the distance apart if the relaxation times are known; this method has been applied to spin-labelled systems as well as to metalloproteins with endogenous paramagnetic centres (pp. 307–312, Examples 3 and 4).

The Types of Information Available from EPR in Biological Systems

Detection and identification

Electron paramagnetic resonance can assist the chemical identification of transient free radicals in enzymic reactions (see p. 307, Example 3), photosynthesis (p. 315, Example 6), mitochondrial respiration, and radiation-damaged systems (p. 318, Example 7), mainly by comparative studies and by observing hyperfine splitting patterns and intensities. Electron paramagnetic resonance intensities, g-values and hyperfine structure can be used to characterize metal atoms in metalloproteins with regard to number, kind, and oxidation state (pp. 305–309, Examples 2 and 3).

Monitoring changes

Electron paramagnetic resonance can be used empirically like NMR to determine concentrations, measure time-courses of reactions or investigate equilibria. Transient effects in the millisecond range can be followed in conjunction with rapid reaction techniques (see pp. 48–54) providing useful information on reaction mechanisms (p. 307, Example 3). The identity and number of molecular species giving EPR spectra must be established if possible; in many studies the EPR signal may be associated with some other change such as a visible absorption change (p. 315, Example 6), or isosbestic points may be observed and be interpreted in the usual way as being due to an equilibrium between two spectral forms (p. 309, Example 4).

Probing the general environment

Electron paramagnetic resonance spectra are sensitive to mobility because of spectral anisotropy (pp. 309–315, Examples 4 and 5), and to polarity and electron delocalization because of hyperfine interactions. In transition-metal complexes the chemical nature and arrangement of ligands around the metal can be investigated by studying g-value anisotropy and hyperfine splittings (pp. 303–307, Examples 1 and 2). Thus, EPR is useful for detecting conformation changes in general.

Kinetic and dynamic information

Molecular motion studies with spin-labels can often be made quantitative when the motion falls into the nanosecond (10^{-9} s) range which is a shorter time-scale than that usually detected by NMR (pp. 309–315, Examples 4 and 5). Spin–spin interactions between paramagnetic centres and EPR relaxation times, can give motional information in principle. Temperature-dependence studies may give thermodynamic information on the various changes.

Structural information: distances and orientations

Electron paramagnetic resonance is better suited than NMR for studying molecular orientations in ordered or semi-ordered systems (single crystals, membranes) because of spectral anisotropy, (p. 303, p. 312, Examples 1 and 5). As in NMR, distances can sometimes be determined from spin–spin interactions (hyperfine nuclear interactions and interactions with other electron spins (pp. 307–312, Examples 3 and 4). Practical difficulties and less well-developed theory limit the utilization of these effects in EPR.

Applications of EPR to Biological Systems

The following selection covers several of the most significant areas of application but is by no means comprehensive. The aim has been to indicate the types of approach used with different biological systems at progressively increasing levels of complexity and give some impression of the types of system which can be profitably studied by EPR. In the interests of space and readability, no attempt has been made to give full details of individual experiments; the reader who develops a working interest in the subject will learn this detail better from practical situations and from the extensive literature on particular topics, although it is hoped that the interest of other readers will have been stimulated sufficiently to consult the brief bibliography at the end of the chapter.

Example 1. Structural studies on haemoproteins

Myoglobin and haemoglobin were among the first biological molecules to be studied by EPR because of their ease of purification and well-known structural features. They can only be studied by EPR in their oxidized (Fe(III)) forms, which may in turn exist in high-spin or low-spin states. High-spin Fe(III) in the haem group of metmyoglobin gives a signal with a large g-value anisotropy which is axially symmetric ($g_{\perp}=6, g_{\parallel}=2$) (p. 288 and Fig. 7.3). From the featureless nature of the spectrum, even at very low temperatures (Fig. 7.11) it might seem surprising at first sight that anything useful can be discovered by EPR, particularly in view of our advanced knowledge of the structure from X-ray diffraction, but by a suitable choice of experimental method a large variety of useful information has in fact been obtained.

First, single crystal rotation studies were performed on myoglobin and the g-value variation was measured as the crystal was rotated about different crystallographic axes. Two interesting facts emerged from these early studies, which were performed before the details of the crystal structure were known: (i) two signals were observed showing the same angular dependence of g, but

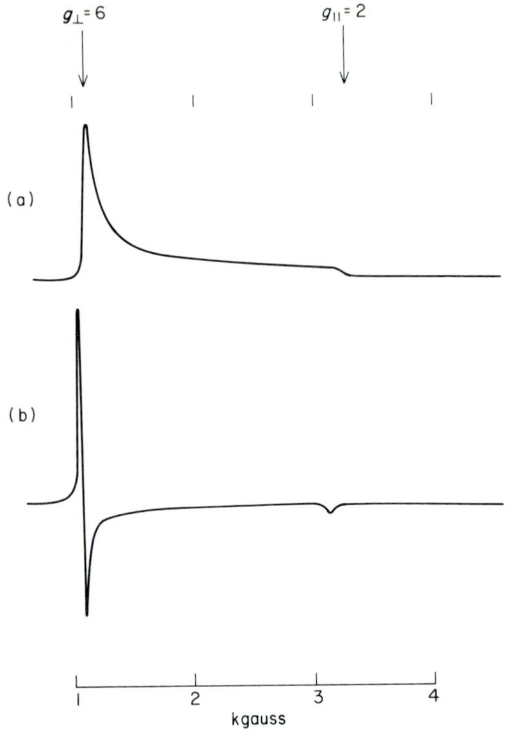

FIG. 7.11 Electron paramagnetic resonance spectrum, 8·9 GHz, of frozen metmyoglobin at 2 K presented as (a) the absorption spectrum and (b) derivative spectrum (from Feher, G., Isaacson, R. A. and Scholes, C. P. (1973). *Ann. N.Y. Acad. Sci.* **222,** 86. Reproduced by permission of the New York Academy of Sciences).

displaced by 46°, indicating that the unit cell contained two molecules with differently orientated haem groups; (ii) the g-value variation observed for rotation about different axes could be used to determine the orientation of the haem planes relative to the crystallographic axes (eqn (7.5)); this information was a useful aid in the early work on the crystal structure. The studies were successfully extended to the more complex case of haemoglobin.

g-Value anisotropy studies have also provided information on the ligands of haem. By a suitable choice of ligand in the 6th coordination position, normally occupied by O_2, a form of myoglobin with low-spin Fe(III) can be obtained. This was found to have g-values much closer to 2 which deviated from axial symmetry ($g_{xx} \neq g_{yy}$). Since the isolated haem group is axially symmetric this meant that one of the ligands in position 5 or 6, or something in the protein structure, was causing asymmetry in the haem plane. The source of asymmetry was found to be a nitrogen p orbital in the histidine coordinated at position 5 (the proximal histidine), and from the anisotropy the orientation of the histidine ring was determined. Other information on the ligands has been provided by double-resonance studies (p. 301) of the nuclear hyperfine interactions. The barren EPR spectrum of Fig. 7.11, which shows no sign of hyperfine interactions, contrasts with the more detailed ^1H or ^{14}N ENDOR spectra obtained under similar conditions. Using ^{14}N ENDOR, the interactions of the haem nitrogens and the histidine nitrogen at position 5 with the electron spin can be clearly seen, and in haemoglobin, mutants lacking this histidine can be distinguished from normal haemoglobin.

Example 2. Other metalloproteins: the characterization of iron–sulphur proteins and copper proteins by EPR

We turn now to cases where EPR has played a significant role in the initial characterization of other groups of metalloproteins, no less vital to metabolic function than haem proteins, but about which comparatively little was known until recently. In early studies on mitochondrial respiration, photosynthesis and nitrogen fixation, the existence of a variety of "reducing factors" had to be inferred from chemical studies, but in most cases their precise molecular identity was not known. Clues as to the nature of some of these reducing factors came from the discovery of a protein involved in bacterial nitrogen fixation which contained Fe in a non-haem form, which was given the name ferredoxin, and the apparently unconnected discovery in mitochondrial particles of an EPR signal with an unusual g-value—a principal absorption at $g \approx 1{\cdot}94$ and a minor absorption at $g \approx 2{\cdot}00$. Similar EPR signals were subsequently found in a variety of intact sub-cellular preparations at very low temperatures under reducing conditions. Typical signals are shown in Fig. 7.12. Chemical analysis revealed the presence of non-haem iron in all these systems (which was only released when the protein structure was disrupted) and also acid-labile sulphur. The bound iron had much less intense visible absorption bands than the haem group in the more abundant haem-proteins (see p. 39), which largely accounted for its later discovery.

The origin of the EPR signal remained a mystery for some time. Its asymmetry and short relaxation time ruled out free radicals as the source. In addition, the signal occurred in a number of systems with no known prosthetic groups, such as flavins, which could give rise to free radicals and in systems with no metal other than Fe present. The g-value ruled out the possibility that the signal was from Fe(III) in haem groups and in any case, under the reducing conditions where the signal occurs, iron is expected to be present as Fe(II). In the low-spin form of Fe(II) there are no unpaired electrons and no EPR signal (Table 7.2); in the high-spin form a much broader signal with a much shorter

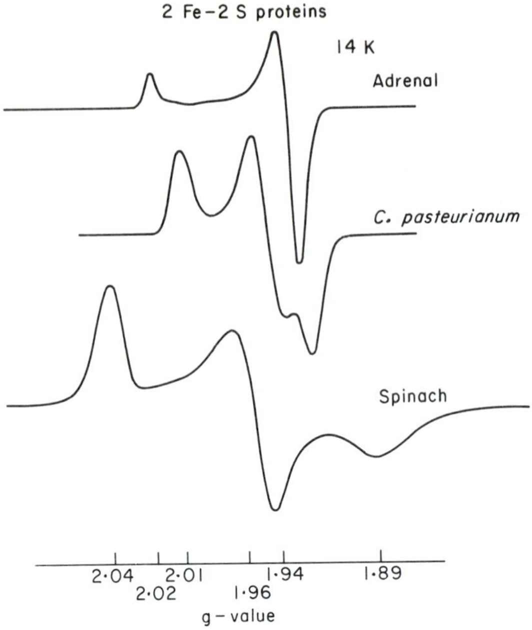

FIG. 7.12 Electron paramagnetic resonance spectra, 9·17 GHz, of reduced Fe—S proteins at 14 K (from Orme-Johnson (1973). "Iron–Sulphur Proteins", Vol. II. (Ed. Lovenberg), Academic Press, New York and London).

relaxation time than that observed would be expected, and thus ordinary ionic Fe^{2+} could not account for the signal. That the signal was in fact from Fe was proved in experiments where microorganisms were grown on a ^{57}Fe source ($I=\frac{1}{2}$), with the result that nuclear hyperfine structure with a large splitting appeared indicating that the unpaired electron was localized largely on Fe-atoms. Similar isotopic substitution experiments with ^{33}S strongly suggested that S was liganded to Fe, and the stoichiometry of Fe-atoms and acid-labile sulphur was consistent with this. Theoretical considerations of the value $g<2{\cdot}0$ and the hyperfine structure led to the conclusion that Fe-atoms must occur in pairs of high-spin Fe(III) coupled to high-spin Fe(II) acting as a single spin unit of $S=\frac{1}{2}$. It thus became clear that there was a large variety of proteins, of which ferredoxins form the largest group, with active centres of the form Fe_2S_2, Fe_4S_4, etc., catalysing a wide range of vital redox reactions, and EPR has played a major role in their characterization.

Electron paramagnetic resonance has played a similar role in the investigation of Cu-proteins which catalyse a large range of oxidative reactions. The best known of these is cytochrome oxidase which catalyses the terminal reaction of mitochondrial respiration, but other Cu-proteins have been more

intensively studied by EPR. Fortunately, the paramagnetic centres in all these proteins have many features in common and information obtained with one enzyme will often be relevant to the catalytic mechanism of others. The general appearance of the EPR spectrum is indicated in Fig. 7.2. Fungal laccase is well studied and catalyses the transfer of electrons from hydroquinones to O_2 which is reduced to H_2O in a four-electron reduction. Chemical analysis showed that four Cu-atoms were present, but quantitative EPR studies indicated that only two were paramagnetic and must therefore be Cu(II) (Table 7.2). These in turn have different EPR spectra and are termed type I and type II Cu(II).

Type I Cu(II) has a narrow nuclear hyperfine splitting suggesting extensive electron delocalization, has an intense blue colour, and is readily reduced by hydroquinones with a loss of EPR signal. Type II Cu(II) has no detectable colour and a large hyperfine splitting (cf. p. 42, Fig. 2.16). It is not reduced by substrates but appears to be involved in inhibitor binding and its EPR spectrum shows hyperfine interactions from inhibitor nuclei. The properties of type I Cu(II) are obviously consistent with a catalytic function, but where do the other Cu-atoms fit into the scheme? Studies with different reducing agents and spectrophotometric and EPR kinetic studies have now established a role for all the Cu-atoms in the electron-transport scheme, with type I Cu as the primary electron acceptor, and this mechanism will probably be useful in understanding other Cu-enzymes. The next example provides a further illustration of the use of rapid kinetic studies (see p. 300) in elucidating the mechanism of an oxidative enzyme.

Example 3. Flavins and flavoenzymes: kinetic studies on xanthine oxidase

We will later see examples of photo-induced or radiation-induced free radicals (pp. 315–319, Examples 6 and 7), but free-radical intermediates also occur in many enzyme systems. Of these, the flavoenzymes are undoubtedly the best-studied group, in which reduction of flavins often proceeds in two one-electron steps with a free-radical intermediate or *semiquinone* according to the general scheme

$$\text{Flavin} \xrightarrow{e^-} (\text{Flavin}\cdot)^- \xrightarrow{e^-,H^+} (\text{Flavin H})^-$$

where the exact protonation state will vary with pH and other conditions. The existence of semiquinone intermediates in redox reactions was first postulated long before the days of biological EPR, but kinetic studies on intermediates using conventional methods such as visible absorption could not easily be made quantitative because of the complexity arising from overlapping bands.

Although EPR could only detect some of the intermediates, the interpretation of results was very much simpler and consequently EPR has contributed significantly to our knowledge of these systems, especially with the use of double resonance techniques for studying hyperfine interactions.

Xanthine oxidase is a flavoenzyme which also contains the paramagnetic metals Mo and Fe. It catalyses the oxidation of xanthine to uric acid and the simultaneous reduction of O_2 to H_2O_2 and is readily purified from milk. It is well suited to EPR, and has probably been more studied by this technique than any other single enzyme. This is largely because it has several different observable paramagnetic centres all involved in the electron transfer process and because its overall catalytic centre activity or turnover number is rather low ($\approx 5\ s^{-1}$), thus facilitating kinetic studies. Chemical analysis of the enzyme indicated that it contained two FAD molecules, eight Fe-atoms and two Mo-atoms per molecule. On addition of reduced substrate, EPR signals were observed from all of these centres and, in the case of Mo, at least two distinct signals were seen. The kinetics of appearance and decay of each signal was investigated in detail by the rapid-freeze quenched-flow method (p. 300) in which the reaction mixture was rapidly frozen in isopentane at $-140°C$ and then analysed by EPR. The signals from Mo(V) were the first to rise and fall and appeared to represent different bound states on the enzyme. By a suitable choice of reducing conditions they could be observed individually. The signals from Fe were the last to decline and had a g-value of 1·94, characteristic of Fe—S centres. On the basis of these studies the electron transfer scheme shown in Fig. 7.13 was postulated. The system has been further elucidated by a variety of other studies, including (i) relaxation rate studies on distances between the paramagnetic centres, (ii) isotopic substitution studies showing that a proton is removed from the 8-position of xanthine to give a protonated form of Mo(V) which then shows hyperfine structure, and (iii) detection of the superoxide radical, $O_2^{\cdot -}$, an intermediate in the final two-electron reduction of O_2 to H_2O_2.

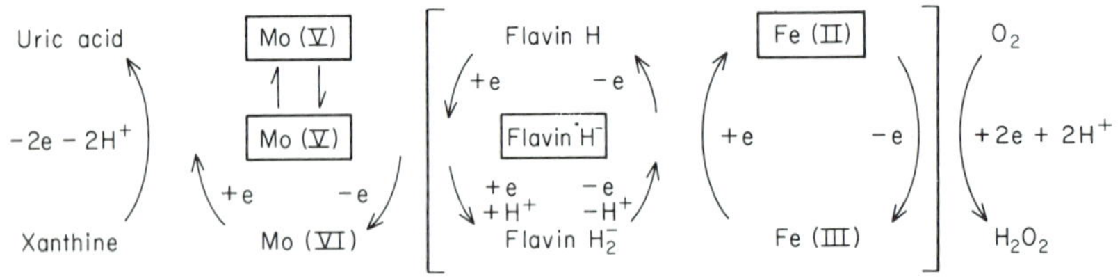

FIG. 7.13 Electron transport in xanthine oxidase as deduced from early EPR kinetic studies. Signals can be observed from both forms of Mo(V), from Fe(II) and from the flavin free radical (shown in boxes). Recent views of the mechanism (e.g. R. Bray, 1975 *In* "The Enzymes", 3rd Edn, Vol. XII. (Ed. P. D. Boyer), Academic Press, London and New York) suggest that part of the sequence (shown in brackets) is not well defined and that all the intramolecular redox centres may be in rapid equilibrium.

The scheme shown in Fig. 7.13, although generally accepted for some time, was recently shown to be inaccurate in a number of respects. The scheme was based essentially on changes following reduction of the fully oxidized enzyme. More recent experiments, including the technically difficult one of obtaining the fully reduced enzyme and then studying changes following oxidation, now appear to show that certain modifications are required. Although the primary interaction of the oxidized enzyme with reducing substrates at the Mo centres as depicted in Fig. 7.13 is essentially correct, the most usual site of interaction of oxidizing substrates with reduced enzyme is FAD rather than Fe. The part of Fig. 7.13 shown in brackets is thus less well defined than originally supposed. Furthermore, the original assumption that the intramolecular electron transfer processes within the enzyme must be slow has now been shown to be unnecessary in explaining the EPR changes. The various redox centres shown in Fig. 7.13 are now thought to be in rapid equilibrium and thus it may be irrelevant to speak of an "electron transfer sequence" in this particular system.

All the examples discussed so far are relevant to an understanding of the central biochemical processes of respiration and photosynthesis since information gained by applying EPR to well-characterized haem proteins, Fe—S proteins, Cu-proteins and flavoproteins may be applied to these more complex processes, both of which involve all four kinds of paramagnetic centre. In addition they emphasize the incompleteness of present knowledge of these processes. For example, xanthine oxidase has been much more intensively studied than most components of mitochondrial respiration and photosynthesis, yet even here we have seen how the picture of electron transport has recently altered. To what extent will details of the electron transport "chains" in respiration and photosynthesis, so beloved of biochemists, ultimately have to be modified?

Example 4. Spin-labelled proteins and interactions of proteins with spin-labelled ligands

The introduction of stable nitroxide spin labels has greatly expanded the scope of EPR in biological systems. Spin-labels can be specially designed for many purposes, e.g. to react covalently with different groups on proteins, to probe dimensions using labels of varying length and shape and to act as analogues to different kinds of biological molecule such as substrates, haptens, phospholipids and steroids. Thus studies can be made very selective, but the corresponding disadvantage of spin-labels is that they inevitably perturb the system to a greater or lesser extent. Spin-labelling should be made specific as far as possible and changes in the biological and molecular properties of the native system upon introduction of the label should ideally be small.

A large number of successful studies have been made on spin-labelled enzymes, although in some cases activity was lost or modified by the labelling. This was not the case with phosphofructokinase, a regulatory enzyme of glycolysis which has also been studied by NMR (p. 269, Example 6). The enzyme has an exceptionally reactive sulphydryl group (one per subunit) which has been specifically modified with the commonly used spin-label 4-(2-iodoacetamido)2,2,6,6-tetramethyl piperidinooxyl:

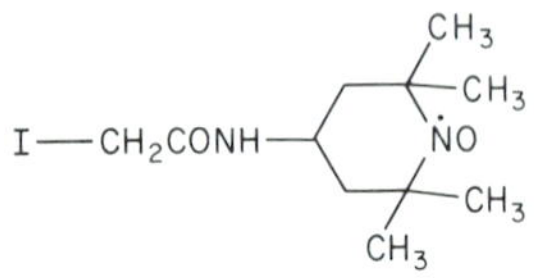

The modified enzyme retains all its usual allosteric kinetic behaviour, i.e. allosteric inhibition by ATP and sigmoidal kinetics with respect to the other substrate, fructose-6-phosphate. Its EPR spectrum is that of a "moderately-immobilized" nitroxide (Fig. 7.5(c) and p. 290) and from it a rotational correlation time of $\tau_R = 4 \times 10^{-9}$ s was calculated. From the molecular weight of the enzyme, which normally exists in a highly polymerized state, it was calculated that the rotational motion of the whole enzyme must be much slower than this, and hence, although the nitroxide is not fully mobile, it must have considerable motion independent of the protein.

More interesting than this is the fact that the EPR spectrum shows components with different mobility even though the labelling is highly specific. By titrating the enzyme with its substrates, fructose-6-phosphate and MgATP, the components can be interconverted. Figure 7.14 shows that there are isosbestic points (see p. 55) and the changes correlate well with the kinetic behaviour which has usually been explained in terms of an allosteric transition between active and inhibited forms. Thus useful non-kinetic evidence for an allosteric equilibrium involving two states has been obtained in this case. Furthermore, when the spin-labelled enzyme was titrated with MnATP instead of MgATP a striking difference in behaviour was observed; the spectrum was strongly quenched to a final intensity only 32% of the corresponding intensity with MgATP, although chemically Mg and Mn behave almost identically. The quenching was due to paramagnetic spin–spin interactions and, from it, an Mn-nitroxide distance on the protein of 1·2 nm was determined. In addition the spin-label caused paramagnetic broadening in the ^{1}H NMR spectrum of bound MgATP, from which the ATP proton–nitroxide distances could be determined. Since the MnATP–proton distances were already known it could be shown that all the distances were self-consistent.

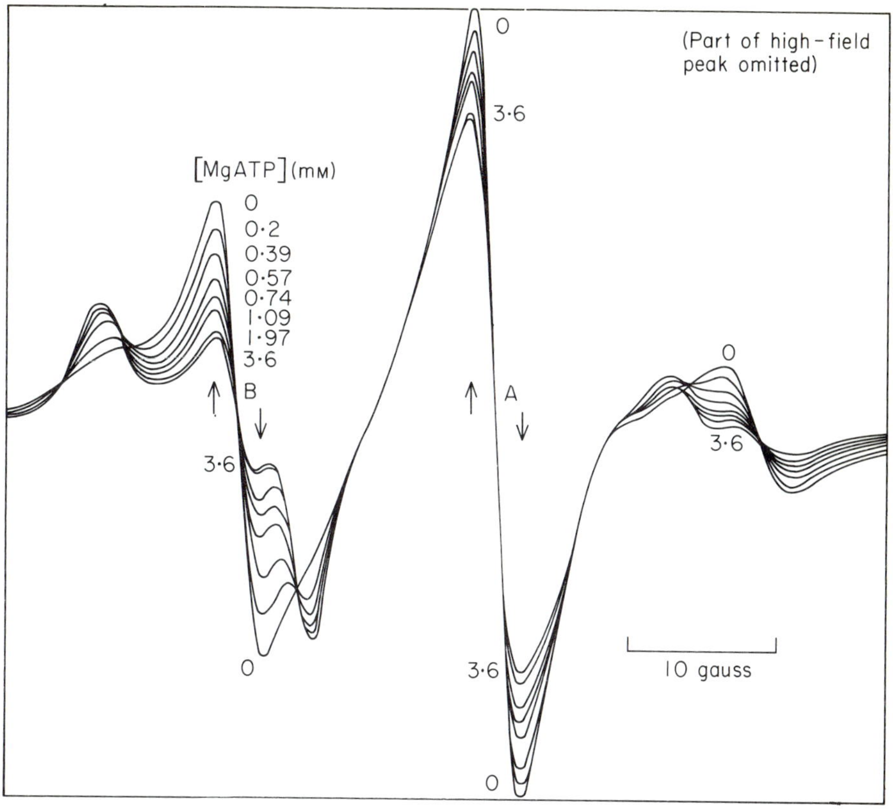

FIG. 7.14 Electron paramagnetic resonance spectra, 9·2 GHz, from spin-labelled phosphofructokinase titrated with its substrate, MgATP, at pH 7·5. The isosbestic points indicate that only two spectral states are present in equilibrium. The titration curve obtained from this data is sigmoidal, indicating cooperative interactions (from Jones, R., Dwek, R. A. and Walker, I. O. (1972). *FEBS Lett.* **26**, 92).

With the above kind of approach there is no guarantee that the labelling site will be of special interest; in phosphofructokinase it appears to be near to a regulatory ATP-binding site. A different approach is to use spin-labelled substrates and study their turnover by EPR, as has been done for chymotrypsin, or use inactive substrate analogues and study their interactions with different groups in the catalytic site by NMR or double resonance methods. Similar studies have been done with immunoglobulins. In the mouse myeloma protein, MOPC 315, spin-labelled dinitrophenyl haptens have been used to perturb NMR spectra (p. 268, Example 5) and give structural information. In addition, the EPR spectra of a series of spin-labelled haptens of graded lengths have been used to probe the geometry and polarity of the binding pocket. These labels were of the general type

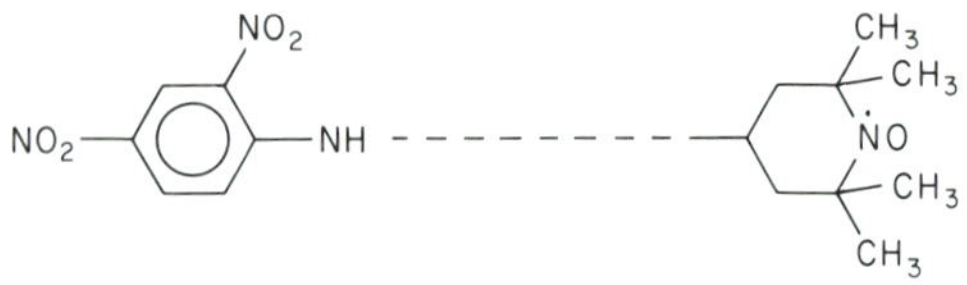

where variations in length were provided by alkyl and other groups. The binding of dinitrophenyl haptens to MOPC 315 is strong so that it was possible to obtain EPR spectra under conditions where essentially all the hapten was bound. Binding specificity is provided largely by the dinitrophenyl group which is rigidly held in the binding site by hydrophobic interactions. The mobility of the nitroxide is determined by the flexibility of the chain, i.e. the number of bonds allowing free rotation, the depth of the binding pocket in relation to the length of the hapten, and the size of the bulky tetramethyl piperidinooxyl group in relation to the lateral dimensions of the binding pocket. By analysing EPR line shapes of a series of bound haptens it was found that the mobility of the nitroxide increased significantly when the distance from the 4-nitro group to the far end of the chain was about 0·9 nm indicating that the bulky end group was then largely free of constraints and that the depth of the binding pocket must be of this order. Information on the lateral dimensions of the pocket was also obtained. An analysis of hyperfine splittings (p. 285) showed that the nitroxide groups of shorter (more immobilized) haptens were located in an apolar environment whereas the longer haptens had splittings characteristic of an aqueous environment as expected. The results are summarized in Fig. 7.15; the whole binding pocket has considerable hydrophobic character with a depth of 1·1–1·2 nm and entrance dimensions of about 0·6 × 0·9 nm.

Example 5. Spin-labelled membranes

This is one of the most important areas of application of the spin-labelling technique. Spin-labels are ideal probes of membrane structure and motion because their EPR spectra are sensitive to polarity and mobility and because different labels can be synthesized, e.g. lipid and steroid spin-labels, which are close analogues of membrane components and provide complementary information on membrane structure. Electron paramagnetic resonance can answer similar questions to NMR concerning membrane structure and function and the types of membrane preparation which can be studied are also similar. Model membranes and orientated multilayers are particularly suitable subjects for EPR studies. They are commonly prepared by simply allowing

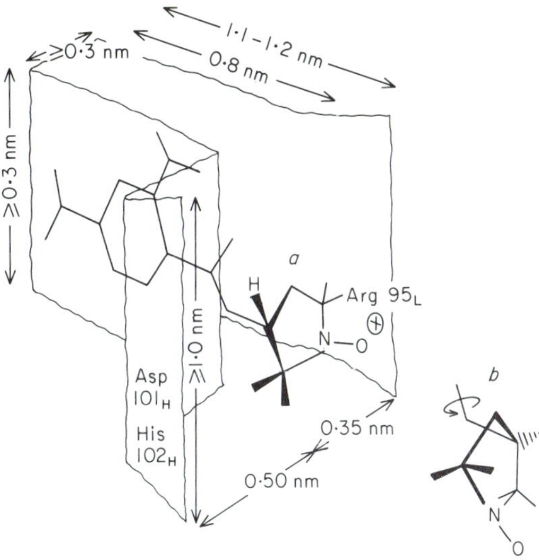

FIG. 7.15 The overall dimensions of the dinitrophenyl binding site on the mouse myeloma protein MOPC 315 as determined by spin-label mapping. The two enantiomers of one of the haptens are shown, only one of which (b) is free to undergo rotation (from Dwek, R. A., Wain-Hobson, S., Dower, S., Gettins, S., Sutton, B., Perkins, S. J. and Givol, D. (1977). *Nature (London)* **266**, 31. Reproduced by permission of Macmillan Journals Ltd).

solutions of the lipid in a suitable organic solvent to evaporate on a glass plate and then hydrating, but a better method in many cases is by repeatedly dipping quartz plates through monomolecular films formed at liquid–air interfaces. A variety of components may be incorporated into model membranes but we shall only discuss a few which occur frequently in biological membranes, i.e. dipalmitoyllecithin (a saturated phospholipid), egg lecithin (a partially unsaturated phospholipid) and cholesterol. The spin-labels commonly used include simple fatty acids and phospholipids with a nitroxide group attached at any required position along the alkyl chain. They may be incorporated into model membranes simply by mixing with the other components before the bilayer is formed; incorporation into biological membranes is done in a variety of ways, usually by allowing the label to enter the bilayer by simple diffusion. Biosynthetic incorporation by feeding the label to the organism concerned is sometimes possible, but as it usually results in metabolism of the label to a greater or lesser extent, it is not commonly used.

Bilayer membranes in general can undergo a variety of phase changes. At low temperatures they usually exist as a gel structure in which the molecules are ordered and both the polar head group and the alkyl chain are relatively rigid. The normal state in which membranes are studied by EPR is the *liquid–crystalline state* present at higher temperatures, where the integrity of the bilayer is still preserved and the polar head groups are still rigid but the apolar region now has a mobile, liquid-like structure. When a series of phospholipids, spin-labelled at various positions along the alkyl chain, are incorporated into simple membranes in their liquid–crystalline state and their EPR spectra are compared, two main features emerge: first, the hyperfine splitting changes according to the depth of the nitroxide group in the bilayer, indicating that the polarity decreases towards the centre of the bilayer; secondly, the order parameter S (eqn (7.7)) decreases towards the interior and a plot of log S against labelling position down the chain gives a straight line. Labels attached near the polar head group have $S \approx 1{\cdot}0$ indicating highly anisotropic motion; the molecules are ordered and can only spin about their long axis. The value of A_{zz} is unaffected by the motion and is equal to $A_{\parallel}$ (Fig. 7.6). Values A_{xx} and A_{yy} are subjected to rapid averaging and the average has the same value as $A_{\perp}$, the splitting in the plane of the membrane. Towards the centre of the bilayer, motion becomes more isotropic with $S \approx 0{\cdot}1$ for fully saturated alkyl chains. This behaviour has been interpreted in terms of two superimposed modes of molecular motion. First, rapid *gauche–trans* interconversions occur about the C—C single bonds and the further down the chain the spin-label is located the more likely it is that the chain and hence the nitroxide A_{zz} direction will be tilted relative to the membrane normal. The fluidity gradient towards the centre of the bilayer is thus partly due to the cumulative effect of these random *gauche–trans* isomerizations. The overall motion in the centre of the bilayer due to this cause is not appreciably different in amplitude from that near the polar head, but it is more random and thus some averaging can now occur between A_{zz} and the other values, giving more isotropic (i.e. disordered) motion. The second mode of motion involves a "precession" of the whole molecule as a rigid stick about an anchor point at the polar head. Here the overall motion increases towards the centre of the bilayer although the inclination of the nitroxide relative to the membrane normal remains constant. This is a typical example of the quantitative information EPR can give on polarity and mobility in membranes, the two properties most basic to their biological functioning.

This kind of study has been made more biologically relevant by including in the model membrane, components such as unsaturated lipids (e.g. egg lecithin) and cholesterol which occur in natural membranes. Membranes containing unsaturated lipids have greater fluidity and a lower *liquid–crystal transition temperature* than membranes containing saturated hydrocarbons; at ordinary temperatures unsaturated bilayers are usually above the transition and have an essentially fluid (but still partially ordered) liquid–crystalline structure, whereas saturated bilayers are below the transition and have a gel structure. The fluidity in unsaturated bilayers is due to the permanent bend introduced into chains by *cis* double bonds, which prevent efficient packing of chains. These mobility effects are well studied by EPR, as are the effects of adding cholesterol in these systems. It is well known that cholesterol condenses partially unsaturated bilayers, i.e. reduces their surface area. Electron paramagnetic resonance studies indicate that order is increased, and cholesterol appears to do this by occupying spaces between chains and reducing the probability of *gauche–trans* interconversions. By contrast, the addition of cholesterol *decreases* the order in saturated bilayers since the structure is already well ordered and the addition of differently shaped molecules can only disrupt this. Thus, EPR studies provide good support for the hypothesis that the biological function of cholesterol is to stabilize membranes against fluidity changes resulting from variation in lipid composition. The effects can be monitored equally well using both phospholipid and steroid spin-labels. Many other examples might be quoted where EPR has been used to monitor the effects of biologically active molecules on membrane structure and mobility. Thus, general and local anaesthetics are thought to act by blocking conductance channels in nerve membranes necessary for transmitting action potentials and EPR studies suggest that they do this by disordering the bilayer structure. The effects are critically dependent on concentration and size of molecule (e.g. chain-length in alcohols) and there is a correlation between the disordering detected by EPR and pharmacological potency.

Finally, another type of motion of considerable importance to the function of biological membranes may be mentioned. This is *lateral diffusion*, the process whereby molecules can diffuse with unusual rapidity in the plane of the bilayer and which may be important in a variety of cell-surface recognition processes such as hormonal recognition of receptor sites, protein–protein interactions, and antibody–antigen interactions. Lateral diffusion rates have been measured by applying spin-labelled lecithins as a concentrated spot on membranes and observing the rate at which the EPR spectral broadening, initially present due to spin–spin interactions, decreases as the labels diffuse apart. Again, it has been found that cholesterol reduces mobility.

Example 6. The primary photochemical events of photosynthesis

Electron paramagnetic resonance can be readily applied to the study of photosynthetic electron transport because free radicals or metalloproteins are involved at almost every step in the process. The widely-accepted "Z-scheme" of photosynthesis in algae and higher plants involving two separate photosystems (I and II) linked by an electron transport sequence (Fig. 7.16) will be familiar to most biochemists, although in reality our detailed knowledge of this is fragmentary compared with mitochondrial electron transport. Electron transport down the potential gradient is spontaneous and is coupled to ATP synthesis and the reduction of NADP. The initial event in both photosystems

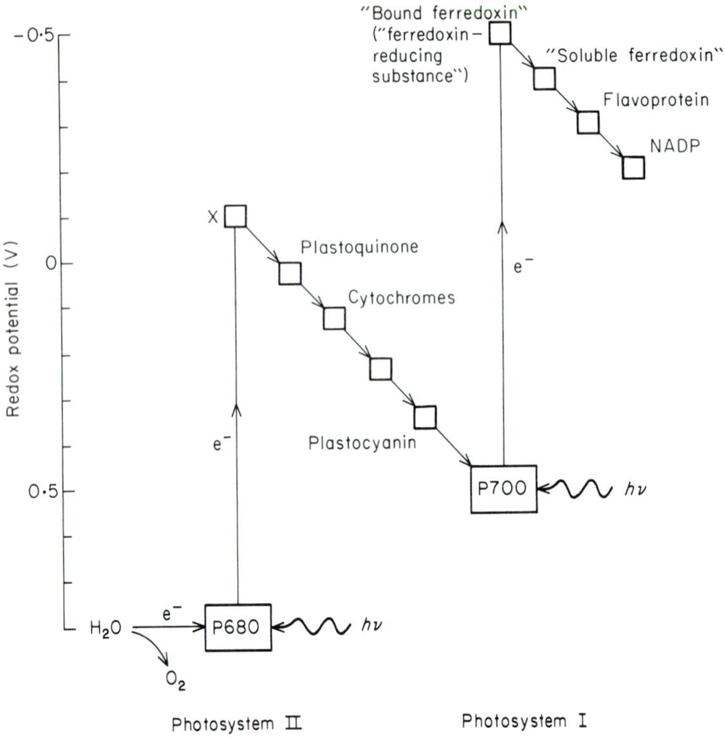

FIG. 7.16 Schematic representation of photosynthetic electron transfer in plants. Free-radical EPR signals can be observed from the primary electron donor in both photosystems I and II, and Fe—S type EPR signals can be observed from electron acceptors at several points in the system.

is the absorption of light quanta by specialized chlorophyll molecules. This results in the loss of an electron, i.e. oxidation, and the chlorophyll is simultaneously bleached. The electron passes to the primary acceptor which is thus reduced and begins the electron transport sequence. The primary donor of photosystem I (P700) is activated by far-red light and the primary donor of photosystem II (P680) by red light.

When intact chloroplast suspensions at room temperature are illuminated, a composite EPR signal from two sources is seen: "signal I" which is narrow with a rapid rise and decay and "signal II" which is broader and complex with a slow decay. In the dark following illumination, signal II persists and can be observed separately. The kinetics of appearance of signal I are unaltered by decreasing the temperature to 77 K or lower, indicating that a physical change rather than a chemical change is involved. Figure 7.17 shows the signals observed at 25 K. The kinetics of signal I correlate exactly with the absorption

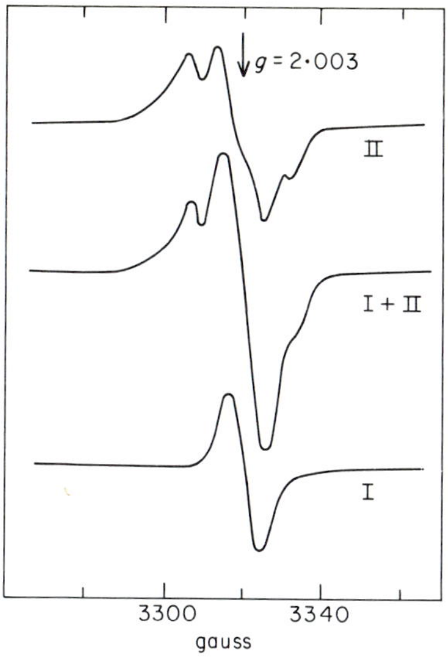

FIG. 7.17 Chloroplast free-radical EPR signals at 25 K. Signal II occurs in dark following illumination, signal (I+II) is a composite signal observed in light, and signal I is the difference (light minus dark) (from Bearden, A. and Malkin, R. (1975). *Quart. Rev. Biophys.* **7**, 131).

change when P700 is oxidized (also temperature independent) and there is a 1 : 1 correspondence between the number of spins produced and the amount of bleached $P700^+$. Several other observations indicate that $P700^+$ is the source of signal I: (i) the action spectra for bleaching and inducing the EPR signal are the same, (ii) signal I and P700 have the same redox potential and (iii) preparations enriched in photosystem I activity have a larger signal. Signal II has similarly been assigned, though somewhat more tentatively, to the primary electron donor of photosystem II (P680) which operates with red light instead of far-red light. The basis of this assignment was that, under conditions where P700 was chemically oxidized and signal I present, far-red light gave no further change but red light induced a new signal (II) indicating that photosystem II was the source. This was confirmed by comparing signal II to the EPR spectrum of oxidized chlorophyll radicals.

In addition to these free-radical signals, EPR signals from a soluble form of ferredoxin (an Fe—S protein) were among the earliest to be detected in photosynthetic systems. Initially, this ferredoxin was proposed as the primary electron acceptor of photosystem I but it became clear from later studies, in which preparations at low temperatures (77 K) were exposed to white light, that another ferredoxin was the primary acceptor. An EPR signal was seen which was neither that of a free radical nor that of the soluble ferredoxin although it was still characteristic of reduced Fe—S proteins ($g = 1{\cdot}86$, $1{\cdot}94$,

2·05). Again, the low temperature precluded chemical processes intervening between the photochemical event and the appearance of the signal, which persisted in systems from which the soluble ferredoxin, had been removed. Thus, a tightly bound form of ferredoxin (P430) is now taken to be the primary acceptor. Far-red light is active in reducing it as expected and the signal is found in chloroplast preparations with only photosystem I activity. The final confirmation that both the primary electron donor and primary electron acceptor of photosystem I had been identified by EPR came from a quantitative correlation between the signals from oxidized $P700^+$ (signal I) and from reduced bound ferredoxin.

Example 7. Studies on irradiated nucleic acids and whole tissues

Free radicals are involved in γ- and X-radiation damage of biological materials, and free radicals and paramagnetic metals can be detected by EPR in both normal and abnormal intact tissue. The study of whole tissues and the connection between free radicals, radiation damage, and carcinogenesis was one of the first applications of EPR in biology because of the obvious clinical relevance. These early studies were limited by the presence of a large number of unidentified EPR signals in whole tissue and by artefacts of sample preparation, but despite this there is much potential for EPR in whole-tissue studies because definite changes have been detected in the early stages of malignancy which could be of diagnostic value. However, it is not possible at present to do whole-body studies and avoid the need for excising tissue.

The most useful findings in γ-irradiated systems have been made with single bases, nucleic acids, amino acids and proteins, and without these the whole-tissue irradiation studies would be practically uninterpretable. Radiation damage of tissues is mostly due to damage of nucleic acids, and it has been shown that the molecular weight of DNA is drastically reduced after irradiation. The effects of irradiating single crystals of bases at low temperatures (particularly thymine and cytosine) have been extensively studied by EPR. Free radicals are produced, and the localization of unpaired electrons can be determined from the hyperfine structure. In many cases a primary anionic radical appears to be formed by reaction with the free electrons generated by irradiation and this is then converted to a hydrogen addition radical, e.g.

$$\mathrm{T} \xrightarrow{e^-} \mathrm{T}^{\cdot -} \xrightarrow{\mathrm{H}^+} \mathrm{TH}^{\cdot}$$

In the earliest EPR studies, which were not carried out at low temperatures, only the second product could be detected. This thymine radical is conspicuous in irradiated whole DNA, and EPR studies also show that long-range transfer of charge can occur over many base-pairs until it is trapped by some component with high electron-affinity. This is important for radiobiology because it means that the absorbed energy may be concentrated in particular regions and the tissue "sensitized" to radiation damage; this could, for instance, render chemical carcinogens more effective because evidence suggests that a common mechanism for their action might be reaction with DNA via free-radical intermediates.

A knowledge of such effects could have application in radiation therapy. It is known that metals can act as electron scavengers and reduce radiation damage in proteins and probably also in nucleic acids. A major cause of protein damage is the breaking of cystine S—S bonds and radical formation, and this is reduced by the presence of metals. In addition, it is known that many malignant tumours have anoxic regions whereas normal tissues have high O_2 levels, and also that radiation damage is increased by the presence of O_2. This tends to defeat the object of radiation therapy, but by the use of suitable exogenous free radicals the situation might be reversed and it might be possible to "sensitize" the anoxic regions to radiation.

General References

Berliner, L. J. (Ed.) (1976). "Spin-labelling: Theory and Applications". Academic Press, New York and London.

Ingram, D. J. E. (1969). "Biological and Biochemical Applications of Electron Spin Resonance". Hilger, London.

Knowles, P. F., Marsh, D. and Rattle, H. W. E. (1976). "Magnetic Resonance of Biomolecules. An introduction to the Theory and Practice of NMR and ESR in Biological Systems". Wiley–Interscience, NewYork.

Swartz, H. M., Bolton, J. R. and Borg, D. C. (1972). "Biological Applications of Electron Spin Resonance". Wiley–Interscience, New York.

8 Atomic Absorption Spectroscopy

E. J. Wood

Introduction

Metallic elements are extremely important in biological systems. Familiar examples of metals which have a key role in living organisms are iron in haemoglobin, magnesium in chlorophyll and calcium in bone. Other metals such as mercury and lead are highly poisonous in comparatively small amounts, indicating that they can interact in a deleterious way with living systems. Therefore, clinicians and toxicologists, as well as scientists working in a wide variety of fields including biochemistry, the pharmaceutical and food-processing industries and pollution control, require methods for determining metals in biological materials, in drugs, in waters, in soils and in foodstuffs. Although there are many chemical techniques that could be used (colorimetry, volumetric and gravimetric analysis), in almost all cases it is likely that atomic absorption spectrophotometry would turn out to be the fastest, most specific, and often the most sensitive method of analysis.

History

The principle of atomic absorption has been known to astronomers for over 150 years. In 1802 Wollaston noticed dark lines in the spectrum of the sun and these were studied more thoroughly by Frauenhofer in 1814. However, the theoretical basis for atomic absorption was really established in 1859 when Bunsen and Kirchhoff showed that the dark D line observed in the solar spectrum corresponded exactly with the yellow line emitted by a flame into which sodium salts had been introduced. The conclusion was that the sun emitted light of practically all wavelengths but that sodium vapour present in the atmosphere surrounding the sun absorbed light at certain wavelengths.

Thus was the technique of atomic absorption spectrometry established, namely *the measurement of the absorption of resonant radiation by free atoms in the gaseous state.* The method was first used in chemistry to measure the concentration of mercury in air. However this was only possible because mercury has a sufficiently high vapour pressure at room temperature and it was not until the 1950s that atomic absorption spectrometry was proposed as a generally applicable method of analysis. Its practical usage had to wait for the development of ways of producing atomic vapour of metals (e.g. by atomizing their compounds in flames) and the manufacture of lamps capable of acting as light sources (the so-called hollow-cathode lamps). As we shall see, the great advantage of the technique compared with other methods is that very little sample preparation or manipulation is required because other metals, on the whole, do not interfere.

Atomic Spectra

In atomic absorption spectrometry we are, as the name implies, concerned with *atomic spectra*, and it is as well to remind ourselves how these arise before we proceed further. In addition, it will help us to compare and contrast with atomic absorption another technique which is commonly used for detecting and measuring the concentrations of elements, namely *emission flame photometry*.

It is a familiar principle that the colours of flames depend on which elements are present in the flame, for example a red flame test for calcium and a green one for barium. In these instances light of certain specific wavelengths is being emitted by thermally excited atoms in the flame. We see a similar phenomenon when certain gases are excited by electrical energy, for example in neon advertising signs.

As we have seen in Chapter 1, such spectra are associated with quantized changes in atomic energy produced by variations in the orbital motions of valence electrons. For any atom there exists an extensive set of possible energy levels, and absorption or emission of radiation results from a transition between particular pairs of levels. Kirchhoff's Law embodies the principle of atomic absorption: "Any species that can be excited to emit radiation at a particular wavelength will also absorb radiation at that wavelength". Hence comes the idea of resonance radiation. Those transitions which are possible are predicted by the so-called "selection rules" of quantum mechanics (see p. 19) and the practical consequence is a unique spectrum for each element. In flame photometry the sample is atomized in a hot flame and atoms of the element are thermally excited to higher energy levels. As they return to the ground state, light of characteristic wavelength(s) is emitted. By selecting a characteristic spectral line and measuring its intensity it is possible to determine the concentration of the element in the flame and hence in the sample being analysed. Flame photometry is still the method of choice for determining certain elements such as sodium and potassium in biological fluids, because of the higher sensitivity for these particular elements. However, for the majority of elements this is not so and on the whole the technique is more difficult to use than atomic absorption for reasons which will become apparent shortly.

The other important feature of atomic spectra is that the transitions appear to be isolated causing the emission of "sharp" lines. Indeed atomic spectra are often called "line spectra", but as we have seen in Chapter 1 the sharpest line has a finite bandwidth. There are numerous well-understood reasons for this, such as "natural" broadening, Doppler broadening, Lorentz broadening, and so on, and this has a number of important consequences in atomic absorption theory and measurement. In emission flame photometry the sharp nature of the lines means that, provided one has a good quality monochromator, it is comparatively easy to isolate (and hence to measure) the light corresponding to a particular transition. Recent developments in the field of multi-element analysis by emission spectrometry have included attempts to use an extremely high resolution device called an echelle spectrometer. A multi-element sample is atomized in an argon "plasma" (which is initiated by arcing a high-voltage spark through argon gas) where the temperature may be as high as 10 000 °C. The emitted light is analysed by an echelle grating which is a special type of diffraction grating with very accurately ruled broad, flat grooves. The resulting spectrum is a two-dimensional pattern of spots of light in a small area, and a bank of photomultipliers at the focal plane measures the intensity of light at each of up to 20 spectral lines simultaneously.

The main problem with the emission method, given good wavelength selection, is that the intensity of emission (which is being measured in order to determine concentration) depends critically on the position and temperature of the flame. (In the method described above, this is overcome by the use of a plasma.) As we shall see, temperature is nowhere near so critical in atomic absorption and in addition, wavelength selection is easier. For convenience some of the advantages and disadvantages of the emission and absorption methods are compared in Table 8.1.

TABLE 8.1 Advantages and disadvantages of emission and atomic absorption spectrophotometry

Emission	
Advantages	Disadvantages
1. No light sources required. 2. For many elements wide choice of spectral lines, often in visible region where measurements are easy. 3. Flexible over a wide concentration range. 4. Qualitative as well as quantitative. 5. Simultaneous multi-element analyses possible	1. Changes in temperature of flame give large changes in intensity of emission. 2. Background emission often a problem. 3. High-quality monochromator essential. 4. Considerable operator skill required for reliable results.

Absorption	
Advantages	Disadvantages
1. Technique simple to set up and operate: easily automated. 2. Double-beam arrangement possible to counteract variations in intensity of light source. 3. Sharp line sources give very good spectral sensitivity. 4. Relatively insensitive to changes in flame temperature	1. Special light source required for each element to be analysed. 2. Many of the useful spectral lines are in the UV where measurement is more difficult. 3. Not easy to make a linear calibration over a wide concentration range. 4. Spectral interferences can occur but are less apparent than with emission and are therefore more likely to cause error.

Principle of Atomic Absorption

In atomic absorption spectrophotometry the characteristic spectrum of an element is passed through a flame in which is present atoms of the same element (the sample to be analysed) and the consequent *reduction* in intensity at a particular wavelength is measured. As it can be shown that the amount of light absorbed in this way depends upon the concentration of the element in the vapour (i.e. the absorbance obeys Beer's law), we have a method for measuring accurately the concentration of a single element in a complicated mixture. Other metal atoms in the flame do not interfere because there is no light of suitable wavelength for them to absorb. By interposing a monochromator between the flame and the detection device, any interfering light, for example, *emission* from the flame itself, is excluded. The problems that present themselves in atomic absorption spectrophotometry are: first, how to produce light of the characteristic wavelength; secondly, how to atomize the sample in the flame, (although of course this problem is shared with emission spectrophotometry); and thirdly, how to monochromate and measure the extent of absorption. A typical layout in an instrument follows this arrangement (see Fig. 8.1), the source of light usually being a so-called "hollow-cathode lamp".

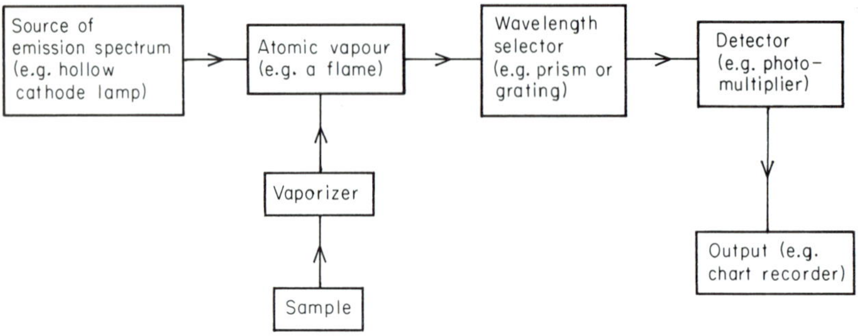

FIG. 8.1 Block diagram of an atomic absorption spectrophotometer.

Theory

Although bandwidths of atomic spectral lines are very narrow (e.g. 0·005 nm), the fact that a spectral line may never quite be represented by a unique wavelength (or frequency) introduces considerable difficulties into the derivation of formulae. This is a major reason for the problems that arise in attempts to perform *absolute* analyses. Usually it is found convenient to keep as many as possible of the instrumental parameters constant and to calibrate with samples of known metal concentration, checking that Beer's law is indeed obeyed over the concentration range being investigated.

Such quantities as the *absorption coefficient*, the *total absorption factor*, and the *absorbance* may be used to represent atomic absorption, but in fact the present widespread use of the technique is based chiefly on the straightforward interpretation of *absorbance* measurements made with sharp line sources. As in conventional UV and visible spectrophotometry (Chapter 2) absorbance is defined by

$$A = \log_{10} \frac{I_0}{I}$$

where I_0 and I are the incident and transmitted intensities at the absorption "cell", usually a flame, but sometimes a graphite furnace or other arrangement.

Instrumental Arrangement

There are of course many ways in which an instrument could be constructed in order to measure the amount of absorption of a certain emission line by the atomic vapour of an element. There are several ways of producing the emission line, several ways of atomizing the sample and several ways of measuring the extent of reduction in light intensity. Figure 8.2 translates the block diagram of Fig. 8.1 into a simple instrumental arrangement. We will now look at each of the components in turn.

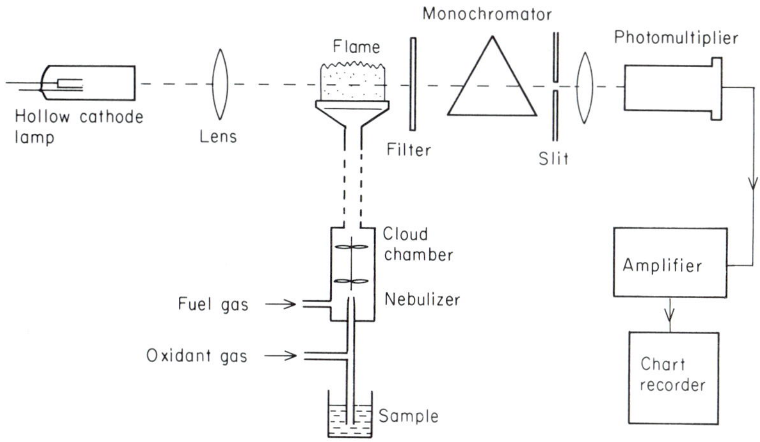

FIG. 8.2 Typical arrangement in a commercial instrument.

The spectral source

The hollow-cathode lamp is, at present, the most widely used light source in atomic absorption instruments, although there are many other systems both possible and actually employed. In the hollow-cathode lamp (Fig. 8.3(a)), the hollow cylindrical cathode contains the metal to be determined, or an alloy of it, and the lamp is filled with an inert gas such as argon or neon at low pressure. When a current is passed this filler gas is positively ionized at the anode and the ions are accelerated towards the cathode. The resulting bombardment of the cathode material causes "sputtering" and formation of atomic vapour which is raised to an excited state. As the atoms return to the ground state the lines characteristic of the metal are emitted (Fig. 8.3(b)).

Several practical technical details need attention in order to obtain the best output intensity coupled with length of life of the lamp. Thus, self-absorption

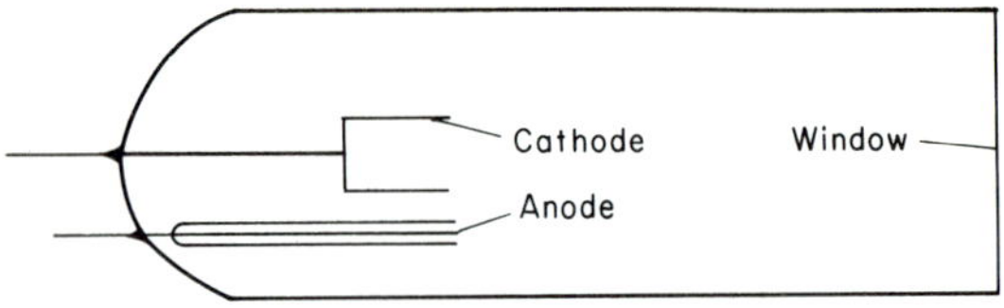
Cathode
Window
Anode

(a)

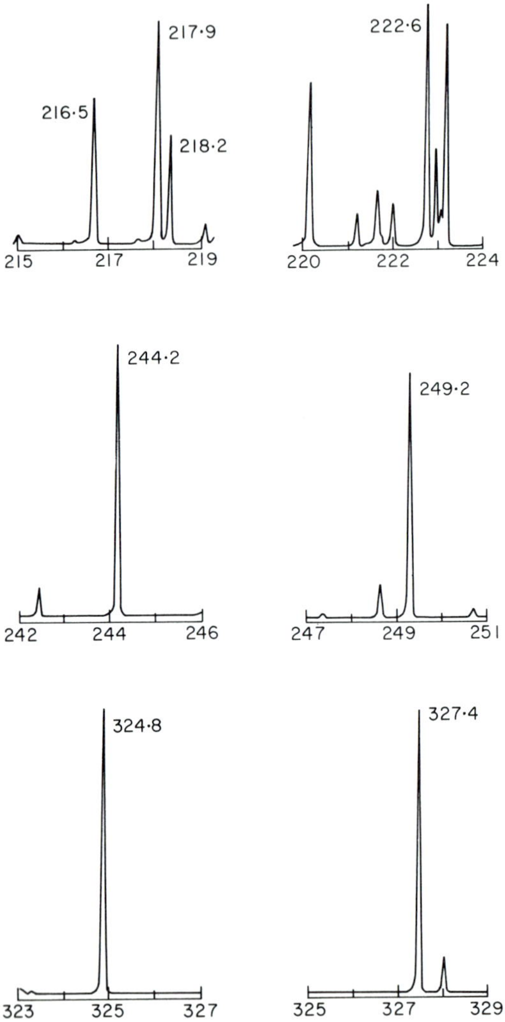
216·5
217·9
218·2
215
217
219
222·6
220
222
224
244·2
242
244
246
249·2
247
249
251
324·8
323
325
327
327·4
325
327
329

(b)

by the atomic vapour is minimized if the lamp is run at the lowest current compatible with adequate light output. Interference by absorption by the filler gas also needs to be considered (see Fig. 8.3(b)). Recent improvements in lamp design have greatly increased the intensity of light emitted. In "high-intensity" lamps, for example, the cathode merely produces an atomic vapour by sputtering, which is excited by a discharge from secondary electrodes. Multi-element lamps are also available in which an alloy of two or more metals is used in the cathode. An alternative solution to the problem of having to have a different lamp for each element is to use demountable lamps. A more recent development is the microwave lamp. This consists of a silica tube containing a small amount of the metal to be analysed and filled with an inert gas at low pressure. This so-called "electrode-less lamp" is excited by microwave frequencies, and the intensity of light emitted is several hundred times that from a conventional hollow-cathode lamp. Such lamps have not as yet found wide application in laboratories dealing with the analysis of biological materials.

Production of atomic vapour

The heart of an atomic absorption spectrophotometer is the flame or furnace where the sample is atomized and through which the light from the spectral source passes. Flame temperature is very critical in emission flame photometry, but it is not so important in atomic absorption. For calcium, for example, at a flame temperature of 2000 K, only one in 10^7 of the atoms present in the flame will be in the excited state (i.e. be capable of emitting light). If the temperature is raised to 3000 K, one in $2{\cdot}5 \times 10^4$ will be excited, with a consequent gain of 400 times in the emission. However there will be a corresponding fall of only 0·004% in the degree of absorption. Thus there is little advantage in atomic absorption in going to the inconvenience of using extremely hot flames, provided that the energy of the flame is sufficient to convert an aerosol to an atomic vapour. The actual flame or furnace temperature is not crucial and indeed at higher temperatures the production of significant ionization actually *reduces* the number of ground state atoms available to absorb the radiation.

What is more important is the *length* of flame traversed by the light beam, since it should be fairly obvious that the amount of absorption depends on the

---

FIG. 8.3 (a). Hollow cathode lamp. If the cathode were copper the filler gas would be neon. As most of the useful emission lines are in the ultraviolet region of the spectrum, the window through which the light emerges would be quartz. (b). Part of the emission spectrum from a copper hollow cathode lamp showing wavelengths in nm. Note the extremely narrow bandwidth: the line at 324·8 nm is that most commonly used. Other lines are present, some of which may be due to the filler gas, and some of the copper lines may be absorbed by the atoms of the filler gas. Thus, in a chromium lamp the chromium line at 357·9 nm may interfere with an argon line at 357·7 nm and therefore neon would be preferred.

volume of atomic vapour traversed by the light beam, as well as the *concentration* of atoms in the flame. Very often, therefore, the light from the hollow-cathode lamp (or other source) is passed through a long narrow flame produced by a slit-burner. An atomizer system (see Fig. 8.2) passes an aerosol containing the element to be analysed from the sample into a flame in which a fuel gas such as acetylene is burned in oxygen or air. All that is required of the flame is that the temperature is such that chemical bonds are broken releasing individual atoms. In this condition they can absorb resonance radiation at the wavelengths characteristic of that element. It is important that the aerosol is homogeneous, because large droplets atomize in the flame at a different rate to small droplets giving variation. Ultrasonic nebulizers have been utilized in an attempt to improve the performance in this respect, but are not in common use.

Although flames are the commonest means of producing atomic vapour, and in some respects the most convenient means in present-day equipment, they are not the only way and certainly not the most efficient way. Often flames are acutally inconvenient for a number of reasons (e.g. direct atomization of solid compared with liquid samples) and it is advantageous to use a "non-flame cell" such as a tubular graphite furnace. This has the additional advantage for the determination of trace amounts of many elements, of achieving higher concentrations of atoms in a smaller confined volume than would be possible in a flame with a nebulizer system, and with the possibility of better temperature control. It must be admitted that with flame-nebulizer systems, over 90% of the sample is wasted, not actually being used in the analysis. A typical graphite furnace that can be used instead of a flame is shown in Fig. 8.4.

Measurement of absorption

When atoms of the sample element are present in the flame or furnace, the amount of light passing through the flame from the hollow-cathode lamp is reduced because of the absorption. The usual way to make a measurement is therefore to record the light intensity before the sample is in the flame, and then observe the reduction in intensity as sample atoms are introduced into the flame. A photomultiplier is the normal way of observing the light intensity, but a monochromator is interposed between flame and photomultiplier tube to select the characteristic spectral line required and to block out any other light arising from the flame by emission, etc. The photomultiplier tube output is amplified and the absorbance is eventually displayed as a meter reading, a chart record or a digital printout. When no sample atoms are present in the flame, the observed intensity actually corresponds to zero absorption but is

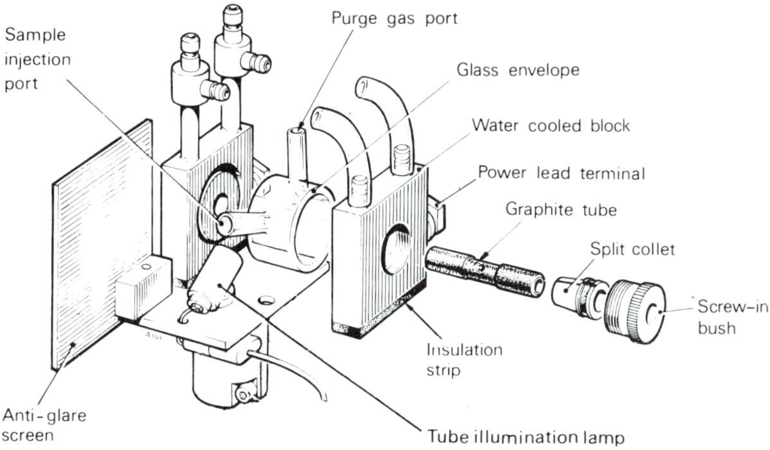

FIG. 8.4 A commercial flameless atomizer (Pye-Unicam). The atomizer unit consists of a graphite tube held between two cooling blocks which act as electrical contacts. The tube is machined from high-purity graphite and, in the centre section, where wall thickness is 0·5 nm, a 2 mm hole has been drilled for sample injection. A protecting atmosphere (Ar or N_2) is maintained around the tube. (Published with the permission of Pye-Unicam Ltd)

often recorded as 100% transmission. When sample atoms are present the absorbance is increased, or the percentage transmission is reduced (Fig. 8.5).

In order to determine the concentration of an element in an unknown solution, the instrument is calibrated by running through the flame pure samples containing known concentrations of the element. When a calibration curve is plotted it can be established that Beer's law is obeyed, and the concentration of the unknown(s) can be read off (Fig. 8.6).

It is obvious that the instrument must be stable whilst measurements are being taken and especially that the output from the hollow-cathode lamp must not change. Similarly, the length and temperature of the flame must not change, nor must the rates of atomization of sample and standard solutions be different. Much of the complexity of modern instruments is aimed at ensuring that this sort of stability is available routinely. Commonly the light beam is "chopped" or modulated before passage through the flame, either by means of a rotating chopper disc or by supplying the hollow-cathode lamp with an alternating voltage. This eliminates light emitted by the flame, because this generates a DC signal at the photomultiplier which can be separated electronically from the AC signal from the hollow-cathode lamp. The more sophisticated modern instruments have a double-beam system (Fig. 8.7) in which light from the lamp is passed alternately through the flame and past the flame, and the *ratio* of the intensities measured. Such an arrangement compensates for drift in lamp intensity, detector sensitivity and electronic gain, giving better precision to analyses.

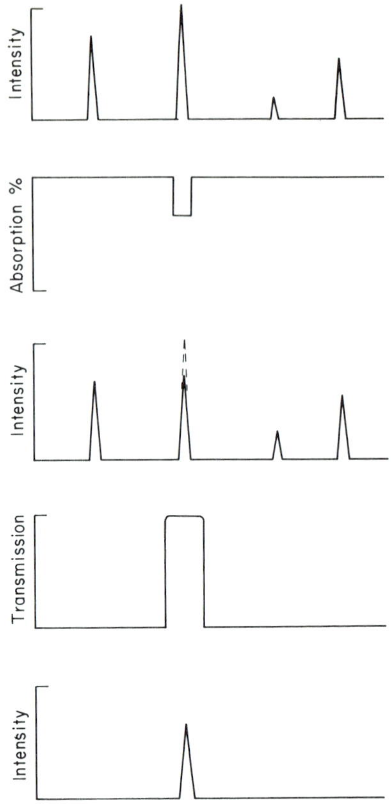

Fig. 8.5 The process of atomic absorption. (i) Hollow-cathode lamp produces emission spectrum of element to be determined; (ii) atomic vapour of sample absorbs energy at the resonance line; (iii) emission spectrum *after* absorption by the sample; (iv) monochromator allows narrow band of wavelengths to pass excluding all but the resonance wavelength; (v) photodetector records all the light reaching it—now in fact only the reduced intensity of the resonance line.

The Sample

Although one of the greatest advantages of atomic absorption spectrophotometry is that it is, on the whole, unnecessary to do much in the way of preparing the sample for analysis, a number of factors need to be considered, otherwise completely erroneous results can be obtained. For example, if one were trying to measure a metal such as iron or copper in serum, it is as well to be aware that serum is a very viscous liquid with a high content of dissolved solids, especially protein. If serum is used undiluted, it will atomize much more

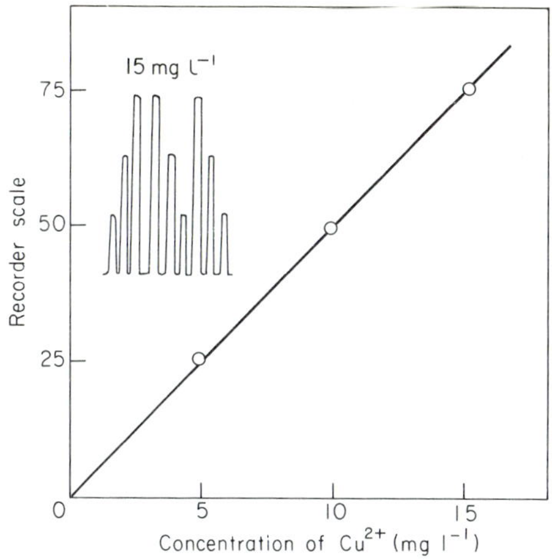

FIG. 8.6 Typical readout and calibration curve in atomic absorption analysis for copper.

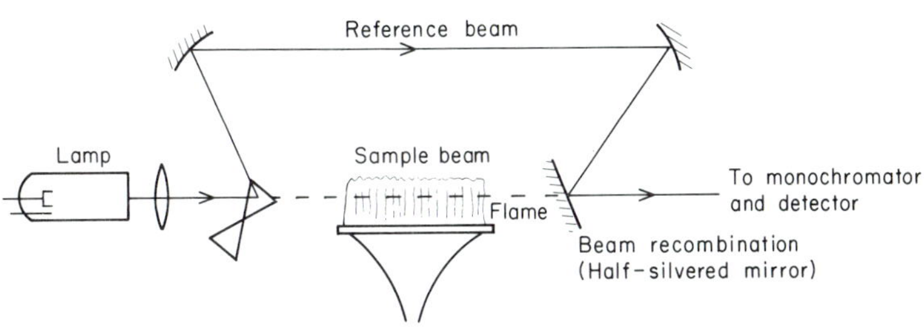

FIG. 8.7 Double-beam optical system. Light from the hollow-cathode lamp is passed alternately through and past the flame. The electronics produce a readout which is the ratio of sample beam intensity to reference beam intensity.

slowly than standard solutions of the metal made up in water. In addition, the protein may precipitate in hot tubes near the burner and cause blockages. Further, minute particles of solid (precipitated) protein, formed actually in the flame, can absorb or scatter light. There are very simple and straightforward solutions to these problems, of course, such as deproteinizing the serum with trichloroacetic acid, but one must first be aware that such problems exist. Therefore, it is essential either to follow standard procedures or to investigate all the variables before accepting the results of an analysis.

The other, and probably most common and serious, source of error is chemical interference which may be due to the composition of the sample or of the flame gases. The presence in the flame of certain elements may result in the formation of stable compounds of the metal to be determined. The outcome of this is that there is a decrease in the number of free atoms which are able to absorb resonance radiation and thus an apparent concentration lower than the true concentration. Examples are the formation of stable phosphates, silicates or aluminates by some metals, especially the alkaline earths. Thus the sensitivity of the method to calcium is lowered in the presence of phosphate. The usual solution to the problem is to add a "releasing agent" in excess which reacts and forms a stable compound with the phosphate but does not otherwise interfere with the analysis. The best-known releasing agent for phosphate is lanthanum. Stable compound formation may also be overcome by using a hotter flame (e.g. nitrous oxide/acetylene, 2950°C, rather than air/acetylene, 2350°C) but the danger at higher temperatures is that the free atoms of the element become excited beyond their ionization potential when the resulting *ion* does not absorb the resonance radiation.

In the determination of metals such as copper, iron and zinc in blood serum it is usual to deproteinize with trichloroacetic acid to avoid viscosity problems, but an alternative which is often used for the determination of calcium, is to dilute the serum in a solution containing lanthanum. Interferences may become a matter of life or death. For example, ever since the discovery that lithium could be successful in treating manic-depressive patients, many laboratories routinely estimate the lithium content of serum. Now the therapeutic level of lithium in the serum (0·6–1·2 mM) is very close to the toxic level (2·0 mM) and wide differences in tolerance to lithium are shown by individual patients. Careful serum lithium level monitoring is therefore mandatory for every patient treated, and reliable analyses are absolutely essential. In this particular case it is very important that interferences of the type mentioned do not give an estimated blood lithium concentration which is *lower* than the true value.

Other problems arise when the metal to be examined is present in very tiny quantities or when it is complexed in the biological material in such a way that it is not easily dissolved, or easily released in the flame. A concentration step may be necessary and this is often achieved by treating the sample with ammonium pyrrolidine dithiocarbamate, a compound which forms complexes with more than 30 elements and which may then be extracted quantitatively into methyl isobutyl ketone. Lead, calcium and mercury have all been estimated in this way and it is obviously important to be able to determine low levels of these cumulative poisons in natural waters and in foodstuffs. (It may also be recalled here that the use of a solid atomization procedure such as a graphite furnace is also a way of "concentrating" the sample.)

When the interference effects from organic constituents are not fully understood or the metals must be released from biological tissues prior to analysis, a [illegible] involving either dry ashing or wet oxidation may be used. Ashing [illegible] onducted at too high a temperature or volatile elements (e.g.

mercury, arsenic) will be lost. Provided that this is taken into account, dry ashing is rapid and the residue may be taken up in dilute HCl and the analysis performed on this. Often, acid extraction procedures are used with plant materials. Complete recovery of copper, zinc, iron, manganese, and chromium from grass, straw, manure and sewage sludge is achieved by boiling samples for 15 min with 6 M HCl. The extracts are cooled and filtered and then used for analysis. Analysis of elements in soils where they may be combined in silicates, etc., demands more vigorous extraction procedures such as treatment with mixtures of hydrofluoric, nitric and perchloric acids. Aluminium, zinc and manganese have been determined in this way and the same extract may also be used for the determination of sodium, potassium, calcium, magnesium and iron after the addition of a "spectroscopic buffer" containing lanthanum.

Range and Sensitivity

Much of the above discussion has been concerned to show why atomic absorption is an attractive method because of its high *specificity*, and not much has been said about the practical consequences of the *sensitivity* of the method. In what follows we consider its application to the determination of a few elements of clinical or biochemical importance in order to illustrate this aspect, and also to help the comparison with emission flame spectrophotometry. Table 8.2 summarizes information on some of these elements.

TABLE 8.2 Atomic absorption spectrophotometry: elements of clinical and biochemical interest

Element	Serum conc. ($\mu g\ l^{-1}$)	Wavelength used (nm)	Absorption[a] sensitivity ($\mu g\ l^{-1}$)	Detection limit[b] Emission ($\mu g\ l^{-1}$)	Detection limit[b] Absorption ($\mu g\ l^{-1}$)
Ca	100 000	422·9	100	0·1	2
Co	10	240·7	200	30	7
Cu	1 000	324·8	100	10	5
Fe	1 000	248·3	100	30	10
K	170 000	766·5	100	0·2	5
Mg	20 000	285·2	10	4	0·3
Na	3 200 000	589·2	40	0·1	5

[a] That amount of the element, expressed as a concentration in the solution, which reduces the optical transmission of the atomic vapour by 1%.
[b] That amount of the element which produces a signal equal to twice the standard deviation of a series of blank determinations (corresponds to 95% confidence level).

under usual condn
error with AA = 1–2%
with precautions 1/10%

Two main groups of metals may be distinguished in living tissues and serum, the major cations (Na, K, Ca, Mg, etc.) and the trace elements (Cu, Fe, etc.). For the former group, the sensitivity is comparatively unimportant, the relevant factors in deciding upon which method to use being ease, lack of interference, and so on. For the latter group the speed of the analysis is much less important than the fact that the analysis is feasible at all. This is where detection limits are of especial interest and it is imperative that the best performance of the instrument should be obtained. In addition, more elaborate pretreatment of the sample may be acceptable.

Thus, sodium and potassium are usually determined by flame emission photometry, not so much because this gives the best detection limit but because of their low excitation energy. Atomic absorption is, in contrast, the method of choice for magnesium which can be determined directly in a 50-fold dilution of serum for example. Indeed, magnesium has the highest sensitivity for atomic absorption and hence the lowest of all detection limits. Copper and zinc in serum are present in sufficient amounts to be determined accurately without dilution, but usually deproteinization is necessary. The technique is less sensitive for iron. Whole blood, of course, presents no difficulties but for serum (normal Fe level, 0·3 ppm) a special technique is required to improve sensitivity. The iron–protein bonds are first broken by treatment with HCl and then the sample is deproteinized with trichloroacetic acid. Acetone is added to the supernatant and the mixture sprayed into the flame.

In an analysis of the concentration of copper, iron, magnesium and zinc in human hair, the washed hair was chopped and digested with nitric acid at room temperature, followed by the addition of perchloric acid and heating to 200°C. The final clear solution was then simply diluted before analysis by atomic absorption spectrophotometry. The mean values obtained for 18 adult males and which illustrate the sensitivity of the method were: Cu, 14·5 $\mu g\ g^{-1}$; Fe, 15·3 $\mu g\ g^{-1}$; Mg, 21·6 $\mu g\ g^{-1}$; and Zn, 178·7 $\mu g\ g^{-1}$.

Another way of increasing the selectivity for a very narrow bandwidth is to use a resonance monochromator instead of a filter, prism or diffraction grating system. In such an instrument, the resonance radiation from a high-intensity lamp passes through the flame in the normal way but then is focussed on to the cathode of a second hollow-cathode lamp of the same element. The radiation excites the atomic vapour in the second lamp to fluoresce and this amount of fluorescence is measured at ninety degrees by means of a photomultiplier tube. Only resonance radiation can excite fluorescence, and therefore the system acts as a highly selective monochromator and one which is very stable.

Finally, it may also be mentioned that after absorption has taken place most of the atoms in the flame return to the ground state without emitting radiation (their energy is lost by collision) but a small proportion rapidly re-emit the light quanta absorbed. This process is called atomic resonance fluorescence. Under the condition normally used in atomic absorption spectrophotometry, resonance fluorescence can be ignored. However, by changing the experimental arrangement to observe the radiation emerging perpendicular to the exciting beam, only atomic fluorescence may be measured. This may be used to determine the concentration of the element in the flame.

Further Reading

General

Christian, G. D. and Feldman, F. J. (1970). "Atomic Absorption Spectrophotometry and its Application to Agriculture, Biology and Medicine", Interscience, New York.

Dawson, J. B. and Heaton, F. W. (1967). "Spectrochemical Analysis of Clinical Material", C. C. Thomas, Springfield, Illinois.

Dulka, J. J. and Risby, T. H. (1976). Ultratrace metals in some environmental and Biological Systems. *Anal. Chem.* **48,** 640A.

Kirkbright, G. F. and Sargent, M. (1974). "Atomic Absorption and Fluorescence Spectroscopy", Academic Press, London and New York.

Price, W. J. (1970). Atomic absorption spectrophotometry. *Process Biochem.* Jan. 1970.

Price, W. J. (1972). "Analytical Atomic Absorption Spectrometry", Heyden, London.

Videotape

A videotaped audiovisual programme on atomic absorption spectrophotometry (principles, instrumental requirements and conditions, sample preparation) is available from Pattison and Co., St. Albans, Hertfordshire, UK.

Articles dealing with the application of the technique

Balazs, N. D. H., Pole, D. J. and Masarei, J. R. (1972). Determination of gold in body fluids by atomic absorption spectrophotometry. *Clin. Chim. Acta* **40,** 213.

Evenson, M. A. and Anderson, C. T. (1975). Ultramicro analysis for copper, cadmium and zinc in human liver tissue by use of atomic absorption spectrophotometry. *Clin. Chem.* **21,** 537.

Hansen, J. L. and Freier, E. F. (1967). The measurement of serum magnesium by atomic absorption spectrophotometry. *Am. J. Med. Technol.* **33,** 1.

Hansen, J. L. (1968). The measurement of serum and urine lithium by atomic absorption spectrophotometry. *Am. J. Med. Technol.* **34,** 1.

Harrison, W. W., Yurachek, J. P. and Benson, C. A. (1969). The determination of trace elements in human hair by atomic absorption spectrophotometry. *Clin. Chim. Acta* **23,** 83.

Trudeau, D. L. and Freier, E. F. (1967). Determination of calcium in urine and serum by atomic absorption spectrophotometry. *Clin. Chem.* **13,** 101.

9 Mass Spectrometry

A. J. Geddes

Introduction and Brief Summary of Applications

The chief uses of mass spectrometry may be briefly summarized as follows:

(i) the determination of the molecular weight of a sample compound;
(ii) the determination of the elemental composition of the compound under investigation or of fragments derived from it;
(iii) the quantitative assessment or detection of trace constituents;
(iv) the elucidation of the chemical structure of a molecule;
(v) the determination of reaction mechanisms.

The major advantage of the technique is its extraordinary sensitivity and accuracy—in favourable cases as little as 10^{-14}–10^{-15} g of sample can provide much useful information. The major disadvantage is the requirement for vaporization of the sample without degradation.

The use of mass spectrometry in biological and clinical investigations has become widespread only during the last ten years or so and has been made possible by a combination of technical advances in instrumentation and design changes appropriate to the solution of biological problems, especially the direct coupling of a mass spectrometer to a gas chromatograph and by the use of knowledge accumulated over the years from studies of a wide range of organic molecules of non-biological origin.

The mass spectrometer performs four essential functions:

(i) the volatilization of the sample molecules if these are not already available in the gaseous phase;
(ii) the production of ions from the vaporized sample molecules;
(iii) the separation of ions according to their mass to charge ratio (m/e) by passage through magnetic and electric fields;
(iv) the recording of the mass spectrum, usually in the form of a graph with m/e values plotted horizontally and ion abundances vertically and, if a computer system is also attached, the automatic calculation and digital output of the mass and abundance of the various ion species.

The initial process of ionization nearly always involves the removal of one electron from the sample molecule to produce the so-called molecular ion (M^+). The mass measurement of this ion immediately gives the molecular weight of the sample compound. In addition, by very accurate mass measurement, the number of each different atomic species in the molecule can be determined. To take a very simple example: the accurate masses of CO and N_2 are 27·994914 and 28·006148 atomic mass units (amu) respectively. Such differences can easily be detected by most mass spectrometers.

Usually the ionization process involves the transfer of excess energy which is sufficient to break one or more bonds in the molecular ion so that a variety of fragment ions and neutral species are produced. The mass measurement of these fragments allows the determination not only of which functional groups are present but also of the way in which they are connected together, i.e. the chemical structure. For example, in the compounds

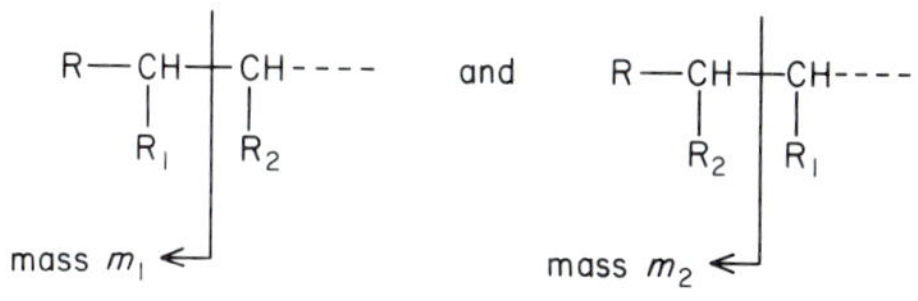

where R_1 and R_2 could be amino and hydroxyl groups respectively, a peak at mass m_1 would indicate that the functional group R_1 was connected via a single CH group to R, whereas if the peak occurred at mass m_2 then it would be R_2 which was so connected to R.

The fragmentations are frequently very complex and numerous and the mass spectrum of even fairly simple compounds can contain peaks at virtually every whole mass number up to the molecular weight. The strongest peaks corresponding to the most abundant ions may result from a completely unexpected fragmentation, while some fragments may be of such low abundance that they are not detected. Rearrangements of the atoms within the fragments can also occur and lead to species which are not components of the original molecule. For example, the ion $C_2H_5^+$ is present in the mass spectrum of

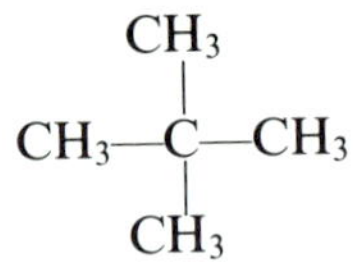

These difficulties should not be a deterrent to use of the technique and are just mentioned as a warning to the potential investigator that a facile interpretation of the spectrum is seldom possible. The known fragmentations from a multitude of chemical moieties are well documented and trial experiments with a series of compounds similar to those which it is proposed to investigate soon brings about a rapid appraisal of the problems likely to be encountered. It is not the purpose here to enter into a detailed discussion of fragmentation mechanisms; there are plenty of excellent books available which are devoted almost entirely to this aspect of mass spectrometry (see end of chapter).

The volatility requirement imposes a mass limit of about 1000 on the molecules which can be investigated directly, but this could hardly be called restrictive. The next generation of mass spectrometers will have a mass limit of 2000–3000. It is common practice to prepare derivatives of the molecule under investigation by the attachment of certain chemical groups which considerably enhance the volatility. Some methods of ionization (see pp. 346–352) are more appropriate than others if the volatility problem is particularly intractable, although these can lead to deficiencies in other areas such as sensitivity and resolution. Biopolymers, of course, present additional problems. They must first be broken down into component parts small enough to satisfy the volatility requirement. As a compensation for this extra effort, the mass spectra of the components can be unusually simple as a result of the fact that the "building blocks" of the biopolymer are connected together in identical fashion and similar fragmentation processes can occur repeatedly along the length of the component parts. It is perhaps worth noting at this point that the amino acid sequence of at least one protein has been completely determined by mass spectrometry and a number of others have been partially sequenced.

In addition to its role in structure determination, mass spectrometry is also used in the detection and identification (by computer file search if necessary) of trace constituents. There are obvious applications here in the fields of clinical and forensic toxicology and in environmental pollution studies. A variety of examples are given later in the chapter but the identification by mass spectrometry of a hundred or so organic compounds in some samples of drinking water, plus an indication of the origin of such pollutants as petroleum products and chlorinated solvents, illustrates the potential of the technique.

Maximum sensitivity is achieved when the energy transfer during the ionization process is low so that only a very few ionic species are produced in any significant abundance and the mass spectrometer is adjusted to look only at these ions instead of trying to cover the whole mass spectrum. A particular application here is the detection of minute amounts of drugs and hormones in various body tissues and fluids. The co-administration of known amounts of non-radioactive isotopically labelled derivatives of the drugs and hormones allows quantitative assays to be made. The advantage of using isotope derivatives is, of course, that they are chemically identical to the parent molecule, but the mass difference allows their separation in the mass spectrometer. Other techniques of similar sensitivity requirc the presence of probes with special fluorescence or absorbance properties or the use of radioactive isotopes. In a similar vein, the use of isotope derivatives can be used to trace metabolic pathways by the detection and identification of the metabolic products. A selection of illustrative examples of these and other applications are given towards the end of the chapter.

Separation of Ions with Different Mass-to-Charge Ratios

The effect of a magnetic field on the motion of charged particles

An ion of mass m, charge e and velocity v, which enters a magnetic field $\boldsymbol{B}$ at right angles to the direction of motion, experiences a force $\boldsymbol{B}ev$ which is at right angles to both the magnetic field and the direction of motion.

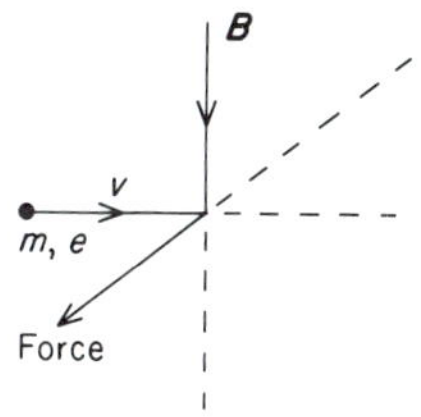

As the ion continues through the field, this force causes it to move in a circle. Balancing the applied force with the centripetal force, we obtain

$$\boldsymbol{B}ev = \frac{mv^2}{R} \text{ or } R = \frac{mv}{e\boldsymbol{B}} \tag{9.1}$$

where R is the radius of the circle of motion. Thus, ions moving with a constant velocity in a constant magnetic field will have a radius of motion dependent on the ratio m/e. Singly charged ions are the most common and in these cases the radius depends on the mass m. In a mass spectrometer the velocity v is imparted to the ions by acceleration through an appropriate electric field.

The single-focussing mass spectrometer

If ions of different mass pass through the entrance slit they can be focussed in turn on the exit slit by allowing the magnetic field to decay from an initial high value which is sufficient for the detection of the heaviest ions. Alternatively the magnetic field can be kept constant and the accelerating voltage V,

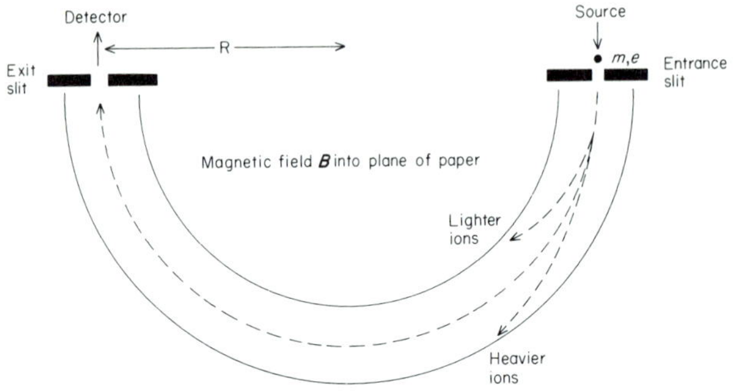

FIG. 9.1 Schematic representation of a single-focussing mass spectrometer

used to give the ions their velocity v, can be altered. Thus, an applied voltage V acting on a charge e will impart an energy eV to the ions. This energy is the kinetic energy $\frac{1}{2}mv^2$ of the ions as they leave the accelerating electric field so that

$$\tfrac{1}{2}mv^2 = eV \text{ or } v^2 = 2V\left(\frac{e}{m}\right) \tag{9.2}$$

If eqn (9.1) is squared to give $R^2 = (m/e)^2 v^2/\boldsymbol{B}^2$ then by substituting for v^2 in eqn (9.2) we obtain

$$R^2 = \left(\frac{m}{e}\right)^2 \frac{2Ve}{\boldsymbol{B}^2 m}$$

so that

$$R = \sqrt{\frac{m2V}{e\boldsymbol{B}^2}} \tag{9.3}$$

Thus the radius of motion can be varied by changing V at constant $\boldsymbol{B}$, or $\boldsymbol{B}$ at constant V. The beam of ions incident on the entrance slit is always slightly divergent due to space charge and other effects and it is usual to mediate this problem by employing a sector magnetic field which refocusses slightly divergent beams (see Fig. 9.2).

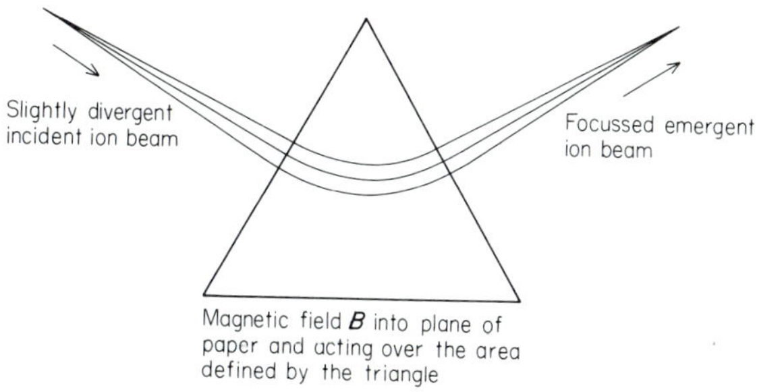

FIG. 9.2 Schematic representation of a sector magnetic field

The double-focussing mass spectrometer

Ions of the same mass incident on the entrance slit to the magnet have slightly different velocities due to differing thermal energies so that the single focussing system shown above does not bring them to a sharp focus at the exit slit. This can be a serious problem when ions of similar mass are required to be separated, but the problem can be overcome by using some kind of velocity or energy selector prior to entry into the magnetic field.

For velocity selection, before entering the main magnetic analysing field $\boldsymbol{B}$, the ion beam passes through crossed electric field $\boldsymbol{X}$ and magnetic field $\boldsymbol{B}'$, as shown in Fig. 9.3.

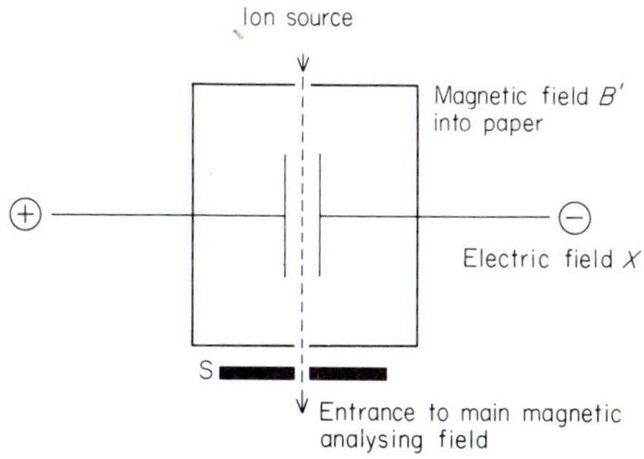

FIG. 9.3 Schematic representation of a velocity selector

The beam can traverse slit S only if the opposite deflections caused by the electric and magnetic fields cancel, i.e. if $\boldsymbol{X}e = \boldsymbol{B}'ev$, i.e. $v = \boldsymbol{X}/\boldsymbol{B}'$. Since $\boldsymbol{X}$ and $\boldsymbol{B}'$ are constant, then v is constant whatever the mass of the ion.

For energy selection, a radial electric field, X', is provided between curved deflecting plates as shown in Fig. 9.4.

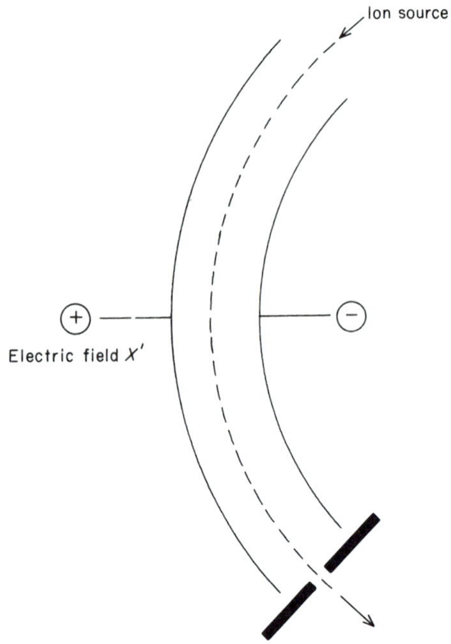

FIG. 9.4 Schematic representation of an electrostatic analyser

The ions are constrained to traverse a circular path of radius R' defined by $X'e = mv^2/R'$. Thus for constant X' and R' a monoenergetic ($\frac{1}{2}mv^2$) beam is produced. In the main analysing field, an ion of mass m and velocity v traverses a circular path of radius R given by eqn (9.1), viz.

$$R = \frac{mV}{e\boldsymbol{B}}$$

Substituting the value of $v = \sqrt{(X'R'e/m)}$ obtained after passage through the electrostatic analyser we obtain

$$R = \frac{m}{e\boldsymbol{B}}\sqrt{\frac{\boldsymbol{X}'R'e}{m}}$$

so that

$$R = \sqrt{\frac{m\boldsymbol{X}'R'}{e\boldsymbol{B}^2}} \tag{9.4}$$

There are a variety of double-focussing systems available depending on whether photographic or electronic detection is used. In the former the ions are brought to focus in a plane whereas in the latter the focus is at a point. Electronic detection would seem to be the more appropriate for a wide range of biological problems.

Resolving power: determination of elemental composition

Two peaks are usually considered to be resolved when the height of the valley between them is not more than 10% of the peak heights.

Considering, for example, compounds with molecular weights less than about 300, the mass spectrometer must be able to resolve masses of 299 and 300. A resolving power rather better than 1 in 300 would be required in practice since, apart from some peaks having an asymmetric shape (for reasons which we will not go into), the atomic weights of the elements are not exactly whole numbers. Thus, taking the atomic weights of H,C,N and O as 1·007825, 12·000000, 14·003074 and 15·994915 respectively, the ion $C_{14}H_{24}N_3O_4^+$, for example, would have a mass of 298·1767. For the assignment of peaks in this mass range, single-focussing mass spectrometers which have resolving powers up to about 3000, are quite adequate.

There are many instances where ions of different elemental compositions have the same mass when this is measured to the nearest whole number. In general, however, their masses will differ and the difference will be less than one part in 3000. For example, the ions $C_{21}H_{29}N_5O_2^+$, $C_{18}H_{31}N_4O_5^+$ and $C_{20}H_{33}N_1O_6^+$ have accurate masses of 383·2320, 383·2293 and 383·2307 respectively. In order to distinguish them and thereby to assign an unambiguous elemental composition to the observed peak, an accuracy of a few parts per million is required. This can only be obtained by a good double-focussing instrument. Apart from this facility of elemental composition analysis, double-focussing instruments are commonly used in a "low-resolution mode" (with a resolving power up to, say, 1000), to allow integral mass assignments for ions with masses up to about 1000.

One of the problems of working at very high resolution is that the widths of all the slits have to be decreased to such an extent that the instrument loses sensitivity and the less abundant ions may not be detected. Moreover, the scan speed has also to be decreased and if only a small quantity of sample material is available, this may have completely evaporated before the scan is complete.

Recording the Spectrum: Mass Measurement

Detectors and recorders

The small ion currents passing through the exit slit of the mass analyser give rise to small ion currents which are detected by an electron multiplier collector. The output is further amplified and fed to a recorder which must have a high-frequency response to cope with a complex and rapidly acquired mass spectrum. In some instruments the recorder contains a mirror galvanometer which reflects a beam of UV light onto sensitive chart paper driven through the recorder during the scan. Several galvanometers of different sensitivities are employed so that suitable peak heights can be obtained on the same recording for the strongest and weakest peaks. The amplified signal is also fed to an oscilloscope so that the output from the mass analyser can be observed during a series of repetitive scans. The oscilloscope traces will indicate the most appropriate time to switch on the chart recorder with a subsequent saving of the expensive UV-sensitive paper.

There is an increasing number of problems where the amount of data acquired is so large that it is not sensible, or even possible, to perform a manual analysis of the recorded spectra. In such cases a computerized data acquisition system is used. The first step is to convert the signal from the electron multiplier (a continuously varying voltage) into digital format using an analogue-to-digital converter (ADC). The digitized recording can be transferred to magnetic tape or disc for subsequent off-line computer processing, i.e. background subtraction, calculation of peak centroid positions, accurate mass determinations (and corresponding elemental composition analysis) by comparison with peaks from a reference compound (see below), etc. Recent developments in ADC and computer hardware have made possible the use of an on-line computer system so that real-time data processing can be achieved. In addition to the immediate feedback of information to the operator, the on-line system will also allow automatic feedback and control of the mass spectrometer operating parameters.

Integral mass measurement

At the low mass end of the spectrum there is usually a number of easily recognizable peaks of known mass due to traces of air in the system. Thus peaks would be expected at 18, 28, 32 and 44 mass units due to the ions H_2O^+, N_2^+, O_2^+ and CO_2^+. Between the significant peaks in the spectrum there is nearly always some intensity, albeit small, around each whole mass number. It is a simple matter then to count the peaks one by one out towards higher mass. The counting can be checked by making sure that the differences in mass between the major peaks make sense in terms of neutral fragments which might be expected to be lost.

Most commonly, the spectrum is output to a chart recorder and the spacing of the peaks decreases logarithmically as the mass increases (because of the nature of the magnetic field decay) so that as higher mass ranges are reached the smaller peaks are less well resolved and the counting becomes more and more inaccurate. However, at a resolving power of about 10 000 using normal scan speeds, peaks can be counted up to a mass of several hundred.

Some instruments have an automatic mass marking facility available whereby a mark is made on the chart at say every 5 or 20 mass units. This is possible because when the magnetic field is allowed to decay, the mass of the ions passing through the slits at a particular time will depend on the field strength and this can be monitored to produce a mass scale. In practice, however, the accuracy in the higher mass ranges is not particularly good.

For convenience and adequate precision in integral mass measurement, use is made of reference compounds which produce a large number of regularly spaced abundant ions of known mass. A common choice is perfluorokerosene where abundant ions are present out to a mass of about 700. Fluorocarbons are not only highly volatile but have the extra advantage that the peaks occur at mass values somewhat lower than the nearest whole number since fluorine is mass deficient ($F = 18{\cdot}998405$ and, for example, the abundant $C_3F_7^+$ ion has a mass of 168·9888) and these peaks are well separated from those derived from biochemical molecules which tend to occur at masses slightly greater than the nearest whole mass number.

Accurate mass measurement

These reference peaks can be used for extremely accurate mass measurement using the so-called "peak-matching" technique.

We have seen from eqn 9.3 that the radius of motion in the main magnetic analyser is given by $R=\sqrt{(m2V/e\boldsymbol{B}^2)}$. Rearranging this to $(m/e = R^2\,\boldsymbol{B}^2/2V)$, if R and $\boldsymbol{B}$ are kept constant, then for singly charged ions $m \propto 1/V$. A value of $V = V_1$ is chosen to allow an unknown ion of mass m_1 to pass through the system. Another value V_2 is chosen to allow a reference ion of accurately known mass to pass through the system. Then

$$\frac{m_1}{m_2} = \frac{V_2}{V_1} \text{ or } m_1 = m_2\frac{V_2}{V_1}$$

The ratio V_2/V_1 can be measured at best to about 3 ppm, hence m_1 is known to this accuracy.

Molecular Weight Determination: The Molecular Ion

There is a need for some caution before interpreting the peak of highest mass as the molecular ion peak because this is frequently not the case. The molecular ion (M^+) may be unstable (especially if the process of ionization has involved a high energy input) and may rapidly fragment so that it is not sufficiently abundant to be detected. Certain types of compound tend to give more abundant molecular ions and these may be the most intense peaks in the mass spectrum. These include aromatic and cyclic molecules. In the latter, at least two bonds must be broken to give ions of lower mass than M^+. However, more often the molecular ion is not the most abundant peak in the spectrum.

There are certain common losses from molecular ions which can aid their identification, e.g. H, H_2, CH_3, O, NH_3, OH, H_2O, CO. Loss of these atoms or groups will give rise to peaks at 1,2,15,16,17,17,18 and 28 mass units lower in the spectrum. The presence of such peaks would tend to confirm the molecular ion while their absence would make the interpretation suspicious. There are no common losses in the mass range 3 to 14 so that the presence of peaks between M-2 and M-15 would indicate an incorrect assignment of the molecular ion. A peak at M-14 would suggest the presence of a homologue, differing by a CH_2 moiety.

Finally, a peak at M + 1 is frequently observed and it will usually be more intense than the peak at M. The $(M+1)^+$ ion arises by the abstraction of a proton from a neutral molecule by the sample. An M + 1 peak can often be recognized since its relative intensity will change with source pressure.

Different Methods of Ionization and Their Particular Applications

Electron impact ionization (EI)

Gaseous samples at atmospheric pressure can be leaked into the source chamber which usually operates at about 10^{-6} torr. Liquid or solid samples can be introduced into the chamber on the tip of a ceramic probe which passes through a vacuum lock (see Fig. 9.5). Solid samples may first be taken up in a small amount of solvent into which the ceramic probe is dipped and the solvent is evaporated off before the probe is inserted into the source chamber. At the low chamber pressure, liquid and solid samples vaporize and this may be encouraged by using a heating element incorporated into the direct insertion probe system.

In the ionization chamber the vapour is bombarded with a stream of electrons accelerated from a hot filament. The positive ions produced on impact of the electron beam with the sample molecules are pushed towards the first accelerator plate by a small potential between this plate and a repeller

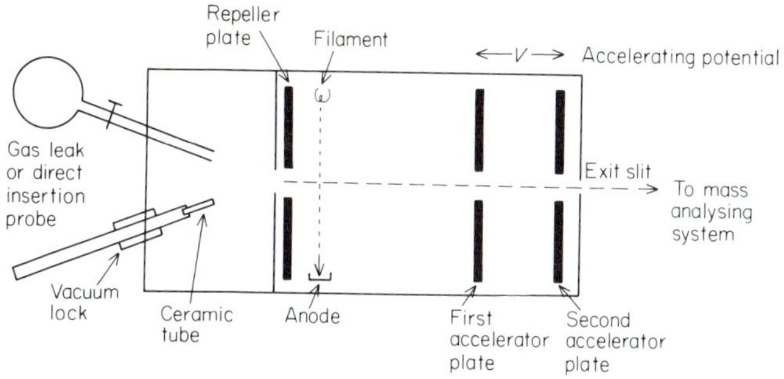

FIG. 9.5 Schematic representation of an ion source chamber

plate. The ions are accelerated through the exit slit towards the mass analysing system by an accelerating voltage of a few kV applied between the first and second acclerator plates.

The energy required to remove an electron from most organic molecules is about 10 eV* (100 kJ mol^{-1}). The efficiency of ionization is low but the energy transfer when ionization occurs is high. The energy of the bombarding electrons is usually in the range 10–70 eV and with the high energy transfer the positive ions first produced have an excess of energy which promotes fragmentation into numerous positive ions of lower mass and neutral fragments. Thus a typical electron ionization (EI) spectrum will show a multitude of peaks but with decreasing intensity towards the high mass end of the spectrum. The value of such a complex spectrum is that it is a record of the fragmentations that have occurred, and a successful analysis and interpretation of the spectrum enables the chemical structure of the sample compound to be elucidated.

Figure 9.6 shows the EI spectrum (above m/e 180) of the diethyl ester derivative of the very successful anti-asthma drug disodium cromoglycate (INTAL®). The disodium salt itself is too involatile for mass spectrometry and has to be derivatized. The chemical structure is shown in the diagram and the bond fissions leading to the significant fragment ions are indicated by vertical dashed lines. The atomic compositions of these fragment ions were confirmed by high resolution measurements.

The loss of small fragments such as water and carbon monoxide is very common for organic molecules and several examples of such losses are shown in the diagram. It is possible to prove that two peaks in the spectrum are related by the loss of a neutral fragment such as H_2O or CO, rather than arising from entirely different parts of the molecule, by the observation of

* An electron-volt (eV) is the energy acquired by an electron when it is accelerated in an electric field with a potential difference of 1 volt.

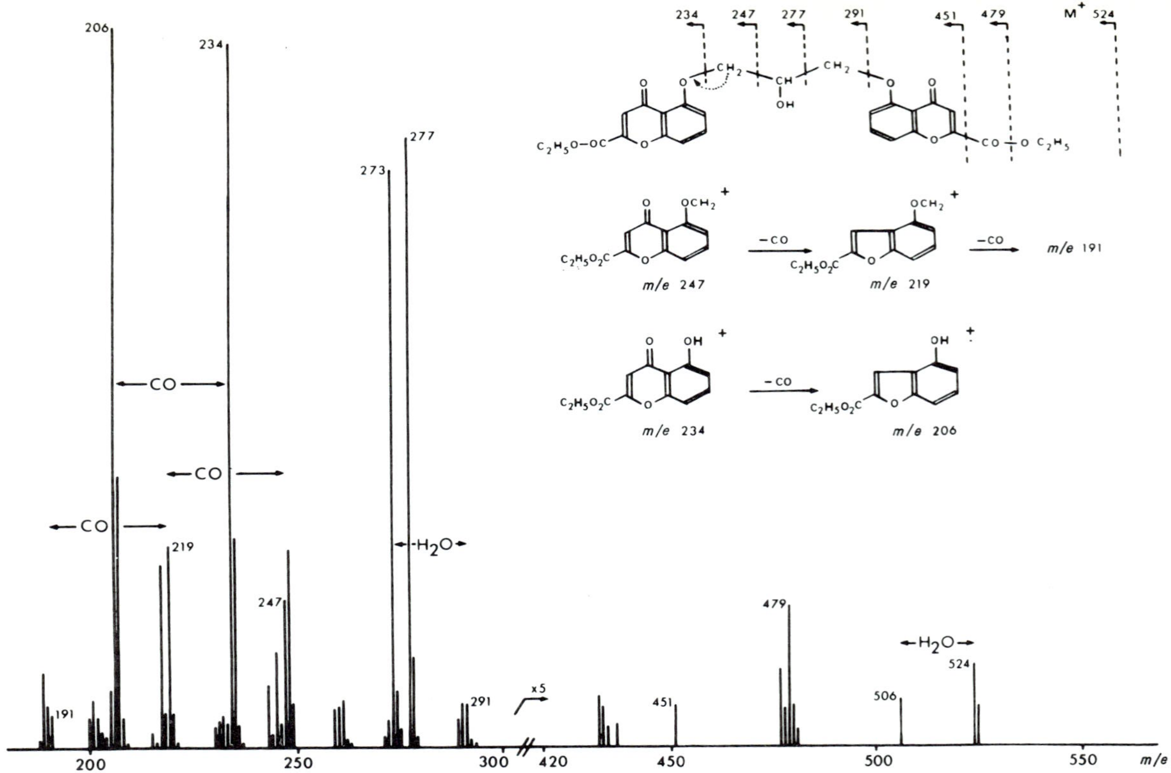

FIG. 9.6 The EI mass spectrum (above *m/e* 180) of the diethyl ester of disodium cromoglycate (INTAL®). Atomic compositions have been confirmed by high-resolution measurements. The losses of H_2O and CO have been confirmed by the presence of metastable peaks (see below). We are grateful to Fisons Ltd for permission to reproduce this spectrum and its interpretation.

so-called "metastable" peaks. The origin, appearance and use of these metastable peaks is discussed in some detail later in the chapter (pp. 358–360).

Apart from illustrating the general appearance of an electron impact mass spectrum, the example chosen here is of special interest since it was the first time that mass spectral data was used as evidence of a molecular structure in the registration of a drug for medicinal use. Usually, such definitive evidence comes from NMR studies but in this case it is clear that the NMR spectrum would not be particularly helpful in providing structural information.

Chemical ionization (CI)

For chemical ionization, the source chamber instrumentation is much the same as that used in EI, but the operating pressure is somewhat higher. The source chamber contains a gas, usually methane, which is ionized by the electron beam. Vaporized sample molecules are present at a low concentration compared to that of the reactant gas and are much more likely to react with an ion from the gas than to be directly ionized by the electron beam.

When methane is used as the reactant gas this is ionized initially to $CH_4^{+\cdot}$, CH_3^+, and $CH_2^{+\cdot}$, and then a series of secondary reactions take place to give ions such as CH_5^+, $C_2H_3^+$, $C_2H_5^+$, $C_3H_5^+$, and $C_3H_7^+$. For example

$$CH_4 + e^- \rightarrow CH_4^{+\cdot} + 2e^-$$
$$CH_4 + CH_4^+ \rightarrow CH_5^+ + CH_3^\cdot$$

The ions CH_5^+ and $C_2H_5^+$ account for about 90% of the total ions produced by electron impact ionization of the methane so that it is predominantly these ions which react with and ionize the sample molecules. Since ion–molecule reactions usually involve low energies, the number of fragment ions produced is relatively small compared with electron impact ionization.

The ion–molecule collision may result in charge transfer alone, giving the molecular ion M^+. Frequently proton transfer occurs, especially if the sample is a strong base:

$$M + CH_5^+ \rightarrow MH^+ + CH_4$$

and the spectrum is dominated by a strong (M + 1) peak. Alkanes and alicyclic compounds (e.g. steroids) frequently have a proton abstracted to give strong (M – 1) peaks, for example *n*-octadecane ionizes according to

$$C_{18}H_{38} + CH_5^+ \rightarrow C_{18}H_{37}^+ + CH_4 + H_2$$

The elimination of water from protonated molecular ions is to be expected when a strong proton donor such as CH_5^+ is used. Thus the methane CI spectrum of codeine, for example, has only four abundant peaks, at (M + 1), M, (M – 1) and (M – 17).

Addition reactions can also take place, particularly with polar molecules, i.e.

$$M + C_2H_5^+ \rightarrow MC_2H_5^+$$
$$M + C_3H_7^+ \rightarrow MC_3H_7^+$$

In these cases, the spectrum will have significant peaks at (M+29) and (M+43).

The EI and methane CI spectra of the *N,N*-dimethyl derivative of phenobarbital are compared in Fig. 9.7 to show the simplicity of the latter and to illustrate some of the reactions discussed above. The EI spectra of barbiturates are usually not very helpful since the molecular ion peak is of low intensity, or not detectable, and elimination of alkyl groups results in more or less identical spectra from different molecules.

There is a particular advantage of CI in gas chromatography–mass spectrometry (GC–MS, see below) since the higher pressures and faster pumping speeds mean that the ion source can take most of the column effluent from the GC, and if the carrier gas in the GC is also used as the reactant gas this may eliminate the need for a molecular separator between the GC and the MS.

Field ionization (FI) and field desorption (FD)

In more recent developments known as field ionization (FI) and field desorption (FD), the ion source consists of an anode (emitter) which may be a thin wire, a sharp edge or blade, or, more recently, 10 μm tungsten wires coated with carbon fibre microneedles about 30 μm long, and a cathode which is a plate with a fine slit or hole according to the shape of the emitter. The two electrodes are close together and a high voltage is applied between them to give a field strength of the order of 10–20 V nm^{-1}. In such a high field the potential field surrounding a sample molecule is distorted to such an extent that when the molecule encounters the emitter an electron can move to a vacant metal orbital in the emitter via a quantum mechanical tunnelling process. The ion produced is immediately accelerated by the field through the hole or slit in the cathode. Because the electrodes are so close together a widely divergent ion beam is produced and some additional focussing is required to get a reasonable ion current through the narrow entrance slit of the mass analysing system. It is difficult to focus the ions which have attained very considerable kinetic energies during acceleration and in practice, the total ion current passing into the mass analyser is about 100-fold less than that normally encountered with EI systems. Thus both resolution and sensitivity tend to be inferior to those attained by EI and CI.

There are two advantages of this method of ionization. First, because there is effectively no transfer of energy to the sample during the ionization process

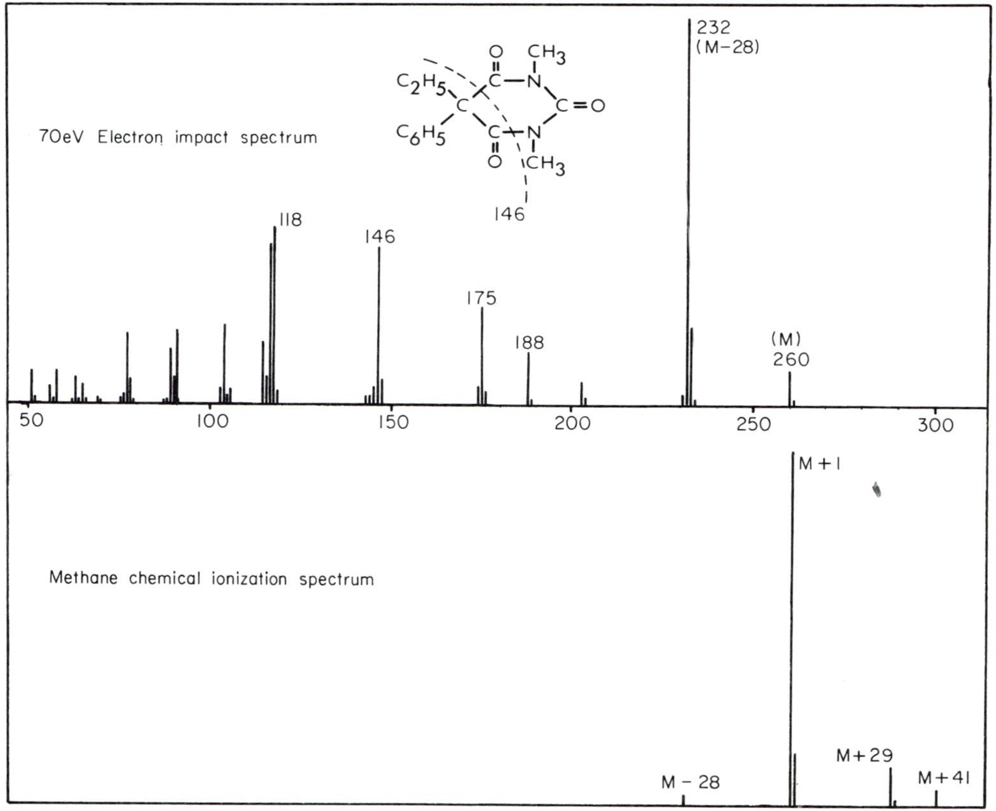

FIG. 9.7 A comparison of the EI and methane CI spectra of the *N*,*N*-dimethyl derivative of phenobarbital. From "Advances in Biochemical Psychopharmacology", Vol 7, E. Costa and B. Holmstedt (Eds), 1973. Reproduced by kind permission of Raven Press, New York.

the abundance of the molecular ion is relatively high. For example, with digitoxin the intensity of the molecular ion peak increases from about 0·003% to 3% of the strongest peak in the spectrum when FI is used instead of EI. Generally the spectra are simple and molecular weights are easily determined. The problems associated with the formation of addition products in CI giving rise to peaks with masses in excess of the molecular ion are not encountered although $(M+1)^+$ peaks can still be observed. Secondly, relatively involatile materials can be studied directly and by depositing the sample in solution or as a suspension directly onto the emitter where it is adsorbed, the requirement to volatilize a sample by heating the tip of an insertion probe is eliminated. This is particularly important with regard to thermally unstable compounds.

The term "field ionization" is used when the sample is introduced by a gas leak or insertion probe. "Field desorption" is reserved for those cases where the sample has been adsorbed directly onto the emitter.

Field ionization–field desorption would seem to have a promising future in the biological field. The simplicity of the spectra will allow analysis of complex mixtures (e.g. of pesticides, antibiotics, tobacco smoke). The ability to tackle samples of low volatility will allow studies of free peptides, carbohydrates, etc. There is a particular application to mixtures of drugs with their corresponding metabolic products, especially when the route along the metabolic pathway leads to compounds of increasing polarity and hence lower volatility.

Figure 9.8 compares the EI and FD spectra of azathioprine, a widely used immunosuppressive agent in clinical organ transplantation. The EI spectrum does not have a molecular ion peak whereas the FD spectrum has essentially just one peak at $M+1$.

Selected Ion Monitoring or Mass Fragmentography

The mass spectrometer can be used not only to provide molecular weight and structural information but also for the detection and quantitative measurement of minute amounts of a known compound.

Detection of very small amounts of sample

In normal operation a complete mass spectrum is produced by allowing the analysing magnetic field to decay so that ions of different mass are successively

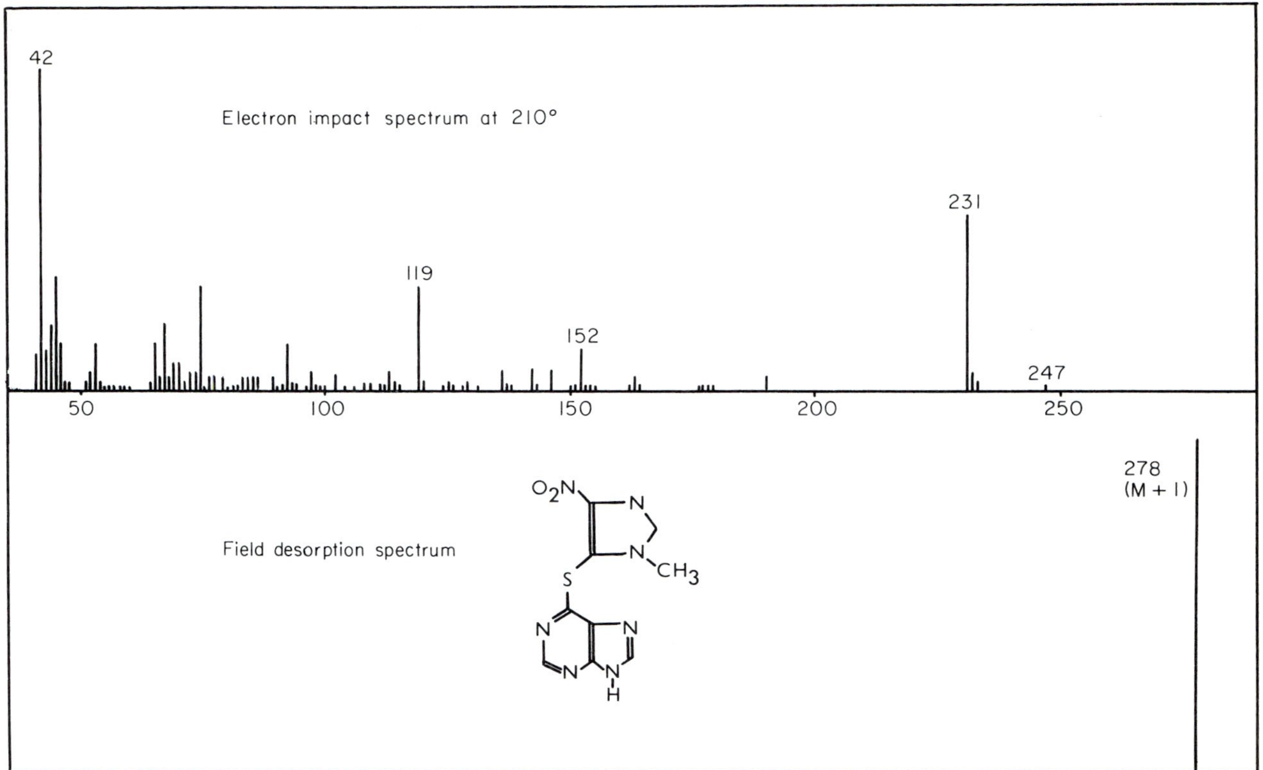

FIG. 9.8 A comparison of the EI and FD spectra of azathioprine. Reprinted with permission from "Advances in Mass Spectrometry in Biochemistry and Medicine", Vol. 1, Alberto Frigerio and Neal Castognoli (Eds), Spectrum Publications, New York, 1975.

brought to focus at the detector. The time spent at each mass value is just a few milliseconds out of a total scan time of, say, 10 seconds. If the fragmentation pattern of the sample is known then the focussing conditions can be preset so that a single (and abundant) ion can be continuously monitored, with a dramatic increase in sensitivity for that ion. With suitable samples under ideal conditions, as little as one femtomole (10^{-15} mol) can be detected, although in day-to-day usage, detection at the picomole (10^{-12} mol) level is more common.

Quantitative assays

Suppose it is required to measure the amount of a particular compound in a small volume of body fluid (e.g. cerebrospinal fluid). To achieve quantitation, a reference compound must be used which ideally has properties identical to the chemical under investigation. The aim is to compensate for variations in yields of extraction and derivative formation and losses due to absorption etc. in the isolation and purification of the required compound. Isotope labelling is the only technique which will produce sufficient identity between reference and sample compounds and of course radioisotope dilution methods and radioimmunoassays are in common usage. A major advantage of using radioisotopes is that the liquid scintillation counters used for detection are sensitive (picogram quantities) and relatively cheap. However, if the problem is to determine the distribution of a drug in human tissue, then it is important to eliminate any possibility of radiation damage. This means that non-radioactive isotopes must be used for labelling, and the only technique that can distinguish a particular compound from its stable isotope labelled derivative is mass spectrometry.

Mass spectrometers are now available that allow focussing conditions for several ions within a limited mass range to be preset. By rapid switching between the accelerating voltages corresponding to the focussing conditions of the ions under investigation, and the use of multichannel detectors, an effectively continuous and simultaneous monitoring of the sample compound and one or more of its stable isotope derivatives can be performed. Comparison of the areas under the peaks corresponding to the sample and the standard gives an accurate quantitative assessment in the pico- to femtomole range. When the mass spectrometer is dedicated to the examination of just a few chosen ions, the technique is referred to as "selected ion monitoring" (SIM), or "mass fragmentography". The term "multiple ion detection" (MID) is becoming less common. The technique is most usefully employed with gas chromatographic separation of the sample prior to analysis by the mass spectrometer.

Gas Chromatography–Mass Spectrometry (GC–MS)

The technique of gas chromatography (GC) is widely used for the separation and detection of very small quantities of organic compounds. Partial characterization may be possible from a knowledge of the column retention times. Mass spectrometry is the most suitable technique for providing definitive structural information on similar quantities of material. A combination of these two techniques, where the effluent from the gas chromatograph is input directly to a fast scanning mass spectrometer, is thus an extremely powerful tool for the complete characterization of components of a mixture when these are present in minute quantities. All compounds which are sufficiently volatile to pass through the GC system will also be sufficiently volatile for mass spectral analysis. However, there are many instances where compounds of interest are not (or cannot be made) volatile enough for GC, although the isolated compound is perfectly suitable for mass spectrometry.

A major problem to be overcome is that of reducing the pressure at the GC outlet (*c.* 1 atm) to a value suitable for the MS inlet (*c.* 10^{-6} torr) with minimum loss of sample and maximum separation of the carrier gas from the sample. A variety of "molecular separators" are available. The one shown in Fig. 9.9 is a two-stage jet separator due to Ryhage.

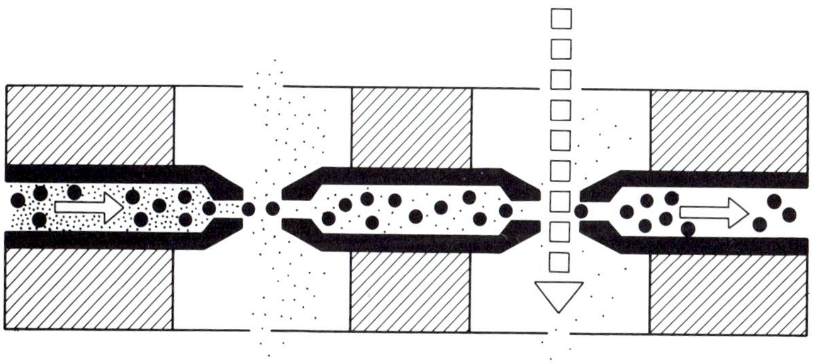

Fig. 9.9 A Ryhage two-stage jet separator. The dashed arrow represents a sliding closure valve which is withdrawn in normal operation but can be operated to prevent expected solvent peaks from entering the MS analyser, or to allow columns to be changed without closing down the system. Reproduced by kind permission of LKB Instruments Ltd.

The lighter carrier gas (He) diffuses away from the first orifice so that a sample-enriched gas passes into the second orifice. A further enrichment occurs between the third and fourth orifices. The maximum enrichment is strongly dependent on the geometrical parameters of the system.

There are various ways of approaching the data collection depending on the information sought. A full mass spectrum can be obtained from each peak passing from the gas chromatograph allowing chemical characterization of the corresponding molecules. Since a number of repetitive MS scans can usually be made during the elution of a single GC peak the mass spectrometer can sometimes resolve compounds which are not resolved by the gas chromatograph. The mass spectrometer can be used for continuous monitoring by repetitive scanning of the GC effluent. The enormous number of mass spectra so obtained have to be input directly into a computer storage device for subsequent analysis, e.g. the calculation of a graph of a specific ion abundance against time (a "mass chromatogram"). The SIM approach is used when prior knowledge is available concerning the components in the GC effluent. A diagrammatic representation of a complete GC–MS system is shown in Fig. 9.10.

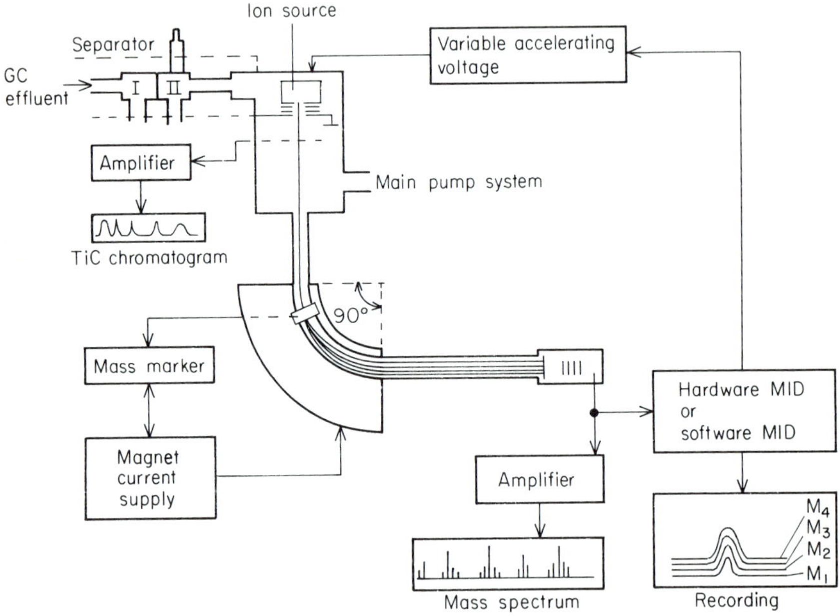

FIG. 9.10 Schematic representation of a GC–MS system. The total ion current (TiC) can be monitored to indicate the most appropriate times to record mass spectra. Reproduced by kind permission of LKB Instruments Ltd.

As an illustration of the use of GC–MS and mass fragmentography, Fig. 9.11 shows the simultaneous determination of dopamine (DA) and its catechol-*O*-methyltransferase product, 4-hydroxy-3-methoxyphenylethylamine (3-MT), in rat brain. The method is particularly useful here since the normal fluorimetric methods for the determination of 3-MT are rather insensitive.

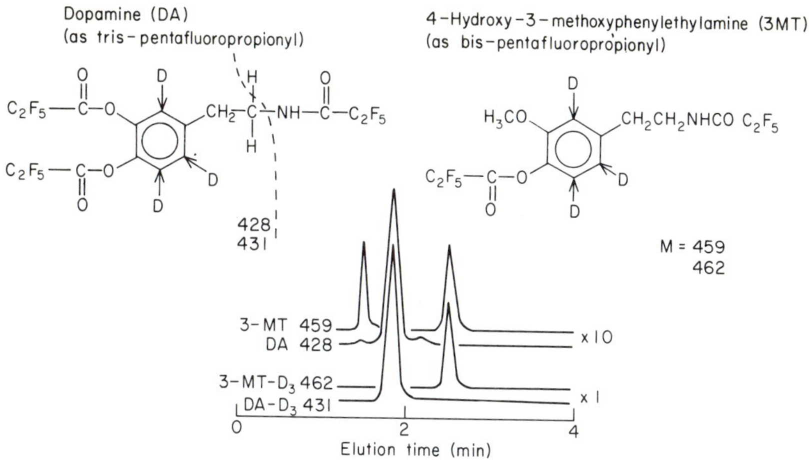

FIG. 9.11 Mass fragmentogram of dopamine (DA) and 4-hydroxy-3-methoxyphenylethylamine (3-MT). Reprinted with permission from "Advances in Mass Spectrometry in Biochemistry and Medicine", Vol. 1, Alberto Frigerio and Neal Castognoli (Eds), Spectrum Publications, New York, 1975.

The compounds were made sufficiently volatile for GC by preparation of their pentafluoropropionyl derivatives. The trideuterated derivatives were used as standards for the mass fragmentogram. The GC–MS/SIM technique allowed determination of the 3-MT level (1·4 nmol g^{-1}) in the olfactory tubercles from one rat brain. The level of 3-MT in striatum was 2·5 nmol g^{-1}.

The Double-Beam Mass Spectrometer

The double-beam mass spectrometer is a fairly recent innovation. Two ion beams are produced by two independent ion sources, pass though a common mass analyser, and are detected by separate collector and amplifier systems (see Fig. 9.12). The spectra of the samples in each source are recorded simultaneously and on identical mass scales.

Such a system allows a convenient, direct comparison of the effects of different ionization potentials and variations in source pressure and temperatures when identical samples are present in each source. The purity of a sample can be determined quickly by direct comparison of its spectrum with that of the same compound of known purity. A particular advantage is in chemical mass marking, since the independent recording of the spectra of a reference and an unknown compound avoids the necessity of having to optimize the

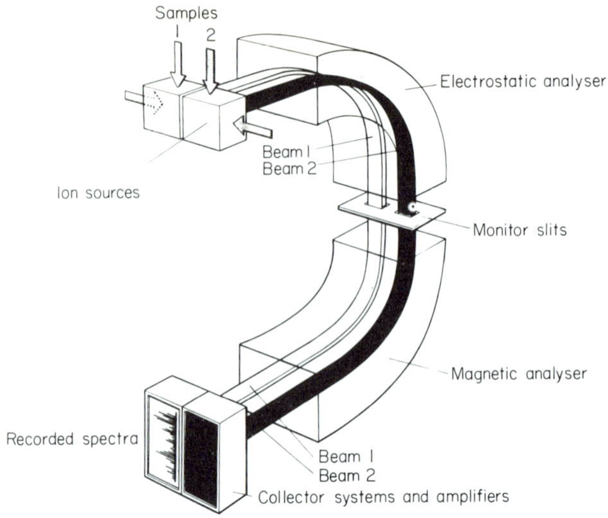

Fig. 9.12 A double-beam mass spectrometer. Reproduced by kind permission of Kratos Ltd (AEI Scientific Instruments).

relative pressures and temperatures of the sample and reference compound. This is a significant advantage in GC–MS work where there are constantly changing concentrations in the GC effluent. Both spectra are recorded back-to-back on the same chart and it is an easy matter to read off the masses of the peaks on the sample spectrum by comparison with the peaks of known mass from the reference compound spectrum.

This method of mass marking is especially useful for spectra produced by chemical ionization. The usual reference compound, perfluorokerosene, does not give an abundance of intense peaks at regular intervals under CI conditions, but with a double-beam mass spectrometer, the reference compound can be ionized by conventional electron impact conditions in one source, while the sample is run under CI conditions in the other source.

Metastable Transitions

Consider the single-focussing instrument first. We have seen from eqn (9.3) that the radius of motion of an ion in the magnetic analyser is given by

$$R=\sqrt{\frac{m2V}{e\boldsymbol{B}^2}}=\frac{1}{e\boldsymbol{B}}\sqrt{2meV}=\frac{1}{e\boldsymbol{B}}\sqrt{2mE}$$

where E is the energy of the ion entering the magnetic field. Suppose that just prior to entry into the magnetic field an ion m_1^+ fragments into an ion m_2^+ and a neutral fragment of mass (m_1-m_2). The energy is partitioned in the ratio of the masses of the product particles so that m_2^+ carries away $(m_2/m_1)E$ and (m_1-m_2) carries away $[(m_1-m_2)/m_1]E$ of the energy of m_1^+. Thus the radius of motion of m_2^+ in the magnetic field is

$$R'=\frac{1}{e\boldsymbol{B}}\sqrt{2m_2\left(\frac{m_2}{m_1}\right)eV}=\frac{1}{e\boldsymbol{B}}\sqrt{\frac{2m_2^2eV}{m_1}}$$

This is the radius of motion of a normal ion of mass m^* produced in the source where $m^*=m_2^2/m_1$ i.e. the fragment ion m_2^+ is recorded at an apparent mass of m^*.

The peak at m^* is known as a "metastable peak". Since the neutral fragment can come off (at extremes) in front of and behind m_2^+, the ion m_2^+ has a small range of kinetic energies giving rise to a broad peak at m^*. In addition to its breadth the metastable peak can be recognized since it does not need to occur near integral mass values and is of low abundance.

For completeness it should be mentioned that if m_1^+ fragments *during* acceleration at say V', it can be shown that the radius of motion of m_2^+ is given by

$$R'=\frac{1}{e\boldsymbol{B}}\sqrt{2m_2\left(\frac{m_2}{m_1}eV'+e(V-V')\right)}$$

Although m^* is known by measurement, m_1 and m_2 are not. However, $m_2=m_1-n$ (where n is an integer) and there are some restrictions on n. For example, common losses are H,H_2,CH_3,H_2O,CO,C_2H_4, CH_3CHCH_2,C_3H_7, etc. with masses of 1,2,15,18,28,28,42,43 respectively. Thus n cannot lie between 2 and 15. For the case where $m^*=m_2^2/m_1$ we can take logs to give

$$\log m_1+\log m^*=2\log m_2$$

or

$$\log m_2=\frac{\log m_1+\log m^*}{2}$$

Constructing three vertical log mass scales as shown in Fig. 9.13, then if one end of a ruler is placed at the observed value of m^*, the ruler will intersect the other scales at "allowed" values of m_1 and m_2. The ruler is rotated until values of m_1 and m_2 are obtained that correspond to reasonably abundant masses in the spectrum. These will always be present and correspond to the cases where m_1^+ has not undergone further fragmentation and where m_1^+ has fragmented into m_2^+ and (m_1-m_2) prior to acceleration. Thus, it is possible to relate two ions in the spectrum via their metastable peak.

In the case of a double-focussing instrument, if fragmentation $m_1^+\rightarrow m_2^+$

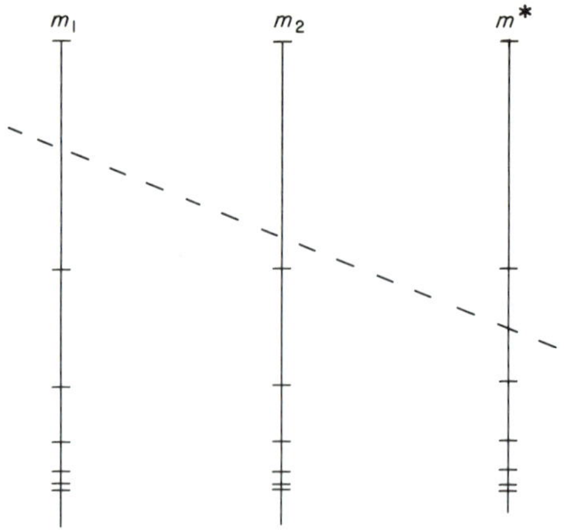

FIG. 9.13 Graphical determination of the masses m_1 and m_2 of two ions related by a metastable peak.

and $(m_1 - m_2)$ occurs after acceleration but before entry to the electrostatic analyser (i.e. the energy selector) then m_2^+ will not be passed since it does not possess the required energy eV.

Metastable refocussing

Suppose the accelerating voltage is set somewhat lower than normal and the magnet current adjusted to pass an ion m_2^+, which has been produced in the source. If the accelerating voltage is increased in the ratio m_1/m_2 then ions m_2^+, produced just prior to the electrostatic analyser, will have the required energy to pass, all other ions now being unable to pass. The procedure then is to scan the accelerating voltage upwards until a peak is recorded; knowledge of m_2 and the ratio m_1/m_2 gives the mass of m_1^+. There are cases where the ion m_2^+ has more than one precursor and the upward scanning of the accelerating voltage allows each precursor to be found in turn.

Examples of the use of metastable peaks are given below in the section on peptide sequence analysis.

Structure Determination by Mass Spectrometry

The rest of this chapter will be concerned with examples of the contribution of mass spectrometry to the solution of problems of biological or clinical importance. An example of the application of the technique to structure determination, that of an unusual amino acid, will now be considered.

The discussion below is not meant to be rigorous but rather to show the sort of approach adopted. The results described were confirmed and aided by chemical syntheses and chemical behaviour studies.

A new amino acid was discovered in the urine of a mentally deficient patient born as a result of a brother–sister mating. On a Beckman 120B automatic amino acid analyser the time of elution lay between that of aspartic acid and threonine. The principal mass spectral peaks (A→G) of the *N*-trifluoracetyl (*N*-TFA) methyl ester derivative are shown in Fig. 9.14. This particular derivatization destroys the zwitterionic nature of the amino acid and substantially increases its volatility

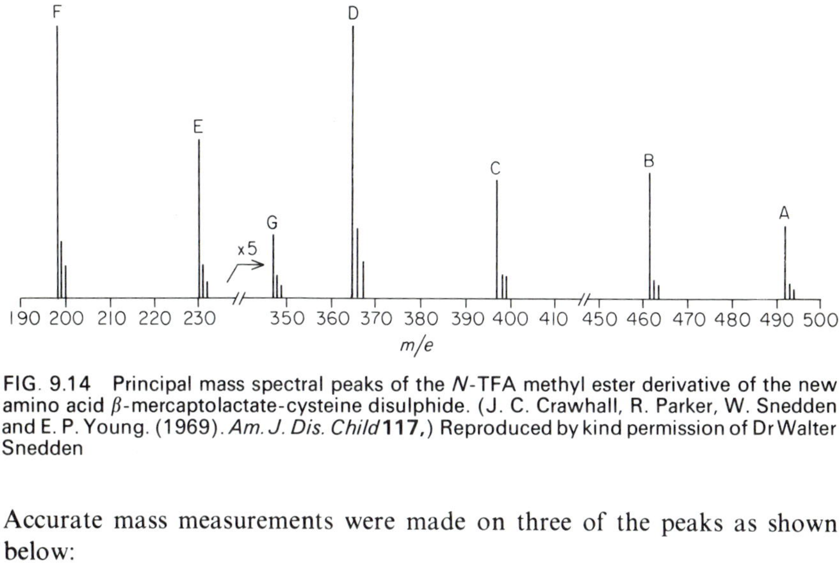

FIG. 9.14 Principal mass spectral peaks of the *N*-TFA methyl ester derivative of the new amino acid β-mercaptolactate-cysteine disulphide. (J. C. Crawhall, R. Parker, W. Snedden and E. P. Young. (1969). *Am. J. Dis. Child* **117**,) Reproduced by kind permission of Dr Walter Snedden

Accurate mass measurements were made on three of the peaks as shown below:

Ion	Mass	Assignment	Theoretical mass
D	365·021 ± 0·001	$C_{10}H_{14}NO_6S_2F_3$	365·0213
E	230·009 ± 0·001	$C_6H_7NO_3SF_3$	230·0098
F	198·039 ± 0·002	$C_6H_7NO_3F_3$	198·0378

Oxidation of the amino acid with performic acid and reduction with dithiothreitol confirmed that a disulphide link was present. Spectra were also taken of the *N*-trifluoracetyl propyl and butyl esters of the amino acid and the only difference between these two derivatives was a shift in mass of 28 amu for ions A,B,C,D and G, and 14 amu for ions E and F. The only corresponding chemical difference between these derivatives is the mass of the alkyl group contained in the carboxyl ester function. Since the propyl and butyl groups differ by 14 amu the results above indicate the presence of two carboxyl groups. (N.B. This process is known as the mass-shift technique and has wider applications.) Ion G occurs 18 amu lower than ion D indicating loss of H_2O from D and suggesting the presence of a hydroxyl group. This was confirmed by infrared spectroscopy.

Since the amino acid contained a disulphide bridge and two carboxyl groups this warranted a comparison with the mass spectrum of the *N*-TFA methyl ester of cystine. This latter spectrum showed ions identical to F and E which are known to be

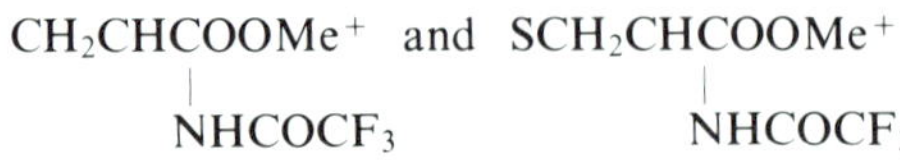

respectively. For the unknown amino acid the mass difference between ions D and E corresponds to $C_4H_7O_3S$. From what has been said above this group of atoms contains an esterified carboxyl group and a hydroxyl group, i.e. $(COOCH_3,OH),S,C_2,H_3$. There would seem to be only two possible structures for ion D, viz.

(i) 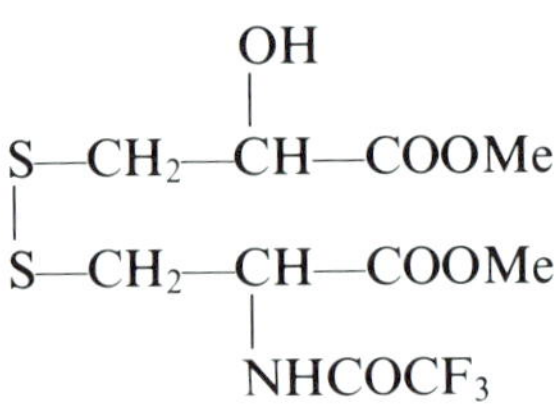 (ii) 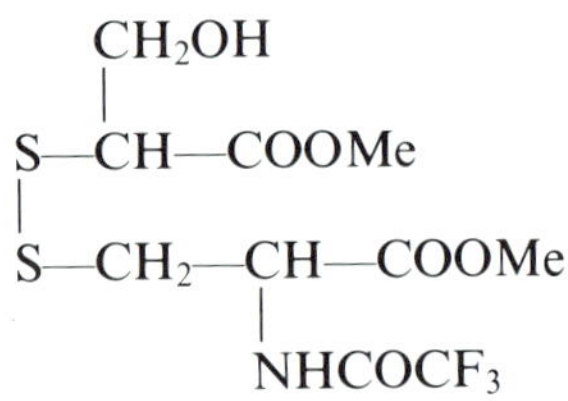

The second was considered unlikely since a strong fragmentation would be expected at the C—CH_2OH bond (indicated by an arrow) giving a fragment ion 31 amu less than D, and this is not observed.

The mass spectrum of the cystine derivative showed the presence of cystine sulphone at 32 amu (O_2) higher than the molecular ion, this artefact probably forming after exposure to air. For the unknown amino acid there is a difference of 32 amu between ions A and B, and C and D with no sign of intermediate fragmentations (i.e. the loss is likely to be O_2 rather than C_2H_8, etc.) so that ion C is probably the sulphone of D, and ion A the sulphone of B.

We are left with the interpretation of ion B. There are no indications of loss of water from B so that it probably does not contain the hydroxyl group. The mass difference between B and D is 96 amu and the implication is that in B the OH group of D has been replaced by $OCOCF_3$ by the derivatization procedure.

Thus a most likely structure for the new amino acid is

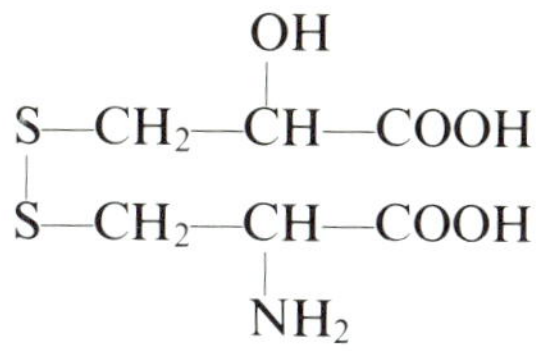

i.e. β-mercaptolactate-cysteine disulphide. Finally, one might add that having identified this compound, further urine samples from similarly mentally deficient patients could be screened to see if its occurrence is typical. On the assumption that the illness is present from childhood, then possible therapeutic treatments for infants may be considered such as dietary restrictions of sulphur-containing amino acids.

The Use of Stable Isotopes as Tracers in Metabolic Studies

It has been mentioned earlier that there is considerable advantage to be gained by using stable isotope labelled compounds for studies *in vivo* since these present no radiation hazard. There are also other advantages; non-radioactive materials are easier to handle, there are no radioactive counterparts of ^{15}N, ^{18}O or ^{17}O which have a sufficiently long halflife to be useful as labels and, in the case of deuterium versus tritium in particular, the smaller mass of the stable isotope reduces any complications which might arise from kinetic effects.

There still remains the problem of having to administer significant amounts of the labelled compound and, in those cases where only a small fraction of the precursors go into the required pathway, the large amounts which have to be administered could produce toxic effects. Sometimes this cannot be avoided, but there are a number of cases where appropriate compounds can be labelled by making the animal breathe $^{18}O_2$ gas, and this has little effect on physiological conditions. Two examples are discussed below.

Catecholamine metabolism

An earlier example showed how it was possible to make simultaneous quantitative assays of dopamine and 4-hydroxy-3-methoxyphenylethylamine (3-MT) in rat brain using mass fragmentography and deuterated analogues. The experiment involved extraction of the compounds from brain tissue and this procedure could not be applied in human studies. However, the main metabolite of dopamine is homovanillic acid (HVA) and this leaves the brain via the cerebrospinal fluid and blood. (Homovanillic acid is derived from dopamine via 3-MT by a monoamine oxidase and via 3,4-dihydroxyphenylacetic acid (DOPAC), by the action of catechol-*O*-methyltransferase. DOPAC itself is formed from dopamine by a monoamine oxidase.) Since the catecholamines are formed by hydroxylases utilizing molecular oxygen, exposure of the animal to $^{18}O_2$ gas produces labelled HVA and hence provides a system for the measurement of dopamine turnover. Such studies were carried out by G. Sedvall, A. Mayevsky, C-G. Fri, B. Sjöquist and D. Samuel (see "Advances in Biochemical Psychopharmacology", Vol. 7, 1973. E. Costa and B. Holmstedt (Eds), Raven Press, New York).

One interesting result to have come out of these studies is proof that increased HVA levels following chlorpromazine and other antipsychotic neuroleptic drug treatments are due to increased dopamine synthesis and turnover and not, as had previously been suggested, due to altered HVA transport from the brain. The latter mechanism would not have produced the observed increased levels of labelled HVA following drug administration.

This type of experiment is applicable not only to dopamine studies but also to compounds such as noradrenaline and 5-HT and, indeed, to any processes involving hydroxylation reactions. Of immediate interest is the study of the effects of monoamine oxidase inhibitors used in the treatment of various neuropsychiatric disorders.

Haem metabolism

In the degradation of haemoglobin two oxygen atoms become incorporated at adjacent rings in the haem group to give bilirubin (see Fig. 9.15)

A variety of intermediate species has been postulated, such as those shown in Fig. 9.16. Several mechanisms are possible depending upon whether the two oxygen atoms come from the same oxygen molecule ("one-molecule" mechanism), one from each of two oxygen molecules ("two-molecule" mechanism), or one from an oxygen molecule and one from a molecule of water ("hydrolytic" mechanism).

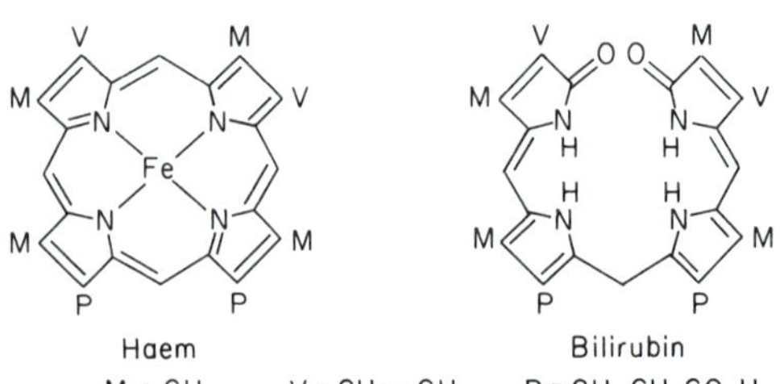

M = CH_3 V = $CH{=}CH_2$ P = $CH_2CH_2CO_2H$

FIG. 9.15 Chemical structures of haem and bilirubin.

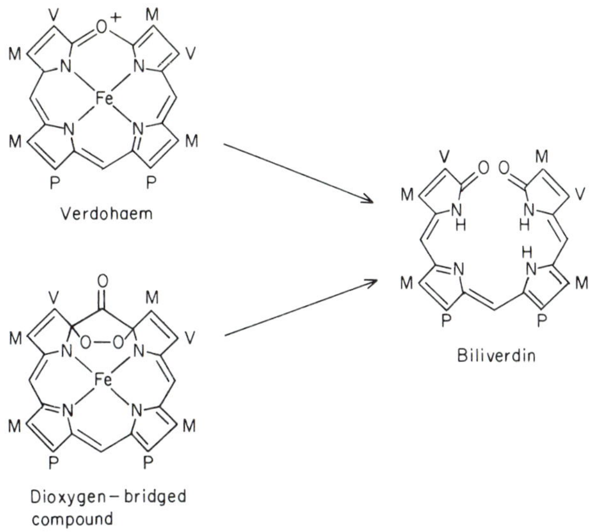

FIG. 9.16 Chemical structures of verdohaem, biliverdin and a dioxygen-bridged compound.

Using a mixture of unlabelled oxygen ($^{16,16}O_2$) and doubly labelled oxygen ($^{18,18}O_2$) gas, the three mechanisms lead to different predictions of ^{18}O incorporation into bilirubin. The molecular ion of bilirubin occurs at *m/e* 584 and peaks at 586 and 588 correspond to the incorporation of one or two heavy oxygen atoms respectively. In the hydrolytic mechanism one oxygen atom is always derived from solvent water so that no peak should occur at *m/e* 588. In the one-molecule mechanism the oxygen atoms incorporated are either both ^{18}O or both ^{16}O so that no peak should occur at *m/e* 586. In the two-molecule mechanism peaks should occur at both *m/e* 586 and 588 with abundances depending on the enrichment of the oxygen gas used in the experiment. Table 9.1 shows the predicted and observed values for 15% $^{18,18}O_2$ for an experiment using living male Wistar rats (see S. B. Brown and R. F. G. J. King. (1975). *Biochem. J.* **150,** 565–567).

TABLE 9.1 Studies on haem metabolism in the rat

m/e	Mol % observed	Mol % predicted: One-molecule mechanism	Two-molecule mechanism	Hydrolytic mechanism
584	70·3	85	72·2	85
586	25·5	0	25·5	15
588	3·5	15	2·3	0

These results clearly indicate that both oxygen atoms incorporated into bilirubin during haem metabolism are derived from molecular oxygen and that they come from two different oxygen molecules. The findings would exclude the direct formation of the intermediates verdohaem and the dioxygen-bridged compound shown in Fig. 9.16, since the conversion of these to bilirubin (via biliverdin) would require in the former, a hydrolytic step and involve an oxygen atom from solvent water, and in the latter, either a hydrolytic step or a one-molecule mechanism.

The Detection and Identification of Small Amounts of Compounds in Body Fluids When Other Methods Have Failed

Melatonin

Melatonin is the compound responsible for the lightening of amphibian skin. There have been suggestions that it is involved in pituitary function and cerebral physiology in mammals. Bio- and radioimmunoassays have failed to provide direct evidence for its existence in man although its biosynthetic enzymes are known to be present in the human pineal gland. Figure 9.17 compares the mass spectra, after gas chromatography, of the indole *N*-trimethylsilyl (*N*-TMS) derivatives of synthetic melatonin and extracts from human plasma and cerebrospinal fluid.

The GC retention times and mass spectra of the synthetic and human extract compounds are identical. These experiments demonstrate unequivocally the existence of circulating melatonin in man. Since there are only two major ions in the spectrum, this compound is eminently suited to quantitative assay by selected ion monitoring (mass fragmentography).

Digoxin

Cardiac glycosides such as digoxin (Fig. 9.18) (from the foxglove *Digitalis lanata*) are widely used in the treatment of congestive heart failure. There is some evidence to suggest that the mechanism of action involves inhibition of an Na^+–K^+ pump and certainly some steroids derived from plants are potent inhibitors of the Na^+–K^+ ATPase and pump, e.g. ouabain. For any drug, it is important to have a urine (and/or circulating fluid) assay system, not only to assess its bioavailability but also to provide information regarding optimum dosage levels for individual patients in cases where there are apparently wide variations in drug absorption and/or rates of conversion to metabolites which may or may not have some degree of activity themselves. In the case of digoxin, radioimmunoassays have been developed which detect both digoxin

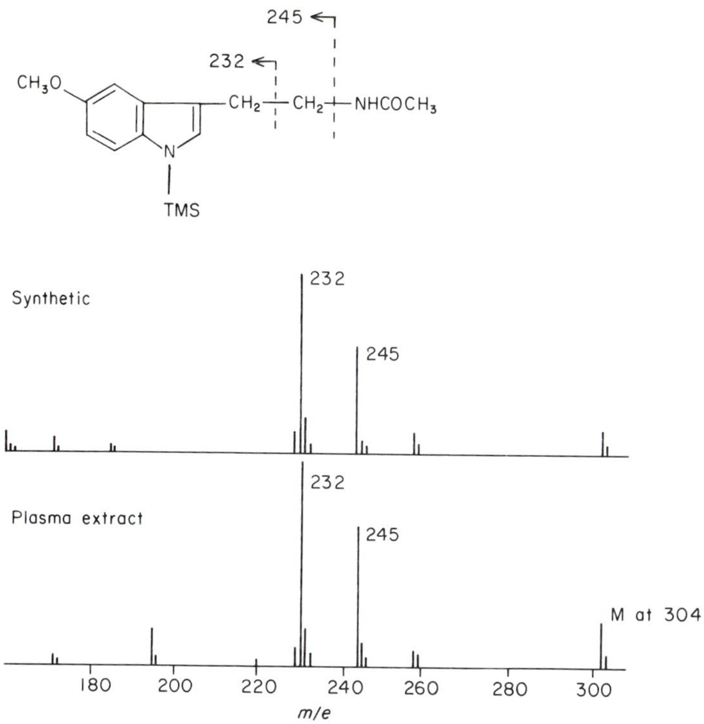

FIG. 9.17 Comparison of the mass spectra of the indole *N*-trimethylsilyl derivatives of synthetic melatonin and extracts from human plasma (I. Smith, P.E. Mullen, R. E. Silman, W. Snedden and B. W. Wilson. (1976). *Nature (London)* **260**.) Reproduced by kind permission of Dr Walter Snedden.

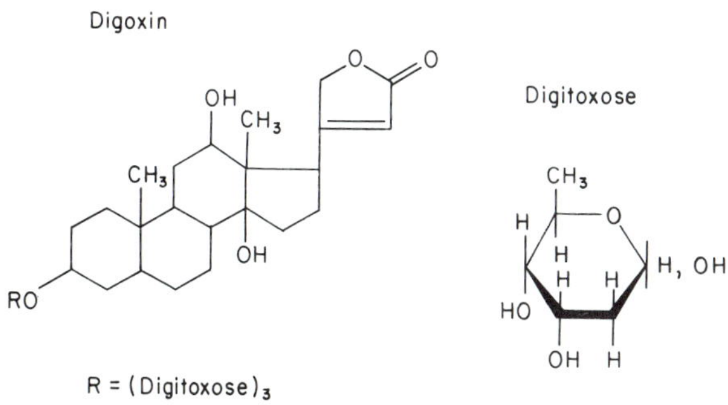

FIG. 9.18 Chemical structure of digoxin. (Details in text from H. Greenwood, W. Snedden, R. P. Hayward and J. Landon, *Clinica Chimica Acta*, (1975). **62**, 213–224).

and the mono- and bis-digitoxosides and digoxigenin metabolites. Mass spectrometric analysis of urine extracts of patients on long-term digoxin therapy has revealed that there is another metabolite, dihydrodigoxin, which is not detected by radioimmunoassay, and moreover that this latter relatively inactive metabolite can account for up to 20% of the orally administered dose of digoxin. Obviously the radioimmunoassay results can sometimes be grossly in error in assessing digoxin and digoxin metabolite levels.

In the mass spectrum of digoxin there are intense peaks at *m/e* 337, 355 and 373, arising from cleavage of the digitoxose residues from the molecule, followed by stepwise loss of two molecules of water. For dihydrodigoxin the corresponding ions occur 2 amu higher. Following prior experiments on digoxin-free urine (to gauge the extent of any other contribution at these *m/e* values) and on urine plus known amounts of digoxin and dihydrodigoxin, it is possible to make quantitative analyses of the amounts of the two compounds present in urine extracts from patients. It turns out that it is sufficient to make measurements at only *m/e* 355 (digoxin) and 357 (dihydrodigoxin).

Sequence Studies on Biopolymers

Peptide and protein sequencing

The volatility of free peptides is limited by their zwitterionic nature, but this is conveniently removed by the preparation of *N*-acyl peptide esters. The mass spectra of these derivatives can be interpreted in terms of splitting of the amide CO—N and C—CO bonds which give rise to amino acyl (Y) and amine (X) fragment ions, with the positive charge being retained on the *N*-terminal side of the molecule.

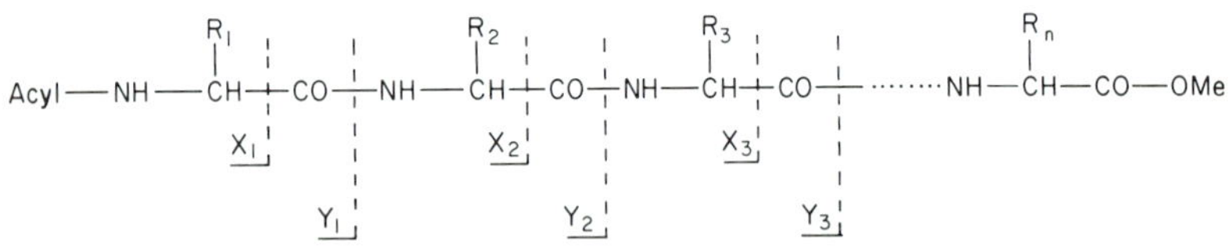

With the exception of Leu and Ile (but see examples below of metastable transitions), each amino acid is uniquely characterized by the mass of its side-chain (R) so that the mass at which the peak corresponding to X_1 (or Y_1) occurs, determines the *N*-terminal amino acid, and successive amino acids are determined by the mass differences X_2-X_1 (or Y_2-Y_1), X_3-X_2 (or Y_3-Y_2) etc. The peaks corresponding to fragmentations at X and Y are referred to as "sequence peaks". There are a number of instances where the sequence peaks are overshadowed by strong peaks arising, for example, by the splitting-off of amino acid side-chains. Such fragmentations can generally be recognized or prevented by prior removal or blocking of the troublesome functional groups. Even so, there are still problems with low volatility and the mass spectra are often quite complicated. The process of *N*,*O*-permethylation of the peptide considerably enhances the volatility (presumably by removing the possibility of hydrogen bonding), and moreover causes a great simplification of the spectra (the amino acids still retain their unique masses after this derivatization, as shown in Table 9.2). The dominant peaks now arise from the amino acyl fragments and the side-chain losses are dramatically reduced. A few minor problems remain but these are gradually being sorted out and the technique can be considered viable for the routine sequencing of peptides containing one to about ten amino acids.

There are a number of advantages of this method over conventional sequencing techniques. The first is that of speed. Imagine the excitement only a few years ago when four or five peptides were being sequenced during the lunch-break (the only time available on the machine for this "novel" application). The second advantage is the ability to cope easily with peptides having a blocked *N*-terminus. For example, the structure of locust adipokinetic hormone pyroglutamyl—Leu—Asn—Phe—Thr—Pro—Asn—Trp—Gly—Thr—NH_2 was determined in large part by mass spectrometric evidence (J. V. Stone, W. Mordue, K. E. Batley and H. R. Morris. (1976). *Nature (London)* **263,** 207–211). *N*-Terminal pyroglutamyl residues give characteristic peaks at *m*/*e* 98 and 126.

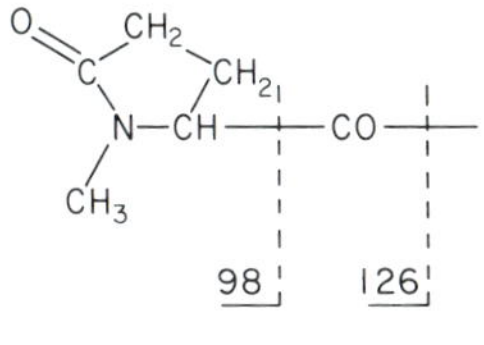

Thirdly, mass spectrometry can cope with a variety of unexpected occurrences as shown by the studies on the recently discovered endogenous opiate enkephalin, isolated from brain. The sequence obtained by conventional (Dansyl–Edman) techniques appeared to be too small for the amino acid analysis and fluorescence measurements seemed to indicate the presence of a novel amino acid. Figure 9.19 shows the mass spectrum obtained from the acetyl permethyl derivative of enkephalin and it is immediately apparent that natural enkephalin is a mixture of two pentapeptides, Tyr—Gly—Gly—Phe—Met and Tyr—Gly—Gly—Phe—Leu, differing only in the *C*-terminal amino acid which may be methionine or leucine in the ratio of about 3 or 4 to 1. The mass spectral interpretation agrees with the amino acid composition results and has been confirmed by synthesis.

TABLE 9.2 To illustrate how each amino acid is uniquely characterized by its mass

Amino acid	Mass of residue	Mass of residue after acetylation and permethylation	As column 3 but residue at *N*-terminus	As column 3 but residue at *C*-terminus
Gly	57	71	114	102
Ala	71	85	128	116
Pro	97	97	140	128
Val	99	113	156	144
Ser	87	115	158	146
Leu/Ile	113	127	170	158
Thr	101	129	172	160
Asp	115	143	186	174
Met	131	145	188	176
Asn	114	156	199	187
Glu	129	157	200	188
Phe	147	161	204	192
His	137	165	208	196
Gln	128	170	213	201
Arg[a]	156	184	227	215
Tyr[b]	163	191(161)	234(161)	222(192)
Lys	128	198	241	229
Trp	186	214	257	245

[a] Arg is converted to ornithine by the derivatization process
[b] Tyr-containing peptides frequently lose a fragment at the *N*-terminal side of the tyrosine residue (see Fig. 9.19)

Fourthly, mass spectrometry is not restricted to dealing with only the twenty common amino acids. Peptides containing new amino acids, or amino acids substituted with lipid or carbohydrate groups, can also be studied. In fact the technique has come of age at just the right time with the recent successes in the isolation of membrane proteins, many of which may be expected to have oligosaccharide branches.

With regard to the complete sequence determination of a protein the rate-limiting step in conventional analysis is the isolation and purification of the component peptides in mixtures obtained by enzymatic digestion. Mass spectrometry allows two ways round this problem. The first involves partial separation by normal methods and then applying mixtures of up to about five peptides onto the direct insertion proble where they are separated by differential thermal evaporation utilizing the different volatilities of the component peptides. The second method is to break the protein down into parts small enough to pass through a gas chromatograph which is coupled directly to a mass spectrometer. These methods are now discussed in turn.

Mixture analysis using the different volatilities of the various components

Consider the four peptides

(a) Leu—Ala—Val—Gly—Phe—Gly
(b) Leu—Gly—Ala—Phe
(c) Leu—Leu—Leu—Gly
(d) Ser—Val—Thr—Pro—Leu—Gly—Ala—Phe

The individual spectra of the acetylated and *N*,*O*-permethylated derivatives are shown in Fig. 9.20. Peptides (a), (b) and (c) can be completely sequenced. For the octapeptide (d) the order of the last two amino acids is not determined, owing to the apparent absence of peptide bond rupture between Ala and Phe.

When the peptides are mixed together a very complex spectrum is to be expected. Figure 9.21 shows how considerable simplification can be achieved by recording just two spectra at different temperatures (250° and 270°). The spectra contain peaks from three and two pepetides of the mixture respectively.

Figure 9.22 shows how simplification of the spectrum from a mixture of peptides (a), (b) and (c) can be achieved by recording a series of spectra at constant temperature and allowing time for the more volatile components to evaporate completely. After 1, 3, 5 and 15 min the spectra show peaks from peptides (c), (c+b), (c+b+a), and (a), respectively. It is easy to appreciate that careful time *and* temperature co-programming could allow more or less complete separation of the component peptides.

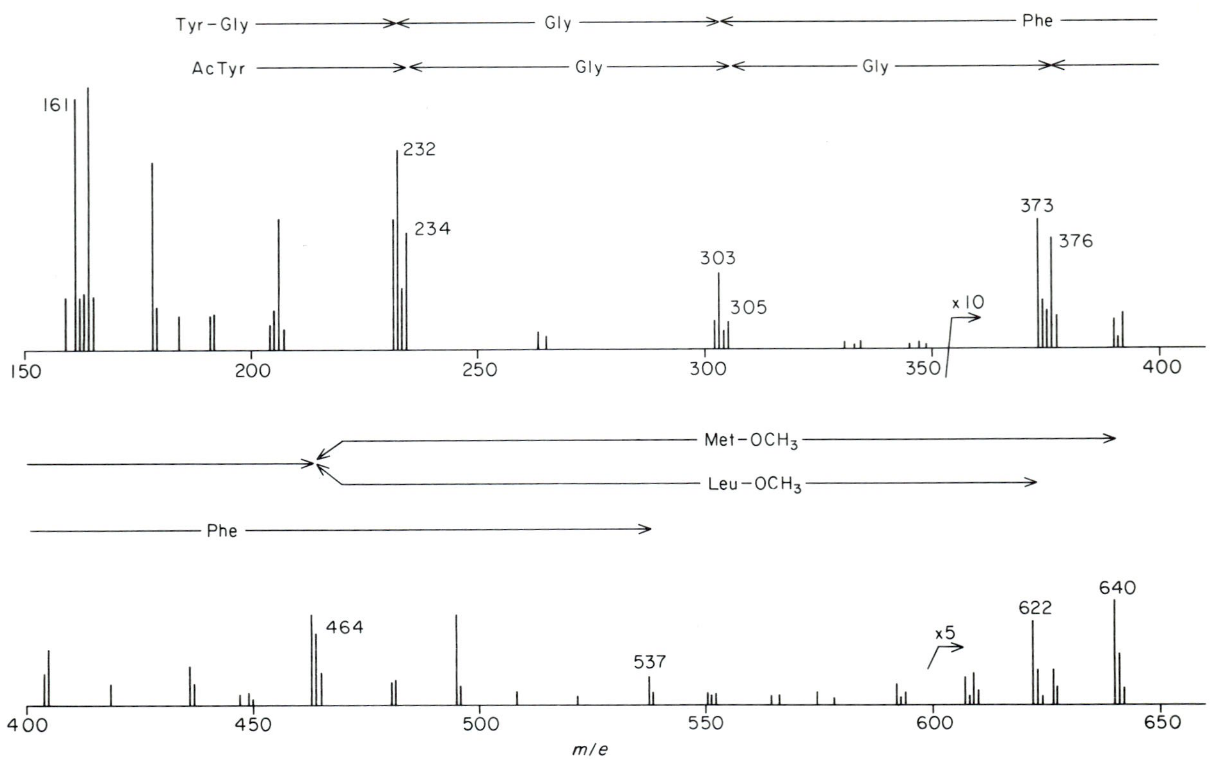
Tyr-Gly
Gly
Phe
AcTyr
Gly
Gly
161
232
234
303
305
x10
373
376
150
200
250
300
350
400
Met-OCH3
Leu-OCH3
Phe
464
537
x5
622
640
400
450
500
550
600
650
m/e

FIG. 9.19 (a)

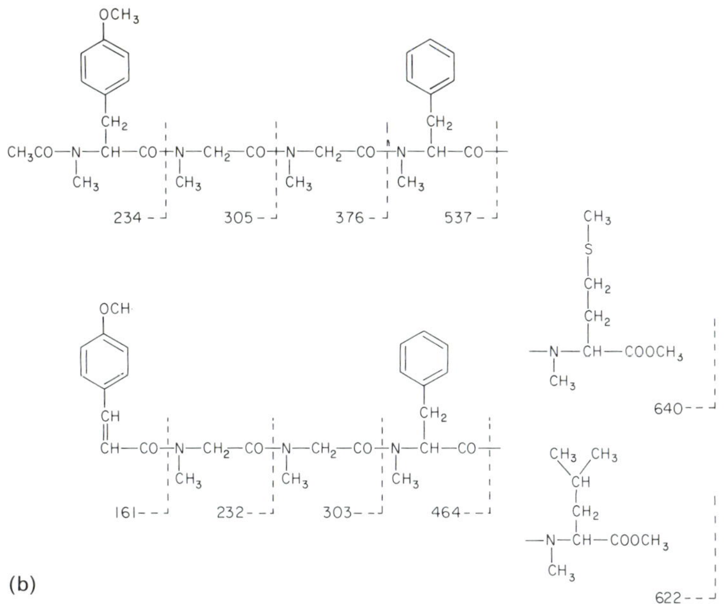

(b)

FIG. 9.19 (a) Mass spectrum of the acetyl permethyl derivative of natural enkephalin. (b) Fragment ions giving rise to the sequence peaks in enkephalin. (J. Hughes, T. W. Smith, J. W. Kosterlitz, L. A. Fothergill, G. A. Morgan and H. R. Morris. (1975). *Nature (London)* **258**, 577–579, Reproduced by kind permission of Dr Howard Morris).

As illustrations of current progress in protein sequencing by this method, it is worth mentioning that about 25% of α-lactalbumin was sequenced using only three consecutive mixture fractions from a Sephadex column effluent, over 200 residues of ribitol dehydrogenase and 170 residues of chloramphenicol transacetylase have been sequenced, and the complete structure of dihydrofolate reductase from *Lactobacillus caseii* has been determined. In the latter two cases, elastase was used for hydrolysis since this enzyme appears to have some size specificity and tends to produce peptides of 2–12 residues which is an ideal range for mass spectrometric work.

Protein sequencing using GC–MS

Trifluoroacetylated *N,O*-permethylated peptides containing from one to about five amino acids are usually sufficiently volatile to pass through a suitable gas chromatograph. The mass spectra of the completely separated peptides are recorded directly and the sequences determined as outlined above. Chemical ionization using isobutane as the reactant gas, with the pressure adjusted to give a good compromise between sequence peaks and

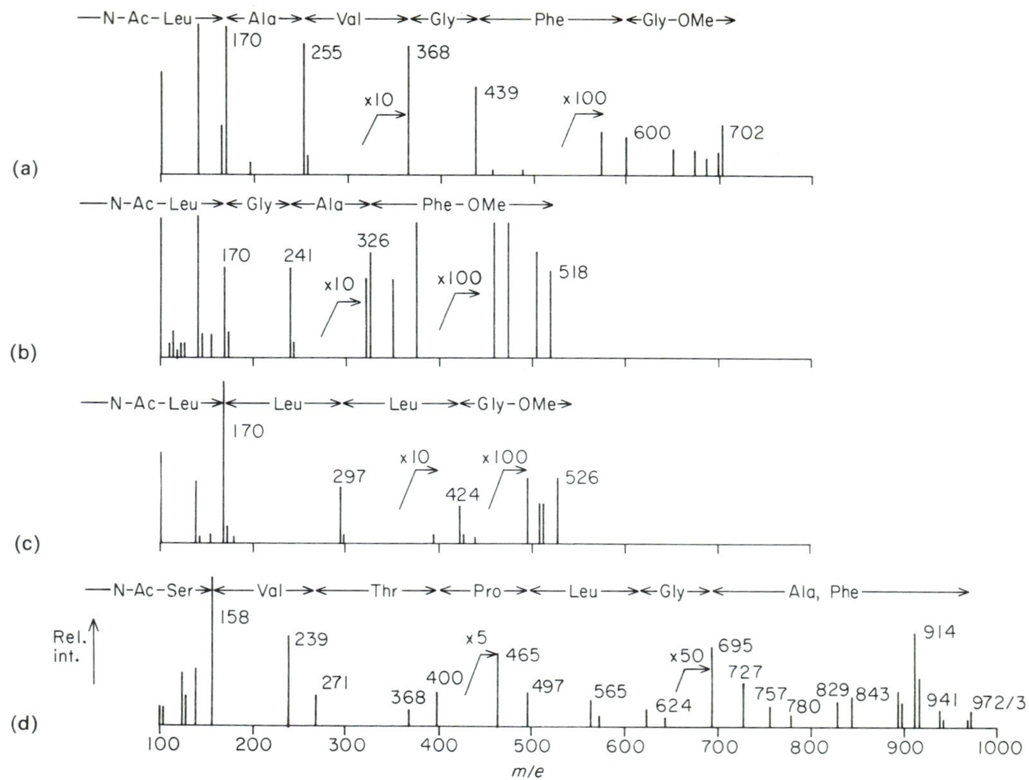

FIG. 9.20 Individual mass spectra of acetyl *N,O*-permethyl derivatives of the peptides (a), (b), (c), and (d) (see text for details).

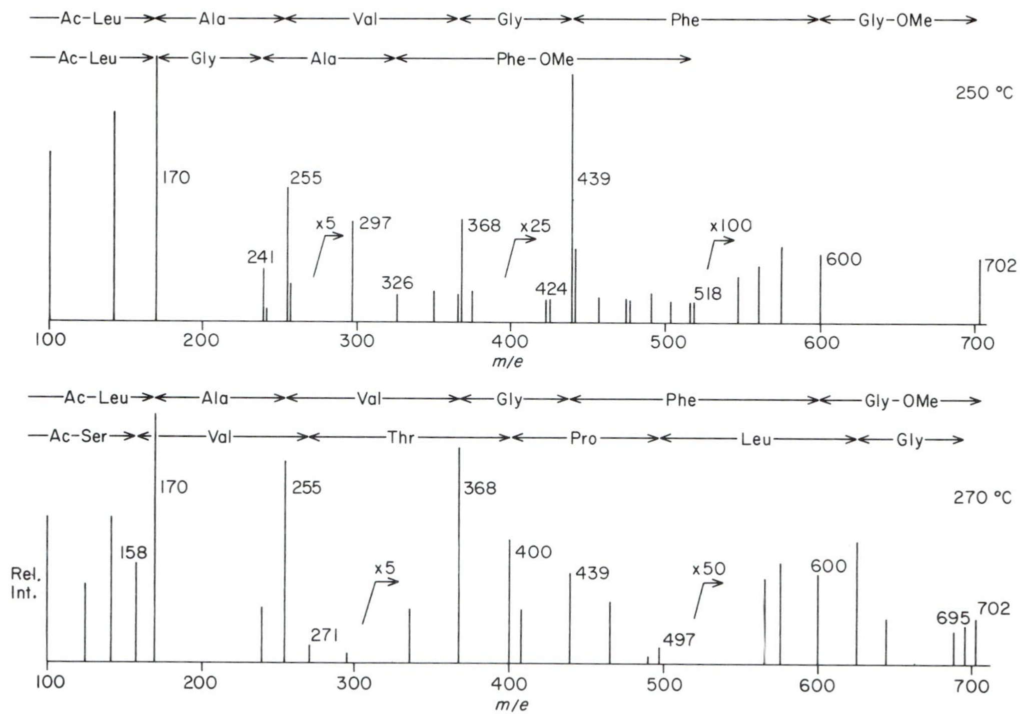

FIG. 9.21 Mass spectra of a mixture of the peptides a,b,c and d at 250°C and 270°C.

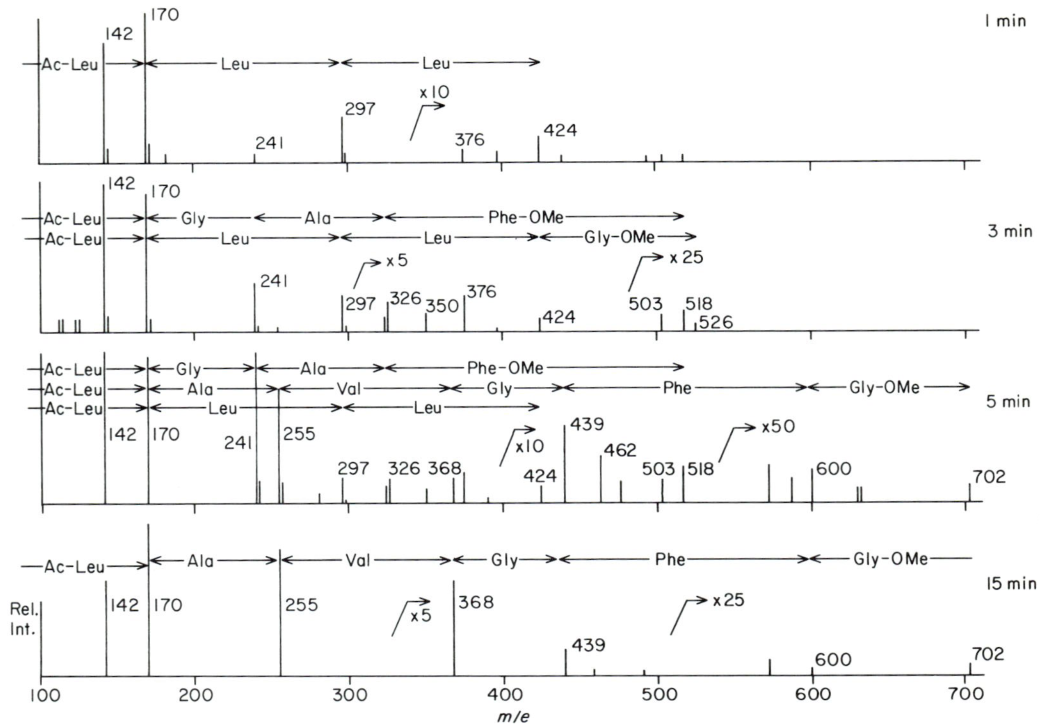

FIG. 9.22 Mass spectra of a mixture of the peptides a,b and c after various time intervals.

molecular ion, seems to offer the most promising results. The volatility requirement of the gas chromatograph means that sequence information has to be derived from peptides somewhat smaller than those that can be applied on the direct insertion probe. However, the ambiguities arising from mixture analysis using the latter approach are not encountered. On the other hand, for a complete protein sequence determination the GC–MS method requires examination of the hydrolysates from a larger number of different enzymatic digests so as to provide sufficient "overlap" information. This may not be particularly disadvantageous with current computerized data processing systems for handling large numbers of spectra. The GC–MS approach is certainly very attractive and offers the possibility of rapid automatic protein sequence determination on very small quantities of material (see, for example, J. D. Priddle, K. Rose and R. E. Offord. (1976). *Biochem. J.* **157,** 777–780). The direct insertion system (EI or FD) with its higher mass range is perhaps more useful for compounds such as the oligopeptide hormones.

Examples of the use of metastable peaks

If an unseparated mixture is being examined then clearly the ability to relate peaks in the spectrum can be of considerable help in deciding which peaks arise from a particular component of the mixture. In addition, of course, the metastable peaks will provide information regarding the fragmentation patterns of the various components. Figure 9.23 shows the mass spectrum of a mixture of two acetylated and permethylated tetrapeptides Gly—Val—Gly—Tyr and Gly—Ala—Gly—Tyr. There is a peak at *m/e* 199 which high-resolution studies showed to be a doublet. The major component is acetyl—MeGly—MeAla ($C_9H_{15}N_2O_3$) but the other component is $C_{10}H_{19}N_2O_2$ corresponding to a loss of CO from acetyl—MeGly—MeVal. This latter fragmentation is confirmed by a metastable peak at *m/e* 174·5 ($199^2/227$).

Another interesting application is the ability to distinguish between leucine and isoleucine, where both compounds have the same mass *and* identical atomic compositions. All oligopeptides containing Leu or Ile (regardless of their position in the sequence) give rise—after perdeuteromethylation—to a significant peak at *m/e* 103 corresponding to ($CD_3\overset{+}{N}H{=}CHC_4H_9$). This ion can fragment further to give peaks at *m/e* 75,69,61,47,35 and 32 but the relative abundances of the products depend on the isomeric form of the parent ion. For Leu-containing peptides, the major fragment from the above ion is at *m/e* 61 (*m** at 36·1) while for Ile-containing peptides, the major fragment is at *m/e* 69 (*m** at 46·2). These metastable peaks are sometimes of sufficient abundance to allow differentiation of Leu and Ile.

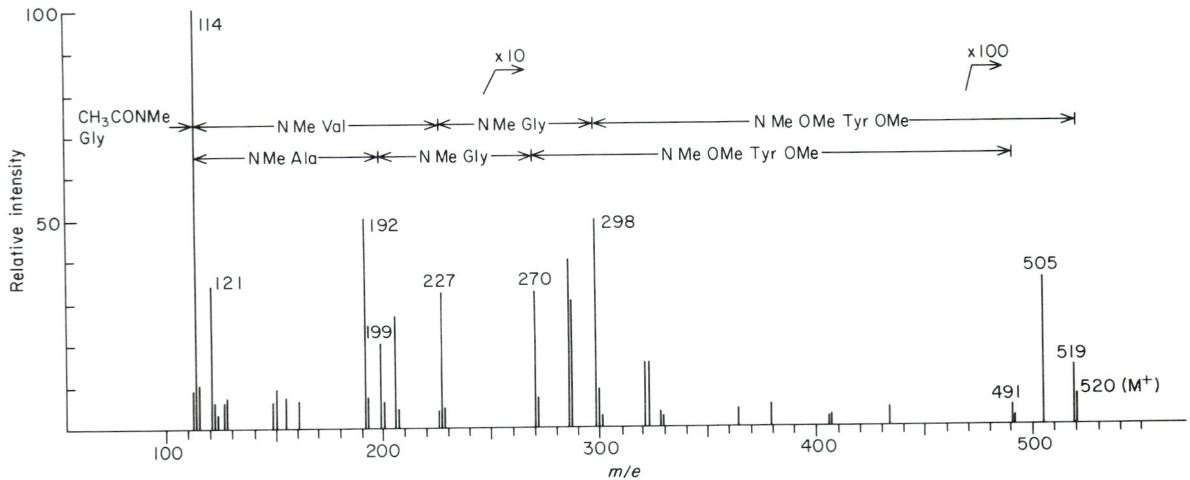

FIG. 9.23 Mass spectrum of a mixture of two acetyl permethyl derivatives of the tetrapeptides detailed in the text. (A. J. Geddes, G. N. Graham, H. R. Morris, F. Lucas, M. Barker and W. A. Wolstenholme. (1969). *Biochem. J.* **114,** 695–702.)

Carbohydrate sequencing

There are a number of problems in carbohydrate sequencing additional to those encountered in peptide sequencing. Carbohydrates require extensive derivatization (to peracetates, pertrimethylsilylethers, etc.) to make them sufficiently volatile. The nature of the glycosidic linkage can vary along the chain, and the monomers themselves are commonly stereoisomers. Mass spectrometry is not well suited to distinguishing stereoisomers although derivatization to suitable structural isomers which give different spectra can provide a means of differentiation.

It seems that the various glycosidic linkages can give rise to characteristic fragmentations but as yet the evidence is thin. Different linkages would appear to produce differences in the ease with which some neutral fragments are lost from certain fragment ions and there would also appear to be differences in the degree of proton transfer during fragmentation.

A promising development has been the incorporation of aromatic units into the terminal sugar. The aromatic component affords stabilization of the molecular ion so that this is present in sufficient abundance to provide a measure of the molecular weight. In a series of acetylated *N*-arylglycosylamine derivatives of tri-, tetra-, and pentasaccharides, the main fragment ions were formed by cleavage of the glycosidic linkages (sequence ions) followed by elimination of acetic acid, ketene, and acetic anhydride. Two series of ions (RS and NS) reflect the sequence from the reducing and non-reducing ends respectively as shown in Fig. 9.24.

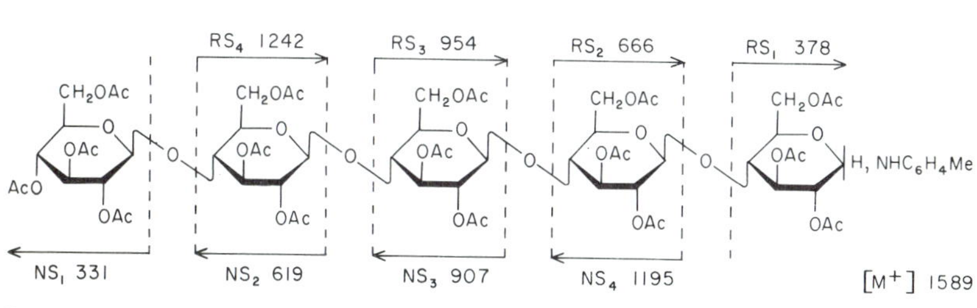

FIG. 9.24 Major fragmentations of acetylated *N*-arylglycosylamine pentasaccharide. (O. S. Chizhov, N. N. Malysheva and N. K. Kochetov, *Carbohydrate Research*, (1973). **28**, 21–29). Reproduced by kind permission of the authors and Elsevier Scientific Publishing Co., Amsterdam.

The different sequences in the isomeric trisaccharides shown in Fig. 9.25 are clearly expressed in the mass spectra. Moreover, the first compound shows a peak at *m*/*e* 377 in addition to the sequence peak at 378 and this is characteristic of a 1→6 linkage.

The tri-, tetra-, and pentasaccharides shown in Fig. 9.26 all have peaks at *m*/*e* 259 and 479 which are characteristic of the 1→2 linkage.

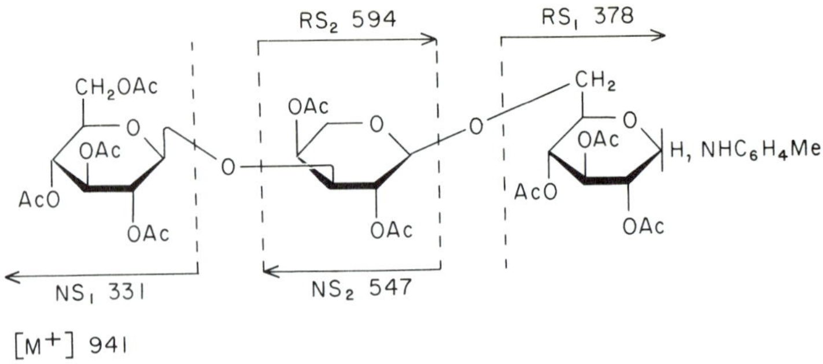

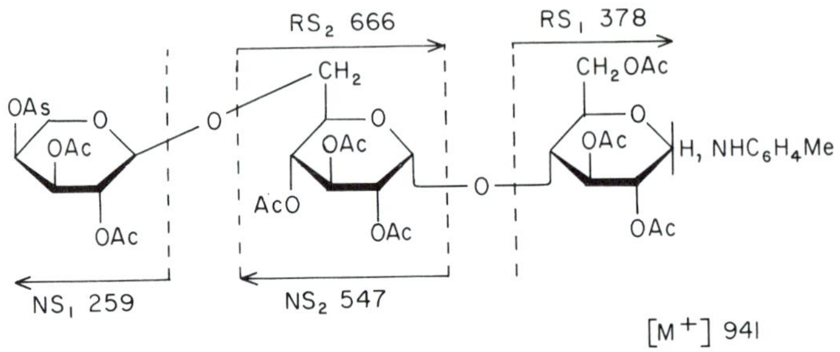

FIG. 9.25 Major fragmentations of two isomeric trisaccharide derivatives (Ref. as for Fig. 9.24).

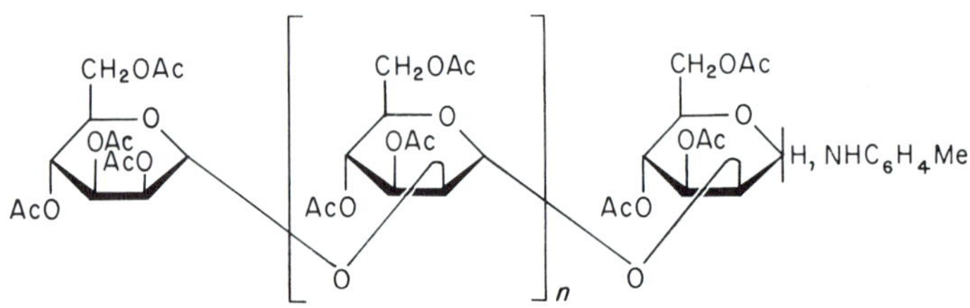

FIG. 9.26 The 1→2 polysaccharide linkages (Ref. as for Fig. 9.24).

The compounds described here are relatively simple but nevertheless the results are encouraging. There have been notable successes in the structure determination of individual compounds of greater complexity but it is clearly going to be an uphill struggle to reach the stage where oligosaccharide sequencing becomes a routine procedure, as is more or less the case with peptides.

Polynucleotides

It is just about possible to sequence dinucleotides using characteristic fragmentations, but extensive derivatization is required, particularly of the phosphate–ester grouping. The increase in mass following derivatization plus the fact that relatively large amounts of material are required for chemical conversion would seem to rule out mass spectrometry for polynucleotide sequencing at the present time. However, much effort has gone into the interpretation of mononucleoside spectra and mass spectrometry probably has a valuable role to play in the identification of unusual bases (transfer RNAs are a good source of these) following conventional chemical degradation.

Rapid Screening for Inborn Errors of Metabolism

There are about 150 well-documented inherited disorders of intermediary metabolism, generally characterized by raised levels of normal metabolites which accumulate in the blood and are excreted in the urine. Well-known examples are phenylketonuria and maple-syrup-urine disease.

Although techniques such as thin-layer chromatography and gas chromatography can detect abnormalities it is clearly an advantage to have an absolute identification of the metabolites which are produced in abnormal amounts. Such positive characterization helps to indicate the primary source of the disorder and the most suitable clinical treatment which should be applied—commonly a dietary restriction. The various disorders can involve the metabolism of steroids, lipids, amino acids, carbohydrates, etc., and examination can be made of plasma, cerebrospinal fluid, amniotic fluid and urine. Gas chromatography–mass spectrometry studies have tended to concentrate on the organic acids which are excreted in the urine since there is such a wide variety of compounds which can be investigated in a single experiment.

Until recently the usual practice was to use the mass spectrometer to characterize a compound detected and isolated by more conventional means, but the stage is rapidly approaching where the GC–MS system can be brought inside the hospital itself, where it could be operated by personnel relatively inexperienced in mass spectral interpretation, and provide the rapid diagnosis essential for the speedy treatment of newborn infants and acutely ill patients. This development has been made possible by the use of on-line minicomputers dedicated to data acquisition and analysis. First of all, a "normal" metabolic profile is established of the volatilizable metabolites—already there is a library collection of several hundred mass spectra of TMS derivatives of compounds in human physiological fluids. The GC–MS metabolic profile from a particular patient is analysed automatically by the computer which then performs a search and comparison with its stored library information and (hopefully) outputs the nature of any abnormality. There is still some rather difficult computer programming to be completed, especially since a number of complications can arise due to different diets and individual metabolic variations and of course to the effects of any drug administration.

Studies of Natural Products of High Volatility but Occurring Only in Minute Amounts

Food aromas

The characteristic aroma of any foodstuff is usually due to the combined effects of a large number of different substances, but in many cases the dominating influence on the aroma arises from just one (or perhaps a few) component(s). This component is known as a "character impact compound" and of course it is an important requirement in the food industry to be able to identify such compounds.

Consider carrots as a typical example. Gas chromatography of a steam distillation solvent extract gives a very complex chromatogram with well over 100 peaks. Figure 9.27 shows such a chromatogram.

Detection of the character impact compounds is performed by "sniffing" the peaks as they are eluted from the column, but in this case the peaks

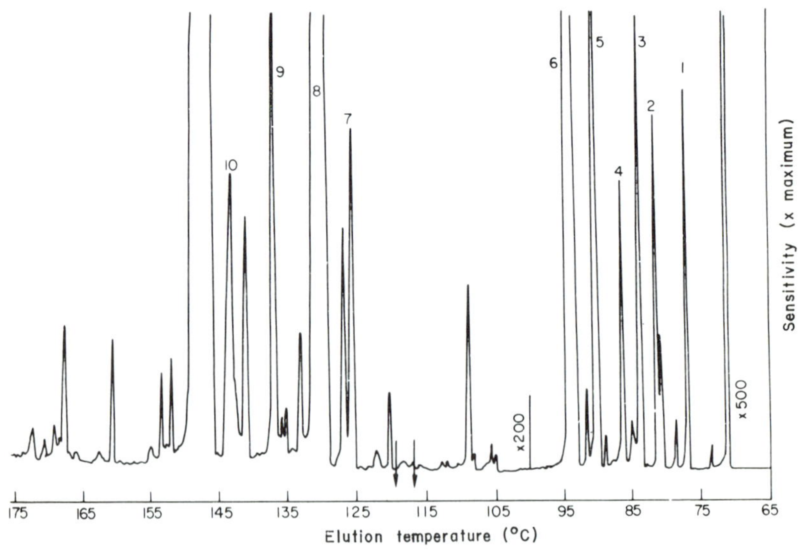

FIG. 9.27 Gas chromatogram of carrot essence (D. A. Cronin and P. Stanton, *J. Sci. Fd Agric.* (1976). **27**, 145–151). Reproduced by kind permission of the authors.

ascribed to the carrot aroma (arrowed in the diagram) were barely detectable above the background and GC–MS studies did not allow identification. However, by using a series of different specially designed and coated columns it was possible to focus on the important region of the eluate and finally to detect a sufficient amount of the character impact compound (Peak C, Fig. 9.28) to run a mass spectrum.

This spectrum is sufficiently close to that of 2-methoxy-3-s-butylpyrazine (see below) to justify the assumption of the identity of the compounds.

Character impact compound		2-Methoxy-3-s-butylpyrazine	
m/e	Rel. Int.	*m/e*	Rel. Int.
152	14	152	12
151	55	151	46
138	87	138	90
137	33	137	50
124	100	124	100
123	20	123	12
105	14	105	14
41	14	41	—

It now appears that 2-methoxy-3-alkylpyrazines are volatile components of over 20 raw vegetables but it remains to be established whether they are all character impact compounds or whether particular aromas are due to a more complex mixture of compounds.

Pheromones

"Semiochemicals" are compounds that are emitted by one organism, travel by diffusion, and cause a behavioural change in a second organism. The term "pheromone" is reserved for those semiochemicals used for *intra*-species communication. They are classified according to the response elicited and the response may be due to a single compound or to a mixture.

The most potent, and the most studied, compounds are the insect airborne sex pheromones used by the female in signalling sexual receptivity; in some cases the threshold concentrations are as low as 200 molecules per ml of air. For moths, many of the sex pheromones are unsaturated straight chain acetates or alcohols with molecular weights in the 100–300 range. These volatile low MW compounds, occurring as "trace" constituents, would seem to be ideal candidates for GC–MS studies. However, it must be said that in a number of cases, the presence of double bonds in the carbon skeleton makes it very difficult, if not impossible, to establish unequivocally the complete structure of the native compound by mass spectrometry alone. Figure 9.29

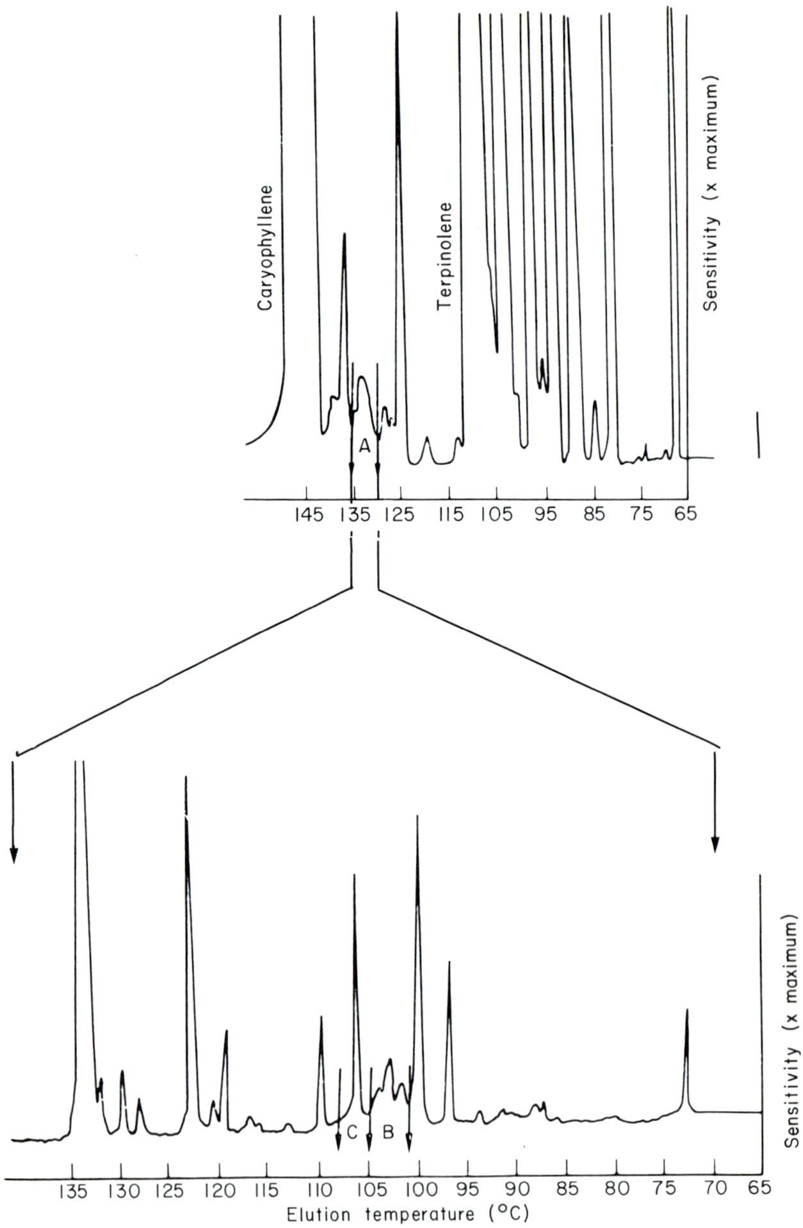

FIG. 9.28 Gas chromatograms of carrot essence illustrating the separation of the character impact compound (Ref. as for Fig. 27).

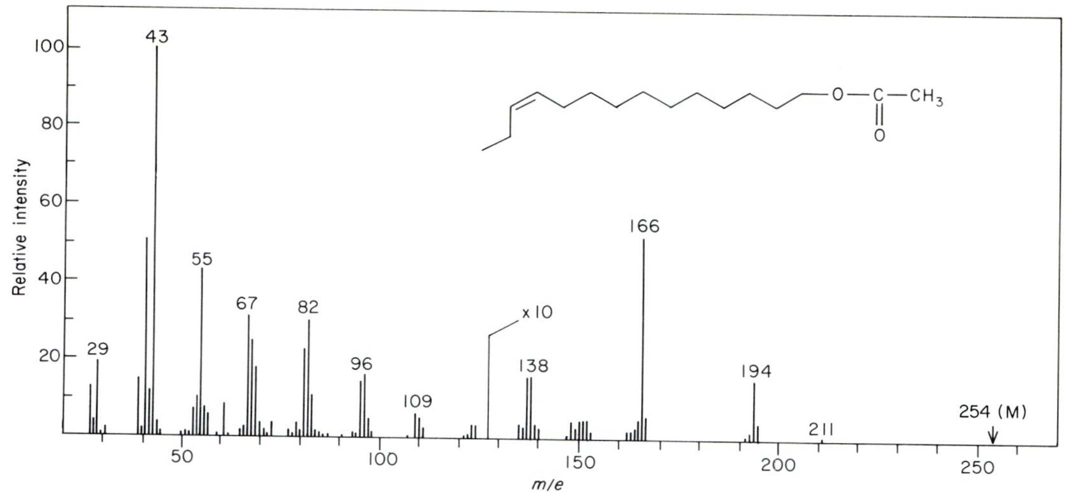

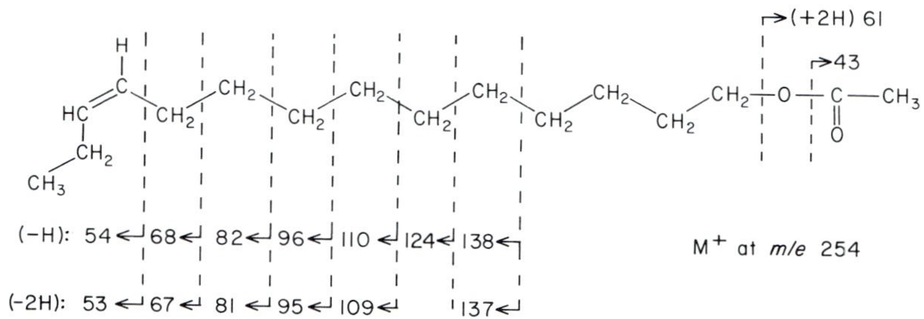

FIG. 9.29 Mass spectrum of *cis*-11-tetradecenyl acetate. (from F. E. Regnier (1972). *In* "Biochemical Applications of Mass Spectrometry", (Ed. George R. Waller), Chapter 26. Wiley, New York.) Reproduced by permission. The sample was kindly provided by Dr W. L. Roelofs.

shows the mass spectrum of *cis*-11-tetradecenyl acetate which is identical to that of the sex pheromone isolated from the red-banded leaf roller moth. Bioassays have shown that the *cis* isomer is about 1000 times as active as the *trans* isomer. In this case the chemical structure is easily determined from peaks arising from a series of fragmentations along the chain and these are also indicated in Fig. 9.29.

Prostaglandins

Prostaglandins (PG) deserve a special section on their own since they represent one of the success stories of the application of mass spectrometry, both in structure determination and quantitative assay. It is perhaps no coincidence that the surge in activity in prostaglandin research in the 1960s coincided with the growth and development of GC–MS techniques.

Prostaglandins are widely distributed in animal tissues, exhibit considerable biological potency, are present at low concentrations (ng g^{-1}) and undergo rapid turnover with halflives sometimes as low as 30 seconds. They have been implicated in an enormous variety of physiological processes, including, for example, gastric secretion and ulcer-healing, blood pressure regulation, bronchial constriction, lipid deposition, platelet aggregation, muscle tone, reproduction and response to inflammation. They are named as derivatives of prostanoic acid, shown below.

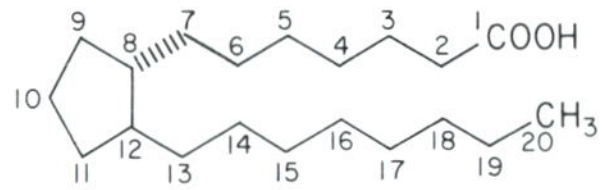

Substituents on the cyclopentane ring are designated α if they are below the plane of the ring (in the molecule shown above) and β if not. Subscripts 1 (often omitted) and 2 are used to denote the number of olefinic double bonds in the two tails. Eight classes (A→H) have now been recognized but the E and F series seem to be the most active and have been the most widely studied. Prostaglandins are synthesized from polyunsaturated fatty acids and Fig. 9.30 shows some of the pathways from arachidonic acid. This diagram is illustrative only and by no means complete; thus, for example, 12 metabolites of $PGF_{2\alpha}$ have been identified so far but only two are shown.

Examples of PG classes A, and D→H are shown in the figure. Classes B and C differ from A in that the double bond in the ring is 8→12 and 11→12 respectively. In Fig. 9.30, "MS" is used to indicate that a mass spectrometric method has been developed for analysis of the compound.

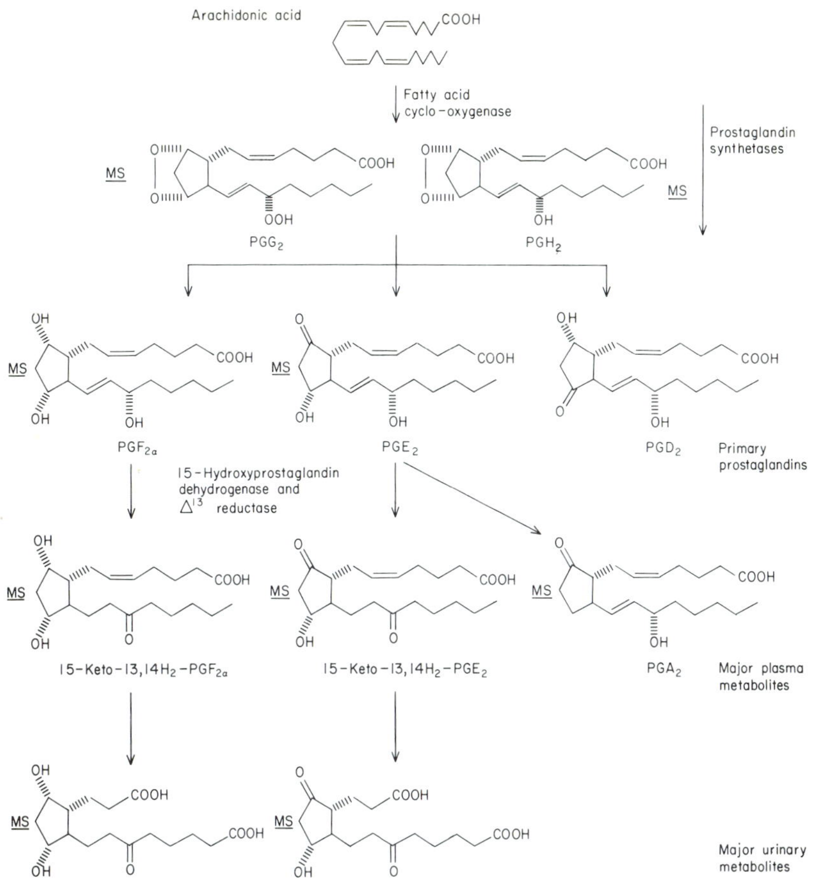

FIG. 9.30 Some of the pathways from arachidonic acid. "MS" indicates that a mass spectrometric method has been developed for analysis of the compound. From "Advances in Prostaglandin and Thromboxane Research", Vol. 1, 1976. B. Samuelsson and R. Paoletti (Eds). Reproduced by kind permission of Raven Press Publishers, New York.

Structure elucidation

The native compounds are polar and the most common methods of increasing volatility are methylation, trimethylsilylation, and the formation of alkyloxime derivatives. The total fragmentation pattern of each type of PG is unique to that type. Thus, for example, the $PGF_{1\alpha}$ derivative undergoes extensive fragmentation of the cyclopentane ring and the PGE_1 derivative gives rise to long chain trimethylsilyl (TMS) fragments and alkoxy group migration. In view of these variations it is more useful here to consider how just a few of the more general structural features may be confirmed.

Consider first those cases where the terminal part of the molecule is that shown in Fig. 9.31(a).

Following trimethylsilylation, the ions shown in Fig. 9.31(b) and (c) appear with compositions confirmed by accurate mass measurement. The atomic composition of the remainder of the molecule may be obtained by accurate mass measurement of the M—C_5H_{11} ion shown in Fig. 9.31(d).

Now consider the major urinary metabolite of $PGF_{2\alpha}$ shown in Fig. 9.31(e) (and see also Fig. 9.30) and its methylated trimethylsilylated derivative. The molecular ion of the derivative occurs at *m/e* 502. The diol is confirmed by a shift of 18 amu to *m/e* 520 in the deuterated (d_9)-TMS derivative. The dicarboxylic acid is confirmed by a shift of 28 amu to *m/e* 530 in the *ethyl* ester derivative (the mass-shift technique, see p. 362). The keto function is established by formation of a methoxime derivative; its location is suggested by the fragment ion produced by the cleavage shown by the dashed line whose mass increases by 14 amu in the ethyl ester but is unchanged in the spectrum of the d_9-TMS derivative.

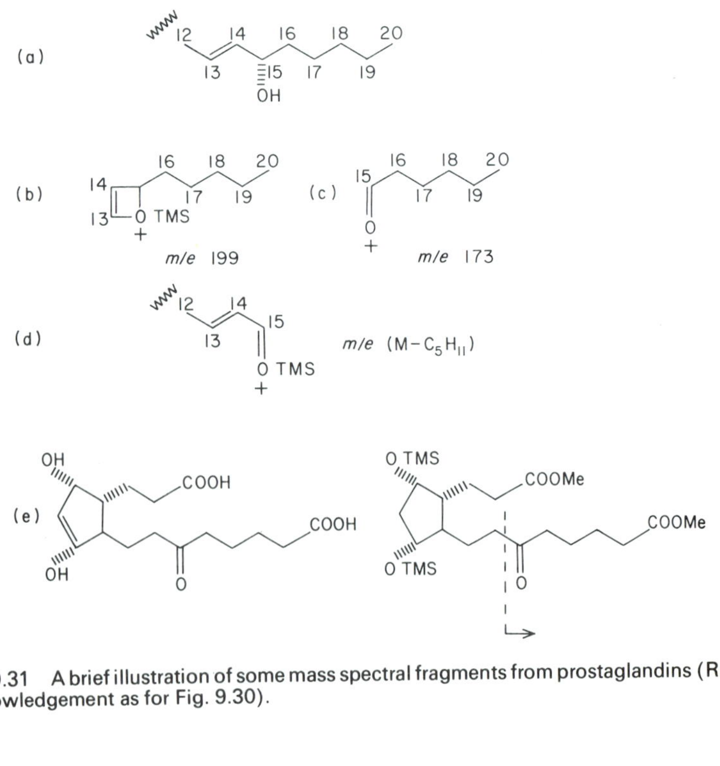

FIG. 9.31 A brief illustration of some mass spectral fragments from prostaglandins (Ref. and acknowledgement as for Fig. 9.30).

An example of metabolic studies

In biosynthetic studies *in vitro* using a mixture of $^{16,16}O_2$ and $^{18,18}O_2$ it was shown in 1965 that the C_9 and C_{11} oxyfunctions came from the same molecule of oxygen since there was no $^{16}O,^{18}O$ contribution in the products (cf. haem metabolism studies discussed on p. 364). It was therefore postulated that endoperoxide compounds (i.e. PGG_2 and PGH_2 in Fig. 9.30) occurred as precursors in the synthesis of the primary prostaglandins. It was not until 1973–74 that the existence of these precursors was directly confirmed by their isolation and structure elucidation.

Monitoring

For monitoring changes in body fluid concentrations an intense peak characteristic of the molecule is required. One way to achieve this is to prepare a derivative which so directs the fragmentation that a single cleavage predominates. A very striking example is the preparation of the methyl ester, oxime, *t*-butyldimethylsilyl (TBDMS) derivative of 15-keto-13, 14-dihydro $PGF_{2\alpha}$ (the major plasma metabolite of $PGF_{2\alpha}$, see Fig. 9.30). The facile loss of the *t*-butyl group gives a peak at *m/e* M-57 which accounts for 95% of the total ionization.

It would be convenient if the primary prostaglandins could be monitored in blood or plasma, but it turns out that their levels are dramatically increased in blood sampling because of platelet biosynthesis. Attention has therefore been focussed on their major plasma metabolites, the 15-keto-dihydro derivatives, and their major urinary metabolites. The results of these monitoring studies include the demonstration of inhibition of PGE synthesis following administration of aspirin-like compounds, of increased PGF synthesis during pregnancy, and of increased levels of 15-keto-13, 14-dihydro $PGF_{2\alpha}$ in peripheral plasma during an allergen-provoked asthmatic attack.

An example of quantitative assay

The $PGF_{2\alpha}$ concentrations in pleural carrageenan-induced inflammatory exudates have been measured at the 5×10^{-14} g level. Figure 9.32 shows the mass spectrum (*m/e* 400–520) of the Me-TMS derivative of $PGF_{2\alpha}$. There is an intense peak at *m/e* 423 corresponding to $(M - C_5H_{11} - TMS - OH)$ which can be calibrated by injection of known amounts of standard $PGF_{2\alpha}$ derivative.

The PGE_2 and $PGF_{2\alpha}$ derivatives are not yet separable by a GC column but fortunately the PGE_2 derivative gives an intense peak at *m/e* 421 with only a small and attributable contribution at *m/e* 423. The PGE_3 derivative could be measured at the 5×10^{-13} g level. There were no non-prostaglandin impurities

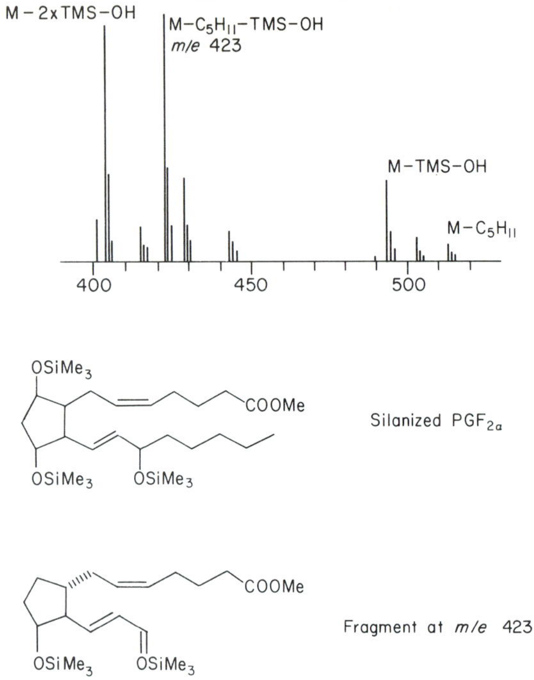

FIG. 9.32 Partial mass spectrum of the Me-TMS derivative of $PGF_{2\alpha}$. (B. W. Wilson, W. Snedden, K. T. Tham and D. A. Willoughby. (1976). *Agents and Actions* **6**, 176–179). Reproduced by kind permission of Dr Walter Snedden.

having the same GC retention time in the exudate so the SIM technique can be used for co-assay of these two prostaglandins. This method is much more sensitive than the current bioassay mainly due to the loss of about 50% of the prostaglandins during the extra purification required for the bioassay.

Suggestions for Further Reading

Books

Budzikiewicz, H., Djerassi, C. and Williams, D. H. (1967). "Mass Spectrometry of Organic Compounds", Holden-Day,

Hill, H. C. (1972). "Introduction to Mass Spectrometry", (2nd edn), Heyden, London.

McLafferty, F. W. (1973). "Interpretation of Mass Spectra", (2nd edn), W. A. Benjamin, Menlo Park, CA.
Millard, B. J. (1978). "Quantitative Mass Spectrometry", Heyden, London.
Waller, G. R. (Ed.) (1972). "Biochemical Applications of Mass Spectrometry", Wiley/Interscience, New York.
Williams, D. H. and Fleming, I. (1973). "Spectroscopic Methods in Organic Chemistry", (2nd edn), McGraw Hill, New York.

Periodicals

Organic Mass Spectrometry, Heyden, London, from 1968.
Advances in Mass Spectrometry in Biochemistry and Medicine, Spectrum Publications, New York, from 1975.

Index